PHOTOMETRIC METHODS IN INORGANIC TRACE ANALYSIS

This book is the revised English version of
Fotometriás nyomelemzési módszerek
published by Műszaki Könyvkiadó, Budapest

Translated by
D. Durham Ph. D.

JOINT EDITION PUBLISHED BY
ELSEVIER SCIENCE PUBLISHERS B. V., AMSTERDAM, THE NETHERLANDS
AND AKADÉMIAI KIADÓ, THE PUBLISHING HOUSE OF THE HUNGARIAN
ACADEMY OF SCIENCES, BUDAPEST, HUNGARY

*The distribution of this book is being handled by the following publishers
for the U.S.A. and Canada*

ELSEVIER SCIENCE PUBLISHING COMPANY, INC.
52 VANDERBILT AVENUE
NEW YORK, NY 10017, U.S.A.

*for the East European countries, Democratic People's Republic of Korea, People's
Republic of Mongolia, Republic of Cuba and the Socialist Republic of Vietnam*

KULTURA, HUNGARIAN FOREIGN TRADING COMPANY
P.O. BOX 149, H-1389 BUDAPEST 62, HUNGARY

for all remaining areas

ELSEVIER SCIENCE PUBLISHERS B.V.
1 MOLENWERF
P.O. BOX 211, 1000 AE AMSTERDAM, THE NETHERLANDS

Library of Congress Cataloging in Publication Data

Upor, Endre.
    Photometric methods in inorganic trace analysis.
    (Wilson and Wilson's Comprehensive analytical chemistry; v. 20)
    Translation of: Fotometriás nyomelemzési módszerek.
    Includes bibliographies and index.
    1. Trace elements—Analysis. 2. Photometry.

I. Mohai, Miklósné. II. Novák, Győző. III. Title.
IV. Series: Comprehensive analytical chemistry; v. 20.
QD75.W75 vol. 20 543s [543'.0852]   84-18741
[QD139.T7]

ISBN 0-444-99588-9 (Vol XX)
ISBN 0-444-41735-4 (Series)

With 9 illustrations and 44 tables

COPYRIGHT © 1985 by AKADÉMIAI KIADÓ, BUDAPEST

Printed in Hungary

Volume XX

PHOTOMETRIC METHODS
IN INORGANIC TRACE ANALYSIS

by

E. UPOR, M. MOHAI and GY. NOVÁK

Mecsek Ore-Mining Enterprise, Pécs, Hungary

Wilson and Wilson's

COMPREHENSIVE ANALYTICAL CHEMISTRY

Edited by

G. SVEHLA, PH.D., D.SC., F.R.S.C.

Reader in Analytical Chemistry
The Queen's University of Belfast

VOLUME XX

PHOTOMETRIC METHODS IN INORGANIC TRACE ANALYSIS

by

E. UPOR, M. MOHAI and GY. NOVÁK

ELSEVIER
AMSTERDAM OXFORD NEW YORK TOKYO
1985

WILSON AND WILSON'S

COMPREHENSIVE ANALYTICAL CHEMISTRY

VOLUMES IN THE SERIES

Separations by Liquid Amalgams
Vacuum Fusion Analysis of Gases in Metals
Electroanalysis in Molten Salts

Contents

Foreword

In *Comprehensive Analytical Chemistry* the aim is to provide a work which, in many instances, should be a self-sufficient reference work; but where this is not possible, it should at least be a starting point for any analytical investigation.

It is hoped to include the widest selection of analytical topics that is possible within the compass of the work, and to give material in sufficient detail to allow it to be used directly, not only by professional analytical chemists, but also by those workers whose use of analytical methods is incidental to their work rather than continual. Where it is not possible to give details of methods, full reference to the pertinent original literature is given.

Volume XX deals with the application of spectrophotometry in inorganic trace analysis, and as such is the continuation of Volume XIX, in which the theoretical and practical aspects of spectrometry were described. The present authors come from an industrial laboratory in southern Hungary, where very often unusual, complex samples have to be analyzed. It is recorded with great regret that one the authors, Dr. Endre Upor, passed away whilst this volume was in production.

Belfast, 25th July, 1984   *G. Svehla*

Preface

Photometric methods of analysis are of great importance, and are likely to remain so for a considerable time. Although the up-to-date procedures requiring large instruments (atomic absorption spectrophotometry, X-ray fluorescence, etc.) are progressively gaining ground, the ever greater demand for material testing means that there is an accompanying increase in the overall number of photometric analyses.

This book sets out to present a survey of the applications of these methods in inorganic trace analysis. Since the activities of the authors are linked primarily to the analysis of rocks and ores, the majority of the methods described in detail for determination of the individual elements are taken from this field. An effort is also made to provide examples from other areas of importance in analytical practice, ranging from metallurgy to biological materials, and from protection of the environment to the investigation of substances of high purity.

The first half of the book reviews the main steps involved in the photometric methods (preparation of the sample and separations to be employed), and in addition gives an account of the experience gained by the authors in developing or improving methods of analysis. It is hoped that this will encourage even those analysts not dealing systematically with research to introduce new procedures or to modify existing ones. An attempt is made to foster the complex-chemical aspect necessary for such work.

Other books appearing in this field treat the theoretical questions of colorimetry and photometry, the correlation of molecular structure and absorbance, the instrumentation, the sources of error and error-calculation, and the reagents used in photometry; accordingly, these topics will not be discussed in detail here.

For reasons of space, tables otherwise necessary for analytical practice, complex stabilities, masking and separation possibilities, etc., are not published. Instead, the reader is referred to the relevant monographs.

The authors are grateful to the Directorate of the Mecsek Ore-Mining Enterprise for moral support and for their aid towards the research and development activity forming the basis of the book. They also express their thanks to their colleagues for their regular help in following the literature, in the elaboration of some of the analytical methods, and in utilizing their experience with regard to application of the methods.

Special thanks are due to Professor K. Burger, who actively supported publication of the earlier, Hungarian version of this book and provided us with much valuable professional advice.

We are very much obliged to Professor J. Inczédy as well, who let us have some professional proposals of great importance for the revised English version of our book.

The Authors

1. *Some questions on the application of (spectro)photometric methods*

The photometric procedures are among the oldest of the trace-analysis methods. They are applied widely and very many of them have proved of great value. Spectrophotometers themselves are relatively cheap instruments, but photometers, which are even simpler instruments, are suitable for the solution of numerous problems, which is very advantageous in smaller laboratories.

Although analytical research demands great experience, a high degree of training is not required for the practical application of photometric analysis.

The attainable sensitivity with a solid sample is 100–0.01 ppm. Appropriate reagents are known for virtually every element and, with the exception of some alkali metals, these can be employed effectively.

1.1. The position of photometric methods in trace analysis

It must be stated that, apart from the analysis of high-purity materials, where its significance is declining, the field of photometric methods is of the highest importance in trace analysis. In a very great number of cases there is no need for separation; if separation is necessary, this can frequently be achieved by simple extraction or, for example, by dissolution via the digestion of solid samples (precipitation separation), which can be regarded as one operational step. It may be noted here that the separations and masking procedures developed in connection with photometric methods of trace analysis can be used without any problem with other techniques too, e.g. spectrochemical methods or neutron activation; hence, the research work relating to this part of the procedure is no longer necessary in these techniques. The importance of spectrophotometric methods has not diminished during the past decade [1a].

Apart from the position occupied in serial analysis, there are certain cases when the application of these methods is required in determinations involving other procedures. Of these, mention should perhaps first be made of the analysis of standard samples. In some methods, the base substance effect or the lack of an exact knowledge of the excitation energy means that the calibration curve can be obtained only with the aid of an external standard (mass-spectrometry, X-ray fluorescence). The spectrophotometric method is generally accepted as the most accurate method for the preparation of the external standard, and remains at present an indispensable procedure.

Spectrophotometry can similarly play the role of a control in serial analyses with other, less accurate methods.

In the analysis of complex samples with unusual compositions, spectrophotometry, and in part atomic absorption, may likewise be employed more advantageously than other methods demanding lengthy preparatory work. The same holds for individual samples not to be analyzed within a series.

The photometric procedures can often be used effectively in combination with other methods. In emission spectrum analysis, for instance, if the need arises for the simultaneous quantitative determination of several elements, the "variable internal standard" may be advantageous [1]. The essence of this is that it is not necessary to add an internal standard to the sample; one or other of the elements already present is instead determined by a different method, e.g. spectrophotometrically, and the base substance effect can be corrected for on the basis of measurement of the line blackening of this element. There are naturally limits to the application of the method (there should not be great differences in the possibilities of excitation of the elements to be determined, and the concentration change should if possible be within two orders of magnitude), but in many cases it is a very favourable one.

As one of the limitations of the photometric methods, it may be mentioned that increase of the sensitivity is in principle restricted by the maximum

4

attainable molar absorbance of $2-5\times10^5$. Only one or a few elements can be determined on one weighed sample. In more difficult trace-analysis tasks it is very difficult to exclude contamination.

The accuracy of photometric methods of trace analysis is generally better than those of other methods. The result is not influenced decisively by the composition of the sample. Data on this are provided by Cook *et al.* [2] and Alimarin [2a]. The determination of four elements ($5\times10^{-4}\%$ Cu, $5\times10^{-4}\%$ Mn, $15\times10^{-4}\%$ Cr and $15\times10^{-4}\%$ Hg) by four methods (spectrophotometry, activation analysis, polarography and spectral analysis) in laboratories in nine countries led to the picture shown in Fig. 1.1.

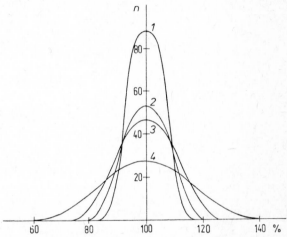

Fig. 1.1. Comparison of the scatters in four different methods of analysis [2, 2a].

1—spectrophotometry; *2*—neutron activation; *3*—polarography; *4*—spectral analysis

In the determination of tantalum in rocks and ores, Lontsykh and Berkovits [3] compared the accuracies of the individual methods in various concentration ranges. They found that the brilliant-green extraction spectrophotometric procedure is of the highest accuracy in every range from 0.0001% to 3.0% (in the lower and higher concentration intervals it is rivalled by spectrochemistry and X-ray fluorescence, respectively).

It is our experience too that the use of spectrophotometric methods is advantageous over several orders of concentration. In the higher concentration interval the ratio of the concentration of the ion to be determined and the concentration of the interfering substances is more favourable, and above a certain limit, therefore, in the best case the result is affected virtually only by the error in the measurement.

The requirements regarding the sensitivity of geochemical research can also generally be met by spectrophotometric methods. Figure 1.2 depicts the sensitivities of the spectrophotometric methods used in our laboratory for rock analysis, compared with the Clark values.

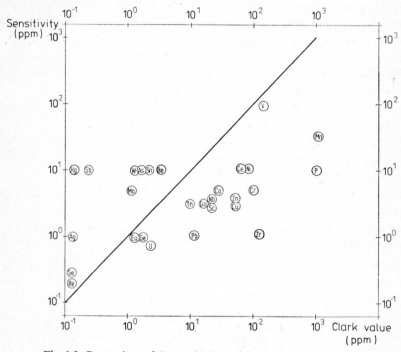

Fig. 1.2. Comparison of the sensitivities of the spectrophotometric methods used for rock analysis in the laboratories of the Mecsek Ore-Mining Enterprise with the Clark values for the elements [4].

A very good auxiliary means of control is not merely to measure the absorbance, but also to record the spectrum, even in serial analyses. Any distortion of the spectrum reveals an error, which may be corrected for in a favourable case on the basis of the course of the curve. In the event of a disturbance-free spectrum, on the other hand, the validity of the result can generally be accepted.

1.2. Further possibilities of application and development of photometric reagents

As regards the application of photometric reagents, the past few decades have seen the emergence of many results which have increased the potential of this method.

Mention must primarily be made of the extension of investigations into mixed ligand complexes, and the analytical utilization of the results [5–10].

The two fundamental types of mixed ligand complexes that are of the greatest importance concerning photometric trace analysis are those in which the central metal ion is linked to the organic base beside the electronegative ligand, or to two electronegative ligands.

In complexes belonging to the former group, ion association is general. Here the metal ion in the anionic complex interacts with an organic cation, most often a ligand of ammonium salt type.

The main advantage of this type of reaction is that in this way even metal complexes with reagents containing strongly hydrophilic groups (SO_3H^-, PO_2H^-, etc.) become extractable.

In mixed ligand complexes the properties do not change in an additive manner. This leads to further advantages. As a consequence of mixed ligand complex formation, for instance, the extractabilities and absorbances of certain ions change to different extents, and overall, therefore, the selectivity and sensitivity of the determination increase. Examples of this are the extraction and photometric determination of the UO_2^{2+}-arsenazo III-diphenylguanidine complex in the presence of otherwise disturbing amounts of Th^{4+} and Ca^{2+} [11], and the extraction of the La^{3+}-arsenazo III complex with the methyltrioctylammonium cation [12]. In the latter case the value of ε_{max} increases by a factor of about 1.5, and Ca^{2+}, which interferes in aqueous medium, is not extracted.

Many new results have emerged in the extension of the application of basic dyestuffs [13]; the conditions of selective extraction have been established for the determination of an ever larger number of anions, primarily metals bound in anionic complexes. Similarly, many authors have studied those complexes in which there are two negative ligands in the inner sphere. The entry of the second ligand is frequently hindered for steric reasons, thereby permitting an increase in selectivity in the determination of certain ions. There is an example of even a 1500-fold increase of the partition coefficient in a complex containing three ligands [14].

There has, of course, been considerable development in the application of mixed ligand complexes of other types too (the use of acidic dyes for the selective extraction of metal ions bound in colourless cationic complexes; ion-pair formation between a complex cation and a complex anion; complexes containing two metal ions; etc.).

Another important, though not so widespread area of development in photometric methods of trace analysis is the production of reagents of higher purity than before. As a consequence of the extensive contamination, the true value of many reagents could not be discerned and hence it did not prove possible to use them in practice.

One of the reasons for the non-uniform composition of reagents is the insufficient purity of the starting compounds, while another is the occurrence of side-reactions in the complicated synthesis and accordingly the presence of undesired decomposition products.

Paper electrophoresis and chromatography have proved to be extremely useful techniques of separation in the study of reagents and certain intermediates [15, 16]. Besides the absorption spectrum, various methods of investigating molecular structure, functional-group analysis and other methods have been utilized in the identification of the products. In this way, products previously stated to be uniform could in many cases be resolved into 6–8 components.

The third most important field in the further development of photometric trace analysis is the work directed towards the discovery of new reagents. This activity has at present reached the stage when not only are reagents containing new functional groups or new heterocycles prepared, but to a certain degree the properties of new reagents and their metal complexes are planned in advance [17–19].

As concerns reagents suitable for the determination of metal ions, there are two fundamental means of increasing their selectivities towards weak-acid anions which exhibit masking action. One is to shift the optimum interval of complex formation in the more acidic direction, while the other is to increase the stability of the complex in the original medium.

In order that the complex formation in the lower pH interval should not be accompanied by a decrease in the stability of the complex (i.e. the structure and character of the functional group should be retained), the acidic dissociation constant of the complex-forming group must be increased. This can be achieved by the introduction of electrophilic substituents (NO_2, Cl, etc.) in the *ortho* or *para* positions.

An increase of 1–2 in the value of pK or pK_{OH} (phenolic OH) decreases the optimum pH of complex formation to a similar extent, and increases the permissible concentration of masking anions by 2–3 orders of magnitude.

The most effective way to increase the stabilities of complexes, often causing a substantial fall in the optimum pH too, is to elevate the number of functional groups.

The higher selectivity of reagents towards cations depends primarily on the acid–base nature of the reagent, on the electron structure of the metal ion, and also on the bonding type and on the complex-forming atoms of the functional groups. In the case of oxygen-containing functional groups, the decrease of the pK_{acid} value is associated with a considerable decrease in the sensitivity of determination of the transition metal ions, whereas there is a slight increase in the case of metals preferably forming covalent bonds [20].

Another means of increasing the selectivity towards cations is to introduce various functional groups into the reagent. A condition for effectiveness is that one of the functional groups should form a coloured complex with the ion to be determined, while the other should bind the interfering ion in a colourless complex. The two donor atoms can naturally not be identical, and the steric arrangement of the two groups must exclude the bonding of some ion to both of them.

A complex-forming group with such a masking action is to be found in iminodiacetic acid. A whole series of such reagents was prepared by Basargin [19], particularly interesting ones among them being polyphenols condensed from the monomer with hexamethylenetetramine.

Another effect will arise if the reagent contains groups involving different donor atoms that form coloured complexes. As demonstrated by the examples of dithizone or oxine, in such cases the selectivity will be poor. It is worthwhile to induce this effect deliberately if it is desired to extract together as many ions as possible, for purposes of enrichment.

There are numerous possibilities for the improvement of the contrast and the sensitivity, primarily by the preparation of new heterocycles, by increasing the number of the various functional groups, by introducing different substituents and by the planned choice of the medium.

From the considerations touched on here and from others that have not been mentioned, a very large number of new reagents have been developed which have further extended the field of use of spectrophotometry. Methods based on the application of these will be presented in this book.

Quantum-chemical calculations too are widely employed for the "directed synthesis" of reagents [21]. However, the research workers who have attained the best results in this field caution against overdependence on their results. Nevertheless, the substituent-dependent variation of some parameter (sensitivity, selectivity) may be calculated well for a given reagent type. Otherwise, however, the basis is provided by systematic experimentation work, naturally setting out from theoretical considerations. Pilipenko [22] has published a good survey of the recently synthesized reagents that are used for analysis. The types, specificities, selectivities and sensitivities of organic photometric reagents were recently reviewed in detail by Sandell and Onishi [23], Korenman [24], Cheng et al. [25] and Burger [26].

In the light of the results outlined in this chapter, the further development of the application of spectrophotometry can be taken as certain.

References

[1] Ahrens, L. H., Taylor, S. R.: *Spectrochemical Analysis*. Pergamon, Oxford, 1961
[1a] Stromberg, A. G., Orient, J. M., Kameneva, T. M.: *Zh. Anal. Khim.* **37**, 2445 (1982)
[2] Cook, G., Crespi, M., Minczewski, J.: *Talanta* **10**, 917 (1963).
[2a] Alimarin, I. P.: *Zh. Anal. Khim.* **18**, 1412 (1963)
[3] Lontsykh, S. V., Berkovits, L. A.: *Zh. Anal. Khim.* **30**, 213 (1975).
[4] Taylor, S. R.: *Geochim. Cosmochim. Acta* **28**, 1273 (1964).
[5] Babko, A. K.: *Talanta* **15**, 271 (1968).
[6] Burger, K.: *Kémiai Közlemények* **36**, 197 (1971).
[7] Pilipenko, A. T., Tannanajko, M. M.: *Talanta* **21**, 501 (1974).
[8] Menkov, A. A.: *Zh. Anal. Khim.* **32**, 1409 (1977).
[9] Tikhonov, V. N.: *Zh. Anal. Khim.* **32**, 1435 (1977).
[10] Koh, T., Aoki, Y., Suzuki, Y.: *Anal. Chem.* **50**, 881 (1978).
[11] Mohai, M.: *Magy. Kém. Folyóirat* **81**, 164 (1975).
[12] Mohai, M., Inczédy, J.: Paper presented at the 12th Transdanubian Analytical Conference, Nagykanizsa, Hungary, 1976.
[13] Lomonosov, S. A., Kozireva, E. N.: *Zh. Anal. Khim.* **30**, 405 (1975).
[14] Pilipenko, A. T.: *Zh. Anal. Khim.* **31**, 220 (1976).
[15] Akimova, T. G.: *Zh. Anal. Khim.* **32**, 1269 (1977).
[16] Savvin, S. B.: *Zh. Anal. Khim.* **33**, 1813 (1978).

[17] Savvin, S. B.: *Zh. Anal. Khim.* **33,** 651 (1978).

[18] West, T. S.: *Analyst* **87,** 630 (1962).

[19] Basargin, N. N.: *Zh. Anal. Khim.* **23,** 1813 (1968).

[20] Martell, A. E., Calvin, M.: *Chemistry of Metal Chelate Compounds.* Prentice Hall, New York 1956.

[21] Gribov, L. A., Savvin, S. B.: *Zh. Anal. Khim.* **33,** 586 (1978).

[22] Pilipenko, A. T.: *Zav. Lab.* **46,** 979 (1980).

[23] Sandell, E. B., Onishi, H.: *Photometric Determination of Traces of Metals.* Wiley, New York 1978.

[24] Korenman, J. M.: *Organicheskie reagenty v neorganicheskom analyze,* Khimiya, Moscow 1980.

[25] Cheng, K. L., Keihei Ueno, Toshiaki Imamura: *Handbook of Organic Analytical Reagents.* CRC Press, Inc., Boca Raton, Florida 1982.

[26] Burger, K.: *Szerves reagensek a fémanalízisben (Organic reagents in metalanalysis).* Műszaki Könyvkiadó, Budapest 1969.

2. Planning, elaboration and control of analytical methods

For the solution of some analytical problem, a choice may generally be made from among several possibilities. In a fortunate case, a method that appears to have proved a successful one may be taken over from another laboratory or from the literature. It frequently happens, however, that no ready procedure is to be found in the literature, or those that are available do not seem satisfactory for some reason.

The planning and elaboration of a new method involve the following main steps: clarification of the demands to be made of the method; choice of an appropriate reagent, i.e. of a coloured compound considered to be suitable as the basis of the determination; planning of the principle of the method; elaboration of the method; investigation of the suitability and accuracy of the method.

2.1. Clarification of the demands to be made of the method

The more essential factors are as follows:

Accuracy: Apart from the means of evaluation, this governs the make-up of the entire method.

Sensitivity: This influences primarily the choice of the reagent and any necessity for concentration. The sensitivity requirements are fixed by the demands in the geological research.

Speed (time required for analysis): This is of importance mainly in the

12

control of technological processes. The speed of the method can often be increased only at the expense of the accuracy.

Simplicity: This depends in part on the level of training of the work force available, and in part on the location of the analysis (the terrain in geological research, the site of production in works-control studies, etc.); in all cases, however, a striving for simplicity will be of advantage in the procedure.

Analysis costs: This is an important question in analyses performed in large series. The costs are primarily influenced by the productivity and by the investment demands.

Investment demands: These depend largely on the type of the measuring instrumentation and on the space requirements (size of the laboratory). Not only the aim, but also the possibilities must frequently be considered before a decision is reached. The older and the most recent instruments have been subjected to critical comparison by Majer and Azzonz [1].

Time to be devoted to elaboration of the method, and dead-line for its introduction: In many cases this forces the choice of the optimum method into the background.

Possibility of further development (increase of accuracy, possibility of automation of the method).

Other requirements: Very varied stipulations are possible in some cases. These govern the structure of the whole method. For example, it is not advisable to use an expensive spectrophotometer in a very corrosive atmosphere, or with reagents creating such an atmosphere; besides the choice of the site and the material of the apparatus, filtration of the air is also necessary in the examination of high-purity substances; because of the danger of fire, it is not permitted to use inflammable organic solvents in certain places.

When the above demands arise simultaneously, it is often difficult to establish the optimum situation. From model investigations Kuznetsov and Kuznetsova [2] conclude that the information theory leads to a misleading result when methods are compared with regard to effectiveness.

When questions of economy are considered, it is important to take into account what other purposes the instrumentation or the entire laboratory may serve. It is absolutely necessary to reckon with the fact that the good quality of the analytical method may result in significant economic advantages. Without a method of suitable accuracy, for instance, it is impossible to assess technological modifications aimed at improving the yield; wastage can be avoided by keeping the microalloys within prescribed limits in special alloys; etc. However, it is usually not possible to define what benefits will accrue from a good method of analysis, nor what benefits will be lost because of an unsatisfactory method.

A good example of the correct approach to this question is to be found in the publication of Scharnbeck [3]. The analytical procedure is treated as an integrated part of the plant operations; for optimization of the costs, attention is paid to the costs of the consequences of analytical error, and also to the decrease in cost attainable through a low analytical error.

When everything is considered, it may be stated that thorough attention to the aims and to the demands made of the method is an indispensable condition of the planning and elaboration of a suitable method.

Ostroumov [3a] has published a survey on the optimization of the analysis of mineral substances.

The limits of application of spectrophotometric methods have been treated in detail by Ackermann [4]. The continuously increasing number of organic ligands employed is an indication that this research area can by no means be regarded as a completed one.

2.2. Choice of the reagent, or the coloured compound on which the determination is based

2.2.1. CHARACTERIZATION OF PHOTOMETRIC REAGENTS

For the assessment of the value primarily of organic reagents, various classifying systems have become widespread during the past decade; these are also suitable for making comparisons [5–7]. It will be seen that this can be performed for all photometric determinations; at worst, not every datum can be given in each case. A very good basis for comparison is the concept of the molar absorbance (ε), which is the absorbance of one mole of the metal complex, measured in a layer 1 cm in thickness:

$$\varepsilon = \frac{A}{l \cdot c}$$

where l is the layer thickness in cm and c is the concentration in mole dm^{-3}.

Naturally, there is no uniformly accepted system in the literature for the comparison of the applicability of reagents. Accordingly, we make use of both source-works and our own practical experience to list the characteristics and requirements in the knowledge of which sufficient information is available.

The wavelength relating to the maximum absorbance of the reagent ($\lambda_{\max\,HL}$). Even more useful is a knowledge of the absorption spectrum.

The wavelength relating to the maximum absorbance of the complex ($\lambda_{\max\,ML_n}$). In the event of stepwise complex formation, the reagent concentration interval employed must also be given.

Contrast, i.e. the difference between $\lambda_{\max\,ML_n}$ and $\lambda_{\max\,HL}$ ($\Delta\lambda$). A large value of $\Delta\lambda$ (>60 nm) is a very valuable reagent property. It increases the accuracy even in objective measurements, while it is particularly important in the visual colorimetric method. It is also critical if the reagent excess does not pass over into the organic phase in the extraction–photometric procedure, e.g. Ga-rhodamine B. For colourless reagents in the visible range (CNS^-, etc.), in a favourable case the value of $\Delta\lambda$ may be even 200–300 nm.

The optimum measurement wavelength (λ_{opt}). The value relating to the maximum of $\Delta A/\Delta\varepsilon$ can be determined from the absorption spectra of the reagent and the complex. In a favourable case:

$$\lambda_{opt} = \lambda_{\max\,ML_n}$$

pH-dependence of light absorption ($\Delta A/\Delta pH$). Some authors take into consideration the absorbance change corresponding to a pH change of 0.5. Better than this is $\Delta A/A\Delta pH$, since in this way we obtain the pH-dependence of the relative absorbance change. From the customary simple graphical plotting (pH vs. A), however, it can be seen that within the optimum pH range of complex formation there is a "plateau" where the absorbance varies within the acceptable limit of error.

The optimum pH range of determination. To a first approximation, this in effect ensures the maximum relative concentration of the complex ion to be determined, and the maximum absorbance relating to this. In extraction–photometric systems the optimum value refers to the initial (possibly equilibrium) pH of the aqueous phase.

The $\Delta A/\Delta pH$ values of different reagents cannot be compared directly. It is clearly not the same requirement to maintain the pH of the solution

15

between 6.0 and 6.5, and between 1.0 and 1.5. It is therefore better to state that the value of the reagent from this aspect increases with the difference $(pH_{opt} - 7)$. It is more advantageous if this is a negative number, i.e. if the determination can be carried out in acidic medium. Besides the greater ease in establishing the pH, it is then simpler to keep the other ions present in a dissolved state.

In the choice of the optimum pH, it is necessary to take into account the interfering components and also the influence of complex formers added to mask these.

Composition of the complex. Instead of the exact composition, it is generally sufficient to know the metal ion : ligand ratio (M : L). In an indirect determination this implies a knowledge of the stoichiometric equation of the process leading to formation of the coloured complex, or the composition of the reaction product. In more complicated cases the latter is possibly unknown.

In the event of stepwise complex formation, it is most useful to plot the individual ionic species as a function of the reagent concentration. This is illustrated in Fig. 2.1 [8], from which it can be seen that at a free NH_3 concentration of 0.1 mole dm^{-3} practically only the tetrammine-copper(II) complex is present in the solution.

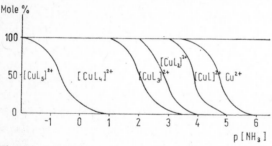

Fig. 2.1. Distribution of copper(II)-ammine complexes as a function of the NH_3 concentration.

The composition of the complex forming the basis of the determination is to be found in the majority of publications describing the properties of reagents, but it is sometimes not given, or in the case of a new reagent it may possibly not be known.

If only one complex species exists in the interval of the reagent concentration, then even in the absence of literature data it is relatively easy to determine the composition of this complex with the Job method or with an auxiliary complex former, e.g. a masking or an extracting agent [9–11]. In a more complicated case (stepwise complex formation or a polynuclear complex), from the complex stability constants alone it is possible to calculate the mole fractions of the individual complexes at some ligand concentration.

Detailed information on the properties of complexes and on the methods of their examination may be obtained from the books by Beck [10], Burger [11a, 11b], Kőrös [11c] and Császár and Bán [11d].

Stability constant of the complex serving as the basis of the determination. For purposes of analytical practice it is enough to know the conditional stability constants. If even these are unknown, the relative complex stability can be determined simply with the method proposed by Kuznetsov [12]. The essence of this is that the corresponding coloured complexes are produced with solutions containing identical molar concentrations of the reagents to be compared. Solutions exhibiting 50% of the initial light absorbance are prepared by the addition of an appropriately chosen auxiliary complex former that behaves as a masking agent. The relative stabilities of the complexes will be proportional to the auxiliary complex-former concentrations necessary for this. This procedure has been used, for instance, to establish that the "stability index" of the UO_2^{2+}-arsenazo III complex is two orders of magnitude higher than that of the UO_2^{2+}-arsenazo II complex [13].

This "stability index" can be of great assistance in the development of a method of trace analysis, even if the composition of the complex is not known. In the event of the existence of a single complex, this too can be determined with the slope method, by interchange of the colour reagent and the auxiliary complex former [14]. If the stability constant of one of the complexes is known, this procedure may be employed in effect to obtain approximate conditional constants.

Constants describing the stepwise dissociation (or protonation) of the reagent. These can be used to calculate the actual concentration of the anion, or the pH necessary for complete dissociation. In many cases these data too may be found from the appropriate diagram (see the section *Composition of the complex* above) or are available in handbooks in the form of α-functions [9].

Difference ($\Delta\varepsilon$) or quotient ($\varepsilon_{ML_n}/\varepsilon_{HL}$) of the molar absorbances of the complex and the reagent. Both the absorbance difference and the quotient are used in the literature, but in themselves neither is sufficient. The actual conditions are controlled not only by the two spectra, but also by the stability of the complex. It is obvious that the reagent concentration necessary for total complexing of the metal ion ($[ML_n] : [M] \gtrsim 10^2$) is inversely proportional to the stability of the complex:

$$[L] \gtrsim \left(\frac{10^2}{\beta ML_n}\right)^{\frac{1}{n}}$$

For practical purposes, therefore, some expression is to be recommended that also takes into account the reagent concentration necessary for the determination. For complexes of low stability, in the case of a small $\Delta\lambda$ value the applicability of the reagent is spoiled to a great extent by a high light absorbance.

Range of validity of Beer's law. Although a photometric (colorimetric) determination can be performed even in the case of deviation from linearity of the dc vs. dA plot, linearity is naturally a condition of greater accuracy. From a practical aspect it is advantageous if this range is fairly wide, for merely by variation of the layer thickness it is then possible to determine samples of different concentrations without dilution or repeated addition of sample.

Temperature-dependence of absorbance. This is of importance in serial analyses; the work is made more difficult by the need for thermostating. In the majority of cases this is not necessary, but it must be checked experimentally. In an unfavourable case, a temperature difference of 5°C may cause an absorbance difference of even several %.

Stability of the colour. The time required for development of the maximum colour intensity and the stability of the resulting colour are additional data that are needed to assess a reagent. The colour is frequently not constant, as a consequence of undesired side-processes (hydrolysis, photochemical reduction, etc.). The rates of the processes are often increased by contaminants; attention must also be paid to these phenomena in establishing disturbing effects.

Lifetime of the reagent. In many cases the solution of the reagent is not sufficiently stable. Thus, it must be prepared freshly or used only for a limited time, and it must be stored under conditions inhibiting decomposition

18

(exclusion of light, application of low temperature). The decomposition may be catalyzed by the impurities in the reagents. Accordingly, the analytical purity of the reagents is important.

Extractabilities of the reagent and the complex. Extraction may have various aims. The most important of these are:

(a) separation from interfering components;
(b) increase of stability of the complex;
(c) removal of excess reagent.

The latter is of great importance if the reagent displays appreciable absorbance at the wavelength of the measurement, e.g. dithizonates. Its importance is fundamental, however, in cases when the absorption spectra of the complex and the reagent are identical; total separation is then a strict requirement. This is the situation with dyestuffs that are triphenylmethane derivatives, e.g. the determination of Ga^{3+} with rhodamine B, or the determination of BO_3^{3-} with crystal violet.

It is therefore necessary to know the distribution of the complex and the reagent in the extraction and photometric procedures depends primarily on the pH and the applied diluent, and possibly on the ionic strength and other factors.

Selectivity. Those ions (compounds) interfering with the use of the reagent must also be listed.

There are various disturbing effects:

(a) increase of the absorbance (due to the colour of the interfering ion itself or to its complex formation);
(b) decrease of the absorbance (due to masking, or possibly to the formation of a polynuclear complex);
(c) decrease of the concentration of the reagent (due to its decomposition).

In establishing interfering effects, it is frequently necessary to take complicated equilibria into consideration. This question will be returned to during the treatment of the methods. Various concentrations are suitable to express the extent of the disturbing effect:

(a) the concentration which is too low to cause any interference at all;

(b) the concentration giving an absorbance, or causing a decrease, equivalent to a certain quantity of the component to be determined.

Not all disturbing effects can be expressed in such units; a list must be given of phenomena such as the breakdown of the reagent, the decomposition of the complex, optical impurity, or any other disturbing factor which does not permit application of the determination.

2.2.2. CLASSIFICATION OF PHOTOMETRIC METHODS ACCORDING TO THE NATURE OF THE ABSORBING COMPONENT

On the basis of the connection between the absorbing ion (molecule) and the component it is desired to determine, the photometric procedures can be classified into direct and indirect methods.

DIRECT DETERMINATIONS

(a) Hydrated (solvated) ions. As a consequence of the low molar absorbances, these are suitable for trace analysis in only a few cases (MnO_4^-, CrO_4^{2-}, Nd^{3+}, Pr^{3+}, etc.).

(b) Complex ions of molecules. These form the basis of the great majority of photometric methods of trace analysis. Numerous forms of subclassification are possible, e.g. the following:

Complexes formed with a colourless ligand. These include most of the inorganic ligands (H_2O_2, CNS^-, NH_3, etc.) and certain of the organic ligands, such as the oxyacids (salicylic acid), oximes (dimethylglyoxime) and phenols. Such complexes are formed by central ions with chromophoric properties (transition metal ions, etc.).

Mixed complexes containing different ligands. The majority of the extracted complexes belong here, e.g. iron(III)-thiocyanate-ethyl ether, but many examples are known in aqueous solution too. The second ligand does not usually have a fundamental effect on the course of the absorption spectrum; its role is primarily to stabilize the coloured complex and to increase the selectivity, for instance by ensuring the extractability [15, 16].

Heteropolyacids. These are very extensively utilized for the determination of some elements (Si: silicomolybdic acid; V: vanadophosphotungstic acid; etc.).

Metal colloids. These are suitable for the determination of some readily reducible elements, if a colloid solution of satisfactory stability and with a characteristic absorption spectrum can be obtained by regulation of the conditions of reduction, e.g. selenium, tellurium. It is customary to use a "sensitizer"; this is another metal ion, e.g. Cu^{2+}, which, besides increasing the sensitivity, regulates the dispersion of the colloid too.

Lakes. These are formed between hydrolyzed polymers of metal ions and coloured complex formers (Al^{3+}-aluminon, Mg^{2+}-titan yellow, etc.). They are non-stoichiometric in composition, they generally form colloids, and a protective colloid is needed to keep them in solution. They do not usually obey Beer's law, and thus these methods are of lower accuracy.

INDIRECT DETERMINATIONS

In contrast with the direct methods, the basis of classification is not the nature of the absorbing compound, but its connection with the ion to be determined.

Decolorization of coloured compounds by means of ligand exchange. This is of use when the ion to be determined does not yield a coloured compound that can be used to advantage, or, even if it does do so, the indirect method is simple. One such procedure is when an anion is determined via the decrease in absorbance of a coloured compound, e.g. determination of F^- with iron(III) thiocyanate, determination of Cl^- with mercury(II) diphenylcarbazone.

Determination of the reagent excess or some ion of it. The classical case among the indirect methods is when the reagent excess is determined. A condition is that the determination should not influence the equilibria, i.e. the ratio of the stability constants should be sufficiently high. An example of this is the determination of a small amount of SO_4^{2-} with Ba^{2+}, where the excess Ba^{2+} can be determined with Orthanil K [17, 17a] or chlorophosphonazo III [17b].

A frequent case is when some element from the reagent is determined in the complex that also contains the ion to be determined. The determination of Si or P by means of molybdenum blue is well known. An analogous possibility, for example, is the solution of Halász *et al.* [18], who in the case of Si(P, Ge, As)-molybdenum-heteropolyacid determine the molybdenum with phenylfluorone. In this way the "analytical molar absorbance" can be increased to several times the original value, possibly by even an order of magnitude, and exceeds the value attainable in the direct methods ($\varepsilon > 2 \times 10^5$).

Catalytic (kinetic) methods. These are based on the concentration-dependent effect of the ion to be determined on the formation or decolorization of some coloured compound. They likewise permit procedures of high sensitivity. As a consequence of the course of the reaction in time, and also the possibility of the presence of other ions that similarly exert a catalytic influence on the process, the practical application involves primarily systems containing H^+ and alkali metal ions (acids, bases, alkali metal salts).

Synthesis of coloured compounds. The Griess–Ilosvay determination of NO_2^- is a well-known method. These procedures are also widely used in the analysis of NH_3 or S^{2-} (determination of NH_3 in the form of indophenol in the presence of phenol and ClO^-; determination of S^{2-} with dimethyl-*p*-phenylenediamine by measurement of the methylene blue formed in the presence of Fe^{3+}).

2.3. Planning and elaboration of photometric methods

2.3.1. INTRODUCTION

We must next examine what parts an analytical method is comprised of, how the individual steps fit together, and what data and information a publication describing a new method or an analytical procedure must contain in an ideal case.

In the sequence of the operations, the elaboration of an analytical method consists of the following processes:

(a) sampling, storage;

(b) sample processing; preliminary treatment of the sample (pulverization, cutting);

(c) decomposition (digestion) of sample, and dissolution of the component to be determined;

(d) separation from interfering compounds; concentration;

(e) establishment of optimum conditions of photometric determination;

(f) compilation of analytical procedure; determination of sensitivity of the procedure;

(g) investigation of accuracy of the method;

22

(h) establishment of other parameters of the method (efficiency, overheads, time requirements, etc.).

However, the sequence of the research work differs from the mechanical sequence of the operations. The sampling and the processing of the sample may be treated separately from this process. The manner of decomposition (digestion) is governed in part by the nature of the sample under examination and by the analytical properties of the component to be determined, but the requirements provided by the further course of the analysis cannot be neglected either. In the analysis of rocks, it must be ensured not only that the component to be determined is dissolved and that the medium is suitable for the further course of the analysis, but also, if possible, that this step should at the same time mean a separation. For instance, alkaline fusion has the advantage that, on aqueous leaching of the melt, VO_3^-, WO_4^{2-}, MoO_4^{2-}, etc. remain in solution, if the alkali concentration is above 1 M. If these ions are being determined, it is no longer necessary to reckon with interfering substances remaining in the hydroxide precipitate; in contrast, these ions can be removed in this way prior to the determination of niobium with thiocyanate, for example.

The central portion of the procedures, however, is the establishment of the optimum conditions for photometric determination and the separation from the disturbing components.

In the elaboration of an analytical method, the properties of the reagent considered to be the best on the basis of literature reports must be checked experimentally, and supplementary studies must be performed with regard to the demands made of the method.

Hence, the most practical sequence in the planning and experimental elaboration of a new analytical method is as follows:

(a) establishment of the attainable sensitivity;

(b) establishment of the optimum conditions for the photometric determination;

(c) establishment of the permitted quantities of interfering substances in the final solution; determination of the necessary separation factors; experimental elaboration of the separations (maskings, etc.);

(d) elaboration of the analytical procedure.

The applied method of decomposition (digestion) has no definite position in this sequence; it is to be fitted into the appropriate section of the experiments in the most logical way.

Questions relating to the sampling and processing, dissolution and separation will be dealt with in later sections. Here we wish primarily to touch only on general questions of the separation, fitting this into the general sequence of thought.

The course of elaboration of the methods is not so mechanical as described in the above points. Even to acquire the information listed here may require preliminary experimentation which would occur only later in the strict sequence.

2.3.2. ESTABLISHMENT OF THE SENSITIVITY ATTAINABLE WITH THE CHOSEN REAGENT

Correlations of use to express the sensitivities of photometric reagents have been mentioned above. These provide a possibility for comparisons, but in the solution of a concrete analytical task it is also necessary to take into account what quantity of sample (what proportion of the solution) can be subjected to photometry in practice.

To decide this to a first approximation, there is no need for an exact knowledge of the optimum conditions for the determination. In addition to the fundamental optical data (ε), however, it is necessary to know the nature of the samples to be investigated, and on this basis the following data:

Quantity of sample available. In certain cases, such as in the examination of flue-dust or biological materials, this may be strongly limited.

Upper limit of sample taken. Even if sufficient sample is available, the upper limit of the sample taken may be limited by the technical possibilities or by the further course of the analysis, above all by the resulting greater amount of interfering substances. In the case of rock samples, for instance, a sample larger than 2–3 g is not generally to be recommended; for plant samples, the ashing or wet destruction of an unlimited amount of sample cannot be solved either; while it is not advisable to take more than 500–1000 cm^3 of water samples. The sample size is often restricted by the high value of the material.

Solution sample to be taken for determination. The quantity of this is restricted by the volume of the reagents. Accordingly, it is advisable to strive to use the most concentrated solutions possible, and to prepare combined reagent solutions. For example, the masking or buffer solution is added together with the photometric reagent.

Final solution volume. Here, naturally, the lowest value must be aimed for. The most important restrictions:

(a) salt precipitation (in the case of evaporation), or a high ion concentration disturbing the determination (in the case of complexes of low stability);

(b) the minimum quantity of solution necessary for the photometric measurement;

(c) the danger of contamination or loss in the concentration operations.

From these data a rough estimate can be made as to whether the selected reagent will satisfy the sensitivity requirements.

As concerns the attainable sensitivity, we take into account measurements on a layer thickness of 5 cm. It is rather difficult to understand the practice referred to in the great majority of methods described in the literature, where a layer thickness of 1 cm is used, or the fact that certain instruments are manufactured without a holder for 5 cm cells. In such cases an appropriate home-made holder is to be recommended (for narrow cells too).

A number of authors publish formulae for calculation of the sensitivity. Adamovich [19] gives

$$\frac{\text{Sample weight [g]}}{\text{Final volume [cm}^3\text{]}} = \frac{A. W._M \times n_M \times A_{min}}{10 \times M\% \text{ in sample} \times \varepsilon_\lambda \times l}$$

where A. W. is atomic weight, n_M is the number of central atoms in the complex; A_{min} is the lower limit of reliable measurement $=0.05$; and l is the layer thickness, in cm.

The advantage of this type of formula is that the sensitivity is expressed as the quotient of the sample weight and the final volume. If the lower limit of reliable measurement is accepted as five times the standard deviation (s_A) [20], the smallest amount that can be determined is

$$M_{min} = m \, (\mu g) = \frac{A. W._M \times n_M \times \text{final volume} \times 5 \times s_A \times 10^3}{\varepsilon_\lambda \times l}$$

For a complex not of high stability $(\beta'_{ML_n}[C_L]^n < 10^2)$, the concentration of the reagent (C_L) and the complex product must also be taken into account:

$$m\,(\mu g)=\frac{5\times10^3\times s_A\times A.\ W._M\times\text{final volume}\times(1+\beta'_{ML_n}[C_L]^n)}{\varepsilon_\lambda l\beta'_{ML_n}[C_L]^n}$$

If these data are known, the applicability of the conceived solution can be decided even without experiment.

2.3.3. ESTABLISHMENT OF OPTIMUM CONDITIONS FOR PHOTOMETRIC DETERMINATION

The sequence of carrying out the experiments cannot be given unambiguously here either. This depends on the degree of detail of the available data, on other analytical experience, and on the complexity of the problem.

DETERMINATION OF THE OPTIMUM pH

This is of great importance particularly in the case of organic chelating agents.

The H^+ and OH^- concentrations act strongly on all equilibria, such as:

(a) equilibrium of formation of hydroxo complex of the metal ion to be determined, or protonation of the anion to be determined;

(b) equilibrium of dissociation (protonation) of the reagent;

(c) equilibrium of formation of hydroxo complex of a disturbing metal ion, or protonation of a disturbing anion;

(d) dissociation of the masking agent added to eliminate the disturbances; equilibria of complex formation with disturbing ions; equilibrium of complex formation with the ion to be determined.

As a resultant of all these effects, changes may occur in the absorbance of the reagent or of the complex to be determined, or in the optical purity of the solution (formation of hydroxide precipitate, etc.).

In extraction–photometric procedures the partition coefficient depends on the pH, while in indirect determinations the fundamental equilibria, the reaction rate, etc. do so. If the optimum pH is examined merely from the aspect of the sensitivity, it is governed in the case of a chelate complex by two opposing requirements. A pH increase exerts an unfavourable effect, diminishing the metal ion concentration as a result of hydrolysis, whereas it also has a favourable effect, for it increases the free anion concentration of the reagent. In the simplest case, the acidity for maximum "yield" of the coloured compound is

$$[H^+]_{opt} = \sqrt{\eta_1 \sigma_1^{-1}}$$

where η_1 is the stepwise dissociation constant of the aqua complex of the metal ion, while σ_1 is the stepwise protonation constant of the reagent.

This pH value can be calculated in the knowledge of the two constants.

As regards the analytical method, the optimum pH is naturally not decided by this value alone.

It is difficult to give advice of general validity as to the sequence in which the above effects should be studied. Naturally, the first step is to establish the pH-dependence of the absorbance of the complex serving as the basis of the determination; the possible interferences should then be considered within the interval suitable for the determination. If the reagent absorbs at the given wavelength, in the examination of the pH vs. A dependence it should not be forgotten that the pH of the reference must be varied in the same way.

In certain methods the acidity (pH) of the solution can be varied within wide limits without any change in the absorbance value. In the phosphotungstic acid determination of vanadium, for example, identical values are obtained in 3 M H_2SO_4 and in 3 M HNO_3 solutions. It is much more frequent, however, that the constancy of the absorbance holds only in a narrow pH interval. If this lies in the range above pH 3.0, use of a buffer solution to keep the desired pH value constant is unavoidable.

In the choice of the buffer, attention must be paid to the appropriate buffer capacity; further, the anion of the buffer solution should not be a ligand forming a complex with the ion to be determined, and it should not yield a precipitate. For this reason, it is frequently not possible to use phosphate buffer, for instance, or even oxyacids (citrate, acetate). Surprising, undesired masking effects sometimes appear. As an example, acetate buffer cannot be used in the determination of calcium with glyoxal-bis(2-hydroxyanil).

In our experience, the buffers recommended for media more acidic than pH 2.0 are of very little practical value. In this range it is more advisable to neutralize the solution approximately, and then to add the calculated quantity of acid.

Abundant data as to the compositions of buffer mixtures necessary to establish various desired pH values are tabulated in manuals [21, 22], but tests must be made to ascertain whether the capacity is sufficient, and whether masking effects or precipitation occur. Attention should also be paid here to the rates of the processes, i.e. to the total time elapsing up to the photometric measurement. One reason for this is that in some cases the development of the colour is slowed down by the buffer solution, even though the latter has no influence on the colour intensity in the equilibrium state. In other cases, however, immediate measurement may be required if the masking effect has not yet been manifested, because of its lower rate.

ESTABLISHMENT OF THE OPTIMUM REAGENT CONCENTRATION

It is desirable to employ the reagent concentration necessary for complete formation ($>99.9\%$) of the coloured complex. With complexes of low stability, this may be foiled by natural restrictions (solubility, or an unattainably high concentration). Nor can an attempt be made to reach this situation if the reagent absorbs at the wavelength of measurement. It may also occur that, with increase of the reagent concentration, above a certain limit, the effects of the disturbing substances which form complexes of lower stability, but which are present in higher concentration, increase more rapidly than the sensitivity of the determination.

It follows from all this that the optimum reagent concentration calculated from the above condition ($K_{ML_n}[L] \gtrless 10^3$) is an upper limit, which must be approached as far as possible without deviating too much from the optimum controlled by the opposing demands due to the sensitivity and the disturbing effects. In the case of complexes of very high stability (Th^{4+} or ZrO^{2+}-arsenazo III), virtually the stoichiometric quantity of reagent is sufficient.

At times the reagent may "mask" other metal ions present in the solution. In determination of the optimum reagent concentration, this too must be checked experimentally.

In certain extraction–photometric procedures, e.g. when dithizone is applied, it is beneficial to add the reagent in small portions, in an effort to extract only the metal complex (the reagent is extracted only subsequently).

It is otherwise common in extraction–photometric procedures that the extraction of the metal ion is not quantitative. Since the calibration plot is obtained in exactly the same way, this does not give rise to error.

In some cases, the macromolecule on which the photometric determination is based is a polymeric compound that has not been exactly defined (Mg-titan yellow, Al-aluminon, etc.). In these situations a protective colloid is generally required. This ensures that a precipitate is not formed during the determination, and that the degree of dispersion of the coloured lake does not change essentially. The most frequently used protective colloids are gelatin, gum arabic and artificial polymers, e.g. polyvinyl derivatives. They are selected experimentally.

SEQUENCE AND TIME OF ADDITION OF REAGENTS

The sequence is normally the addition of the reducing agent or the reagent needed to mask the disturbing substances, followed by the adjustment of the pH, and then the development of the colour. The sequence of the last two steps is sometimes reversed. The most suitable procedure must be ascertained experimentally.

Tests must also be carried out to establish the length of the waiting period after the addition of the various reagents; these tests will demonstrate the time required for development of the colour, or in other cases the need for immediate measurement.

RECORDING OF THE CALIBRATION PLOT;
CHOICE OF THE REFERENCE AND THE MODE OF MEASUREMENT

Various procedures are possible as concerns the reference:

1. Pure solvent (distilled water). This can be used if the reagent does not absorb at the wavelength of measurement, or (e.g. in extraction–photometric methods) it is quite certain that the reagent does not pass into the organic phase.

2. A solution containing the reagent(s). If the reagent does absorb at the wavelength of measurement, but otherwise only the chromophore to be determined does so, this technique must be chosen. Apart from the colour-forming reagent, the other reagents may possibly be neglected.

If the absorbance of the reagent depends to a considerable extent on its concentration or on the pH of the solution, it is advisable to determine the correct value by means of many measurements, but to use pure solvent for the measurement. Naturally, the absorbance corresponding to the reagent solution is subtracted from the measured value.

3. The organic solvent taken through the course of the analysis (without reagent). This procedure is to be used in extraction–photometric methods if the sample contains impurities which are indifferent to the reagent, but which pass into the organic phase, where they absorb.

4. A solution of the sample without reagent is used if the original "basic colour", or that remaining even after addition of the masking agent, does not vary on addition of the reagent. This is a frequently applied technique in industrial or biological analyses.

In fact, one case of this (often used in metallurgy, for example) is when, in the recording of the calibration plot, the absorbing main components of known concentrations are added to the reference and to every member of the series.

5. The examined solution after decolorization of the complex to be determined. This may be applied if disturbing substances are present which yield absorbance with the reagent, these complexes not decomposing on masking of the ion to be determined. The method has the advantage that it also corrects for the "basic colour". This is the solution in the determination of the lanthanides with arsenazo III, for instance, to eliminate the disturbing effect of the small amount of calcium remaining after the separation (hexametaphosphate is used as complex former). A technical simplification is to use the customary reference and to perform the measurement in this way after measurement of the absorbance of the complex and decolorization.

6. In indirect determinations based on decolorization, the reference is the coloured complex itself, e.g. mercury–diphenylcarbazone for the determination of chloride. Naturally, in the measurement the two cells must be transposed compared to the usual positions.

7. A solution of appropriate concentration of the compound to be mea-

sured can be used in differential spectrophotometry. This is a relatively rare case in trace analysis, but if Beer's law holds within wide limits it may be applied with a view to increasing the accuracy.

2.4. Types of disturbing effects and possibilities for their elimination

2.4.1. CLASSIFICATION OF DISTURBING EFFECTS

One type of disturbing effect is optical impurity under the conditions of the determination, e.g. turbidity, precipitate formation. The cause of this may be the formation of a precipitate by interaction between the cations and anions (OH^-, HPO_4^{2-}, etc.) present; separating out due to the poor solubility of the reagent or the metal complexes; or the development of an emulsion from the organic and aqueous phases in contact.

Optical inhomogeneity and undesired absorbance is similarly caused by the evolution of gas, e.g. O_2 originating from the decomposition of H_2O_2.

Undesired redox processes may likewise lead to disturbing effects: decolorization of the reagent (reduction of arsenazo III and other azo dyes by Fe^{2+} in the presence of a complex former of Fe^{3+}), or in contrast the formation of coloured oxidation products (isothiopercyanic acid from thiocyanate).

"Masking" of the reagent by binding into a colourless complex, e.g. formation of thoron complex with Al^{3+}. This decreases the concentration of the reagent and may cause a negative error, the extent of this depending on the concentration and stability conditions. In an unfavourable case, a polynuclear complex containing the ion to be determined is formed.

Masking of the ion to be determined by a complex-forming ligand. This causes a negative error.

A disturbing effect is also produced by a foreign ion (molecule) that absorbs at the wavelength of the determination.

The main cases:

(a) a species indifferent to the reagent, which is added with the sample to be examined or with the reagents;
(b) a disturbing ion that forms a coloured complex with the reagent;
(c) a polynuclear complex which contains the disturbing ion in addition to the ion to be determined.

2.4.2. MEANS OF ELIMINATION OF THE DISTURBING EFFECT

If the disturbing effect is additive (in both directions!), up to a certain limit it can be compensated for on measurement. This is the situation, for instance, in the case of an indifferent absorbing disturbing substance, as discussed above.

In the event of a negative error, i.e. the partial masking of the ion to be determined, an internal standard method may be employed. The result is then not calculated from the calibration plot, but by proportion from the two absorbance values. Conditions of application are the validity of Beer's law, and a not too extensive degree of masking (at most 50%).

If the disturbing effect is a considerable one, is non-additive, is non-reproducible, or is manifested in the form of optical impurity, another means of solution must be sought. The most important possibilities are as follows:

In the case of optical impurity, if this is not too extensive, the following possibilities arise:

(a) if the turbidity is caused by dispersion of the aqueous phase in the organic phase, the solution can be diluted with pure solvent; alternatively, it may be possible to use an additive making the two phases miscible, or to add drying agents; an example of the latter is the addition of a little ethanol after extraction of the UO_2^{2+}-arsenazo III-diphenylguanidine complex with butanol, or the addition of Na_2SO_4 (anhydrous) in chloroform solutions of oxine;

(b) in certain cases, filtration too is successful, but it must be checked that this does not diminish the absorbance of the solution itself;

(c) the disturbing effect may also be avoided by making measurements at two wavelengths, if there is a spectral interval in which only the species causing the interference absorbs.

If the disturbance is caused by *ions forming complexes with the reagent,* between certain concentration limits and in the event of a sufficiently high ratio of the complex stability constants, it is of help to modify the reagent concentration. If the disturbing species gives a colourless complex, it may be beneficial to increase the reagent concentration. In the case of a coloured complex, on the other hand, formation of the disturbing complex may be suppressed by decreasing the reagent concentration (at the expense of the sensitivity of the determination). This solution is used to reduce the effect of titanium in the extraction–thiocyanate determination of niobium, for instance.

Primarily with chelate-forming organic reagents, one of the means of regulating the actual reagent concentration is not via the addition of the reagent, but via the variation of the pH of the solution.

If the disturbing effect is a considerable one, the possible solutions listed so far are not successful. The following possibilities may then be used:

Addition of a buffer cation. If a cation can be found which reacts with the reagent to form a complex that does not absorb at the wavelength of the determination, and if the stability of this complex is intermediate between the stabilities of the complexes of the cation to be determined and the interfering cation, i.e.

$$K_{ML} > K_{ML_{buffer}} > K_{ML_{interfering}}$$

then the interfering cation can be expelled from the complex. This solution is relatively widespread in certain extraction–photometric methods, e.g. determination of Cu^{2+} with $Pb(DDTC)_2$ instead of with sodium diethyl-dithiocarbamate, but its applicability is limited.

Binding of a masking anion (or one which interferes in some other form). This is the simplest, but at the same time a restricted possibility in a few cases. As examples from practice, mention may be made of the binding of fluoride with boric acid or with Al^{3+}, and the binding of nitrate with urea.

Masking of a disturbing cation. This is one of the most important and most general solutions. It is a precondition, of course, that the masking ligand should form a more stable complex with the interfering ion, and a less stable complex with the ion to be determined, than does the reagent itself. In the knowledge of the complex stabilities, it is possible to calculate the necessary concentration of the masking reagent and the maximum permissible amount of it which does not yet cause an essential weakening of the colour.

Compounds used for masking are primarily inorganic ligands forming colourless complexes (mono- and polyphosphates, fluoride, sulphide, etc.), carboxylic acids (oxalic acid), oxyacids and thioacids (tartaric acid, sulphosalicylic acid), OH-containing compounds (triethanolamine) and aminopolycarboxylic acids (complexones). On the basis of the complex-forming properties of the individual ions, it can be decided to a first approximation what types of compounds may come into consideration. For instance, for ions forming fairly stable complexes with OH (Zr^{4+}, Al^{3+}, etc.) oxyacids may primarily be considered, while for those forming more stable complexes with S (Zn^{2+}, Hg^{2+}, etc.) the choice may fall rather on sulphide or thio compounds.

Certain information as to which cations can be masked in the presence of a given cation is provided by the electronic structures, which govern the complex-forming properties.

Masking can be applied to particular advantage in extraction–photometric methods, where, if the ion to be determined has a sufficiently high partition coefficient, the organic phase can be washed with the solution containing the complex former.

In experiments aimed at the elaboration of the method, unpleasant surprises, too, may await the research worker. It may happen, for instance, that a mixed complex is formed, one of the ligands of which is in itself an "innocent" masking agent. Thus, the stability of the Ca^{2+}–arsenazo III complex, and hence its absorbance too, are increased by the addition of phosphate. It similarly occurs, fortunately not often, that a polynuclear complex containing the ion to be determined is formed with the masking ligand, although this is ineffective in the absence of the second cation. Good accounts of the masking effects and the possibilities of their elimination are given by Perrin [23], Skorokhod [24] and Lyalikov [25].

Variation of the valency of the disturbing ion. Although this can be applied only in the case of variable valency, it is nevertheless a method of great importance. As an example, the reduction of Fe^{3+}, an ion which interferes in very many determinations, permits the measurement of numerous components even in the presence of iron. Fe^{2+} yields stable complexes with only a small group of ligands, primarily heterocyclic compounds containing N—N bonds. The aim may likewise be the reduction of anions ($Cr_2O_7^{2-}$, MnO_4^-) which disturb the determination by causing oxidative decomposition. It is rarer for a disturbing ion in a lower oxidation state (e.g. Ti^{3+}) to have to be oxidized.

Metals (Zn), metal amalgams, metal ions with low redox potentials (Sn^{2+}, Ti^{3+}), inorganic anions (SO_3^{2-}, I^-) and other reducing agents (hydroxylamine, hydrazine, ascorbic acid) may be used for these reductions.

Sometimes, the reduction necessary for the determination is achieved with the colour-forming reagent itself, as in the determination of Mo(VI) with dithiol, the first step here being the reduction of Mo(VI) to Mo(V).

A set of suitable reagents can be selected on the basis of the standard redox potentials, but here too the most reasonable solution should be sought experimentally. In certain cases unexpected disturbances occur: if ascorbic acid is used for the reduction, the redox potential of the Fe^{2+}–Fe^{3+} system is lowered to such an extent that the azo dyes, e.g. arsenazo III, are decomposed via their reduction. If measurements are made at a wavelength lower than 400 nm, the use of ascorbic acid, which is otherwise favourable in many cases, is to be avoided, because of a reaction product which increases the absorbance.

At times, the further course of the determination demands the binding of the residual reducing or oxidizing agent. This occasionally makes the resulting equilibrium a labile one, and hence the determination becomes uncertain. This may be exemplified by the virtual impossibility of maintaining the oxidation state of Ce^{4+} in low concentration in acidic medium.

2.4.3. SEPARATION FROM DISTURBING SUBSTANCES

This is the ultimate method for the elimination of interfering effects but with samples with complex compositions it is generally unavoidable. It must often be applied together with one or other, or possibly several, of the solutions listed in the foregoing section.

During the elaboration of such separations, it is fundamental to determine the permissible amounts of the interfering substances, to calculate the necessary separation factors, and to study the loss of the ion to be determined.

Permissible amounts of interfering substances. These amounts may be investigated in several steps, and they may be expressed in various ways, the most useful being as follows:

(a) the amount of the interfering substance giving the same absorbance as a certain quantity (generally 1 μg) of the ion to be determined;

(b) the amount which causes a definite relative error (e.g. $\pm 10\%$) in the presence of a certain concentration of the ion to be determined;

(c) the amount resulting in an absorbance (error) 2–3 times the standard deviation;

(d) the amount not yet causing a demonstrable error.

In practice, all of these modes of expression can be used well. In trace-analysis practice, a value of $\pm 10\%$ is generally accepted for the permissible extent of the disturbing effect. In the planning of the method, however, the starting-point must be a stricter requirement than this.

It is far from satisfactory that in many publications the permissible amount is not referred to the quantity of the component to be determined, or that the extent of the interfering effect is not indicated quantitatively ("interferes in large amounts", etc.).

It is advantageous to give the permissible amounts of disturbing substances when the properties of the reagent are reported; it is to be understood here that no masking is involved. For the determination of the necessary separation factors, however, the basis taken must be the amount which is permissible for masking, or for change of oxidation state. Similar care must be taken in considering the greatest amounts of the individual disturbing substances that probably occur in the samples. In the case of samples of variable composition, taking only the lower level as basis may result in the limited suitability of the method. Elaboration of the procedure is also made more difficult if the disturbances are overestimated. In rock analysis, for example, the main components are taken into consideration on the basis of the expected composition, while the starting values for the trace elements may be 10 times the Clark values [20].

Calculation of the separation factor. The separation factor (Q) is the quotient of the partition coefficients of the ion to be determined and the interfering ion. The required separation factor is also governed by the ratio of the initial concentration of the interfering ion and its permitted concentration in the determination. It is obvious that the desired factor can not arise as the arbitrary quotient of two values. If the loss-free separation of the ion to be determined is desired ($D_1 \geqq 100$), this value governs the separation factor of the interfering ion too:

$$Q = \text{constant} = \frac{D_1}{D_2} \geqq \frac{100}{D_2}$$

36

(If, for instance, the ratio of the initial and the permitted concentrations of the interfering ion is 10^4, the separation factor must be $\geq 10^6$. Naturally, this cannot arise from either of the quotients $10^1/10^{-5}$ or $10^8/10^2$; in the former case a loss occurs of the ion to be determined, while in the latter case there is no separation.)

It is worthwhile to consider the following points in connection with the elaboration of the separation techniques:

(a) The partition coefficients (primarily those of the component to be determined and the main interfering component) should be established with standard solutions and with artificial mixtures.

(b) If possible, the poorer phase should be analyzed to ascertain the values, regardless of whether the loss of the element to be determined or the concentration of the interfering ion remaining after the separation is being measured.

If the opportunity arises, the possibility of radioactive isotopic labelling should be utilized in these investigations, including checks with authentic samples [26–27a]. This makes it unnecessary to elaborate a method for the determination of all interfering elements; it greatly increases the research output and the reliability of the conclusions. Since the quantity of radioactive substance required for one experiment is 4–400 Bq (10^{-4}–10^{-2} µCi), this work does not necessitate a separate isotope laboratory.

In many cases the separation may also mean that the component to be determined is concentrated (collection of the trace element in a precipitate from a large solution volume; extraction with a low organic: aqueous phase ratio; binding on an ion-exchange resin from a large quantity of solution; etc.). If enrichment is a necessity, an effort should be made to carry out the separation at the same time [28, 29].

With samples of complex and variable composition, e.g. ore samples, it may happen that the separation from some interfering substance will not be satisfactory for a small proportion of the samples. In such cases this limitation must be indicated in the description of the method.

No matter whether a new method is being developed or the most appropriate-seeming procedure is being selected from the literature, there is a comparatively simple and reliable possibility for checking the losses arising in the separation. The course of this checking could well run as follows:

(a) overall performance of the analytical procedure with standard solutions ("artificial mixtures"), without the component to be determined too;

(b) overall performance of the analysis with a real sample not containing the component to be determined (or containing it in a concentration below the sensitivity of the determination); the possibility is given for this in many cases, for instance with rock samples;

(c) overall performance of the analysis with a real sample, with addition of a known quantity of the component to be determined (if no "empty" sample is available, then this procedure can be carried out with another sample instead).

2.5. Compilation of the analytical procedure

The various authors give different procedures to describe the analytical method. The most exacting of these is that recommended by the IUPAC [30]. This recommendation is very accurate, but its lengthiness means that in practice it is rarely possible to publish all of the data. In the case of a well-known reagent, it is sufficient if reference is made merely to the most important data relating to the method. These are the following:

(a) the wavelengths of maximum absorbance of the reagent and the complex; the wavelength of optimum measurement;
(b) the optimum pH interval;
(c) the molar absorbance (ε) of the complex;
(d) the range of validity of Beer's law;
(e) the stability of the colour.

Naturally, the description of some method involves much other information too. In our view, it is recommended that the analytical procedure should have the following structure:

A brief description of the principle of the method, including the first three data mentioned above, together with the main sources of interference, and the means of eliminating these.

Analytical procedure. When many different reagents are used, it is best to begin with a list of the necessary reagents. If the stock solution needed for recording of the calibration plot is not prepared by simple dissolution, this should definitely be detailed. This is the situation in the case of metal ions that undergo hydrolysis in acidic medium; here, the solution composition (auxiliary complex former, acidity of the solution) necessary to maintain the salts in solution and to ensure the monomeric state must certainly be given. A niobium stock solution, for example, can be prepared by fusing Nb_2O_5 with alkali metal hydroxide or by acidic dissolution of $NbOCl_3$; in both cases, however, the addition of tartaric acid is indispensable. Mention may be made in this section of any apparatus requirements probably in excess of those generally available in the laboratory (special stills, etc.).

The analytical procedure must naturally be completely unambiguous, and as concerns the variable parameters (sample weight, volume, addition of certain reagents, heating time, etc.) the limiting values must be given, e.g. "highest sample weight 2 g"; "boiling for at least 5 minutes"; etc. Where the pH of the solutions plays an important role in the course of the analysis, the desired pH value must be stated; it is not enough to indicate the buffer solution to be employed.

Apart from publication of the ε value, the description of the method must contain the attainable sensitivity and the data of the calibration plot, though it is not necessary to illustrate the latter graphically. It must be stated, however, whether the determination unconditionally demands a spectrophotometer, or whether some other means of evaluation is also suitable. Finally, the error of the method must be reported in some way; it is most advantageous to give this in relative % for each value interval (in the event of a more detailed description, the standard deviation may be given, or the confidence limits $= \bar{x} \pm ts/\sqrt{n}$; in the latter, $\bar{x}$ is the arithmetical mean value, s is the standard deviation, n is the number of measurements, and t is the Student factor, the value of which is to be found in Table 2.1).

Notes: We consider that mention should be made here of the critical factors which must be controlled strictly in the interest of the accuracy of the method, and also the limitations of the method and the possibilities of its extension.

As an example, special mention may be made that the acid concentration must be kept strictly constant in the stage of extraction into tributyl phosphate from 6.0 M HCl in the determination of uranium(VI) with arsenazo III; a lower acidity leads to the loss of uranium, while a higher acidity causes an increasing extraction of Th^{4+}. Similarly, the non-exactness

39

TABLE 2.1.

Values of the Student factor *(t)* for estimation of the 95% and 99% confidence intervals for a low number of data

No. of data	Student factor (t)	
	95%	99%
2	12.71	63.66
3	4.30	9.92
4	3.18	5.84
5	2.78	4.60
10	2.26	3.25
∞	1.96	2.58

of the reagent concentration in the extraction–thiocyanate determination of niobium results in a lower absorbance, or in a positive error due to the extraction of titanium(IV).

In connection with methods elaborated for rock analysis, it should be stated whether these can be used for the analysis of various ores without any change; otherwise, all slight modifications should be specified. It should also be mentioned whether the procedure permits only a limited amount of some interfering species.

Special specifications relating to sampling, storage or pretreatment must be detailed here, e.g. the storage of biological samples in a refrigerator.

Experience bearing on the output and time requirements of the method similarly form part of the description of the method.

Finally, mention should be made of all other experience which will facilitate the work. Illustrative examples are the precautions necessary to exclude contamination in the analysis of high-purity materials, the possibility of re-use of certain reagents, the means of their regeneration, etc.

As noted above, there is no uniformly accepted practice for the compilation of the analytical specifications [31–36].

It is our opinion that, within the limits of unambiguousness and clarity, and also the permitted scope, the analytical procedure must contain all those data which will allow reliable work and the recognition of the typical errors. In publications describing new methods, all data (partition coefficients, conditional stability constants, etc.) and research experience must be reported which provide credible evidence of the suitability of the methods.

References

[1] Majer, J. R., Azzonz, A. S. P.: *Talanta* **27**, 549 (1980).

[2] Kuznetsov, Yu. N., Kuznetsova, G. A.: *Zh. Anal. Khim.* **29**, 1041 (1974).

[3] Scharnbeck, C.: *Chem. Techn.* **22**, 166 (1970).

[3a] Ostroumov, G. V.: *Zav. Lab.* **48**, 2, 26 (1982).

[4] Ackermann, G.: *Kémiai Közlemények* **45**, 293 (1976).

[5] Kozlicka, M.: *Chem. Analyticzna* **15**, 683 (1970).

[6] Patrovski, V.: *Chem. Listy* **65**, 1121 (1971).

[7] *Tables of Spectrophotometric Absorption Data of Compounds used for the Colorimetric Determination of Elements.* IUPAC, Butterworths, London 1963.

[8] Bjerrum, J.: *Metal Ammine Formation in Aqueous Solution.* Haase, Copenhagen 1957.

[9] Inczédy, J.: *Analytical Applications of Complex Equilibria.* Ellis-Horwood, London 1976.

[10] Beck, M.: *Chemistry of Complex Equilibria.* Van Nostrand, London 1970.

[11] Specker, H., Kettrup, A.: *Angew. Chem.* **81**, 888 (1969).

[11a] Burger, K.: *Szerves reagensek a fémanalízisben (Organic reagents in metalanalysis).* Műszaki Könyvkiadó, Budapest 1969.

[11b] Burger, K.: *Coordination Chemistry. Experimental Methods.* Butterworths, London 1973.

[11c] Kőrös, E.: *Molekulakomplexek (Molecular complexes).* Akadémiai Kiadó, Budapest 1975.

[11d] Császár, J., Bán, M.: *Optikai színkép, ligandumtér-elmélet, komplex szerkezet (Optical spectrum, ligandfield-theory, complex structure).* Akadémiai Kiadó, Budapest 1972.

[12] Kuznetsov, V. I.: *Zh. Anal. Khim.* **14**, 7 (1959).

[13] Savvin, S. B.: *Organicheskie reagenty gruppy arzenazo III.* Atomizdat, Moscow 1971.

[14] Upor, E.: *Kémiai Közlemények* **34**, 45 (1970).

[15] Burger, K.: *Kémiai Közlemények* **36**, 197 (1971).

[16] Pilipenko, A. T., Tananajko, M. M.: *Talanta* **21**, 501 (1974).

[17] Savvin, S. B., Akimova, T. G., Dedkova, V. P.: *Organicheskie reagenty dlya opredeleniya Ba²⁺ i Sr²⁺.* Nauka, Moscow 1971.

[17a] Dedkova, V. P., Akimova, T. G., Savvin, S. B.: *Zh. Anal. Khim.* **36**, 1358 (1981).

[17b] Busev, A. J., Pryakhina, V. M., Boyko, A. J., Bigma, V. V., Pampushko, T. A., Shcherbina, A. A.: *Zav. Lab.* **48**, 10, 20 (1982).

[18] Halász, A., Pungor, E. Polyák, K.: *Magy. Kém. Folyóirat* **76**, 539 (1970).

[19] Adamovich, L. P.: *Ratsionalnye priemi sostavleniya analiticheskikh propisei.* Kharkov University, 1973.

[20] Bulatov, M. I.: *Zh. Anal. Khim.* **30**, 5 (1975).

[21] Pungor, E. (ed.): *Analitikai kémiai kislexikon (Encyclopedia of analytical chemistry).* Műszaki Könyvkiadó, Budapest 1978.

[22] Sipos, B.: *Vegyipari Táblázatok (Tables for the Chemical Industry)*. Műszaki Könyvkiadó, Budapest 1980.

[23] Perrin, D. D.: *Masking and Demasking of Chemical Reactions.* Wiley, New York 1970.

[24] Skorokhod, O. R.: *Khimicheskii analiz.* B. G. U., Minsk, 1980.

[25] Lyalikov, Yu. S.: *Theoretical Foundations of Modern Chemical Analysis.* Mir, Moscow 1980.

[26] Reynolds, S. A., Leddicotte, G. W.: *Nucleonics* **21,** 128 (1963).

[27] Upor, E., Mohai, M., Nagy, Gy., Szalay, J., Vados, I.: *Proc. Anal. Chem. Conf. Budapest, 1966,* Vol. II, p. 276.

[27a] Kirsh, M., Bilimovich, G. N. (ed.): *Novye metody radioanaliticheskoy khimii.* Energoizdat, Moscow, 1982.

[28] Delayette-Mills, M., Karm, L., Janauer, G. E., Ping-Kwan Chan: *Anal. Chim. Acta* **124,** 365 (1981).

[29] Broekaert, J. A. C.: *Anal. Chim. Acta* **124,** 421 (1981).

[30] IUPAC Information Bulletin No. 44, 1975.

[31] Umland, F., Janssen, A., Thierig, D., Wünsch, G.: *Theorie und praktische Anwendung von Komplexbildnern.* Akademische Verlag, Frankfurt-am-Main 1971.

[32] Fries, J., Getrost, H.: *Organische Reagenzien für die Spurenanalyse.* E. Merck, Darmstadt 1975.

[33] Busev, A. I., Tiptsova, V. G., Ivanov, V. M.: *Rukovodstvo po analiticheskoi khimii redkikh elementov.* Khimiya, Moscow 1978.

[34] Koch, O. G., Koch-Dedie, G. A.: *Handbuch der Spurenanalyse,* I–II. Springer, Berlin 1974.

[35] Holzbecher, Z., Divis, L., Kral, M., Sucha, L., Vlacil, F.: *Handbook of Organic Reagents in Inorganic Chemistry.* Ellis-Horwood, London 1976.

[36] Marczenko, Z.: *Spectrophotometric Determination of Elements.* Ellis-Horwood, London 1976.

3. Methods suitable for separation of interferences, and concentration possibilities

There are cases when the composition of the sample to be examined is favourable, and not very selective reagents may be applied, without separation. This is the situation, for example, in the investigation of certain chemicals (alkali compounds, mineral acids, etc.), piped water or various biological meterials. Another group of samples, pure metals or metal compounds, contain practically only one main component; in a favourable case this does not interfere with the determination. Even if separation must then be performed, in general only one component need be removed. Finally, in very many cases we are faced with a sample of complex composition, when several interfering effects must be overcome (rocks, ores, alloys, various hydrometallurgical samples). This is the maximum requirement; often, the solution is provided only by a series of separations.

The separation methods used at present may be classified as follows:

Precipitation separation. This technique includes the inorganic ion-exchangers, and also ion and froth flotation. Since they are similarly based on partition between solid and liquid phases, crystallization and even selective dissolution also belong here.

Extraction ("liquid extraction"). Besides the traditional procedure based on partition between two mutually immiscible liquid phases, this technique covers homogeneous-phase and three-phase extraction too.

The materials known in some of the literature as "liquid ion-exchangers" must likewise be classified here, of course.

Ion-exchange and chromatographic methods. These include artificial ion-exchange resin procedures, the various types of chromatography, and paper impregnated with an ion-exchanger.

43

Mention may be made here of thin-layer chromatography, although this might also be included among the inorganic ion-exchangers, because of its chemical nature.

Distillation methods. Among others, microdiffusion and sublimation belong here.

Electrochemical separations

Inczédy [1a] uses a similar classification in his discussion of the possibilities of separating the elements.

Even from this listing, it can be seen that certain methods can be classified in different groups as a consequence of their borderline character. For instance, it is customary to refer to the organic precipitating agents as "solid extracting agents", while one group of them are suitable for froth flotation too. Some authors use the expression amalgam extraction for cementation; reversed-phase partition chromatography also features in thin-layer chromatography.

3.1. Precipitation separation

This is the oldest used method of separation in analytical practice. Accordingly, numerous procedures are known which have been "inherited" without the dependence on the conditions ever having been thoroughly clarified. On the other hand, just because it is an old technique, many analysts do not consider its application in up-to-date analysis in general to be appropriate. A role was played in the development of this conception by the over emphasized morphological attitude, which slowed the complex-chemical study of the phenomena (coprecipitation, "adsorption", etc.) accompanying precipitate formation.

In the determination of trace elements there are three main cases of precipitation separation:

(a) selective dissolution;

(b) separation of the trace element by trace-collecting precipitation (on a carrier);

(c) maintenance of the trace element in solution while certain interfering substances are precipitated; if the component to be precipitated is similarly of low concentration, here too it may be necessary to add a carrier.

The aim of the precipitation in case (b) may also be concentration.

44

Selective dissolution.

The aim here is clearly the total dissolution of the component to be determined, whereas the disturbing components, or at least a proportion of them, are left undissolved.

This method is very advantageous in certain cases. In rock analysis, for example, it is generally customary in the determination of uranium to apply dissolution with $HCl + H_2O_2$; in this way, the bulk of the interfering titanium or zirconium, among others, is left undissolved.

The method is of general importance for purposes of phase analysis. For instance, the determination of uranium(VI) in the presence of uranium(IV) in rocks is based on dissolution of the former with Na_2CO_3 solution, an inert atmosphere being used to prevent oxidation of uranium(IV). Complete dissolution of uranium(VI) is promoted by application of an anion-exchange resin which influences the equilibrium [1]. In soil analysis, trace elements which can be taken up by plants, which are soluble in organic acid solutions, or which undergo exchange with alkali metal ions, are determined after dissolution in such acid solutions, e.g. citric acid.

For the dissolution of the contaminants from a matrix metal more positive than hydrogen, non-oxidizing acids may be used; as an example, 8 M HCl is applicable for the dissolution of Al, Mn, Mg, Na, Zn, etc. from gallium [2].

Precipitation of trace elements on a carrier. There is no uniformly accepted view as to the mechanism of precipitation with a carrier (coprecipitation). We consider that the viewpoint of Chuiko is the most realistic. He proves with experimental data that coprecipitation is a chemical (complex-chemical) process that can be described by the law of mass action [3, 4], while the Freundlich and Langmuir correlations, earlier termed adsorption isotherms, also describe such processes. From the isotherm itself, of course, exact conclusions cannot be drawn as to the character of the mechanism.

It is certain at least that primarily those precipitates are suitable as carriers, which are chemically and structurally similar to the corresponding precipitate of the trace element to be separated.

Chuiko [4] classifies the suitable trace-collecting precipitates (collectors) on the basis of the properties of the ion to be separated (Table 3.1).

There are other important cases in analytical practice:

Precipitation with oxalate. Poorly-soluble metal ions may feature here as

TABLE 3.1

Carriers used in the precipitation of trace elements

Nature of ion to be separated	Compounds used as collectors
Alkali and alkaline earth metal ions	Isomorphous cocrystallization with salts
Metal ions forming poorly soluble hydroxides and basic salts	Hydroxides, phosphates, fluorides
Ions giving hydroxides and sulphides poorly soluble in weakly alkaline medium	The same, and sulphides in weakly alkaline medium
Ions giving insoluble hydroxides, and sulphide precipitates in acidic medium	Hydroxides, sulphides in acidic medium, phosphates, fluorides
Metal ions of variable valency yielding acidic oxides in the higher oxidation states	Frequently hydroxides, in the lower oxidation states sulphides, phosphates, fluorides
Readily hydrolyzed metal ions (Bi^{3+}, Sb^{5+}, Sn^{4+}, etc.)	$MnO_2.nH_2O$, sulphides, hydroxides
Metal ions easily reduced to elemental state (Se, Au, Pt)	Mercury, tellurium, sometimes sulphides

ions to be separated and carriers alike. Precipitation with CaC_2O_4 is of particular importance in the determination of Th^{4+}, Sc^{3+} and Ln^{3+}.

Precipitation of metal halides

Precipitation of iodates. This is important in the analysis of thorium and cerium.

Precipitation of sulphates. This finds application in the determination of Pb^{2+}, for example.

As already referred to in the introduction, coprecipitation in a given system depends to only a slight extent on some factors which are otherwise important in precipitation separation, e.g. the following: the quantity of the carrier, the ionic strength of the solution, the temperature of precipitation, the sequence of addition of the reagents, and the concentration of the trace element.

The procedure can also be carried out with partial precipitation of the carrier; this means at the same time that its efficiency does not depend on the solubility product of the macrocomponent.

This technique is particularly advantageous when, following dissolution of the sample, the component to be determined passes into the solid phase on crystallization of a small proportion of the matrix, and in this way can be concentrated substantially. The partition is controlled above all by the pH (pOH) of the solution; in all cases, therefore, the pH-dependence curve of the sorption must be established, especially when precipitation is carried out with metal hydroxides containing the OH group as a complex-forming ligand.

Partition between the two phases may be influenced by complex formers present in the solution. For Cu^{2+} or Ag^+, which form ammine complexes, for example, the pH-dependence of their precipitation on $Fe(OH)_3$ will be different in NaOH solution from that in NH_4OH solution.

For the subsequent course of the analysis, the component to be determined must, after precipitation on the carrier, again be dissolved up. In the majority of cases this is achieved merely by simple dissolution, e.g. in acid, but at times the precipitate must be ignited (Ln^{3+} precipitated with CaC_2O_4), or even digested (phosphates, fluorides). These difficulties too must be taken into consideration when the carrier is chosen.

Organic precipitating agents

These have long been known in many cases for the precipitation of macro-components, e.g. the precipitation of tungsten with tannin, or that of magnesium with oxine. In this case, however, we are concerned with precipitation from very dilute solutions, without addition of a metal ion carrier. One of the pioneers in this subject, Kuznetsov, reviews the following mechanisms of precipitation [5]:

Normal salt formation. A metal ion bound in a cationic or anionic complex is precipitated by a heavy organic anion or cation and a neutral additive, e.g. the precipitation of [Tl(I)-*o*-phenanthroline]$^+$ with methyl orange.

Electrically neutral complex formation. If the complex is insoluble in water, e.g. dithizonates, dithiocarbamates, it may be precipitated with an inactive additive (2,4-dinitroaniline, β-naphthol, etc.). If the complex is soluble in water, e.g. arsenazo III complexes, a large organic cation too is necessary in the precipitation (triphenylmethane dyes, for example, to bind hydrophilic SO_3 groups in this case).

Precipitation of a colloid or hydrolyzed polynuclear cation with tannin + basic dyes. In spite of the insufficiently widespread nature of the application of organic precipitating agents, and also other difficulties (problems of preparing pure reagents; difficulties in filtering certain precipitates; possible losses during ignition of the precipitate; etc.), this procedure has the following well-exploitable advantages:

(a) applicability to very dilute solutions and to water of high salt content, e.g. even 0.01 µg/dm³ uranium can be precipitated with oxine via the addition of an inactive precipitating agent;

(b) the excess of a water-soluble complex former is not precipitated;

(c) it can also be made suitable for the precipitation of trace elements in groups by the combination of several reagents; this can be utilized well in water analysis, for instance;

(d) in certain cases photometry may be carried out directly after dissolution of the precipitate in an organic solvent.

Separation by precipitation of interfering ions

The basis of the method is that the component to be determined in the system in question is kept in solution as a simple or complex ion, while the interfering substances are precipitated.

Even if the equilibria governing the solubility indicate that the ion to be determined should remain in solution, an undesired loss may occur because of its coprecipitation.

The following points must be borne in mind when consideration is given to the causes of losses in precipitation separation and the possibilities of avoiding these.

Partition between the precipitate and solution phases is controlled by involved complex-chemical equilibria, in which the functional groups of the polymeric precipitate also behave as ligands. The partition is governed by the relative stabilities of the sorption complex and the complex ions (possibly simple ions) in the solution.

The scope of this book does not permit a detailed treatment of the mechanism of coprecipitation, but this topic must be discussed briefly. In the earlier view, which is still generally widespread even now, great significance was ascribed to the morphological factors in the adsorption on the surface, and also to the electrokinetic potential. Occlusion and mixed crystal formation were generally accepted concepts; adsorption was differentiated from coprecipitation.

The modern conception explains the phenomena from the aspect of coordination chemistry; ion-exchange processes are regarded as fundamental and proved. This conception has the great merit, for instance, that the opposite charge is not considered a criterion in the precipitation; charge identity too is permitted by the logical extension of copolymerization.

At the same time, all this is not in contradiction with the possibility of speaking of a "solid solution", for example.

Naturally, this theory does not deny the laws of homogeneous and heterogeneous partition found for crystalline precipitates, the concept of molecular sorption, the dependence of the stability of colloid systems on the charge density of the ions, the lyotropic series of ions and other regularities, and it does not make their application unnecessary. However, it does eliminate much formality, which inhibited the recognition of complex-chemical correlations.

The background for the treatment of this question is to be found primarily in the works of Chuiko, Skorokhod, Kolarik, Novikov, Kraus *et al.*, Egorov, and Amphlett [3, 4, 6-12].

The extent of sorption (coprecipitation) is influenced to different degrees by the individual factors.

The pH of the solution. This controls the equilibria in the solution (hydrolysis; complex formation with other ligands), and also influences the behaviour of the functional groups in the carrier precipitate, primarily that of the OH group [13]. It follows from this that this is the parameter mainly governing the losses. For instance, the oxyanions of metals of variable valency in the highest oxidation state (VO_3^-, $Cr_2O_7^{2-}$, etc.) or metal ions forming soluble anionic hydroxo complexes (Al^{3+}, Zn^{2+}, etc.) are precipitated quantitatively on metal hydroxide precipitates at lower pH (< 10), whereas at higher pH (> 13) they remain in solution, disregarding a slight loss. The influence of the pH in the acidic range is less in the case of carriers not containing OH groups. Thus, the precipitation of Mn^{2+} or Zn^{2+} on CaC_2O_4 does not vary in the interval pH 0.7-3.0.

Complex-forming ligands possibly present in the solution (NH_3, CO_3^{2-}, etc.) likewise ensure that the metal ion, e.g. Cu^{2+} or UO_2^{2+}, is kept in solution only if the pH is optimum and the stability of the complex is sufficiently high. Otherwise, the complex ion decomposes to an extent in accordance with the relative stabilities, and the metal ion passes partially or completely into the precipitate [14]. Thus, the low-stability lanthanide carbonates decompose in the presence of $Fe(OH)_3$ [15].

The partition depends to some extent on the concentration of the dissolved ion. The cause of this is that the degree of polymerization is not the same in the two phases (the solution generally contains monomer). For a given system the correlation can be described by the Freundlich or the Langmuir isotherm.

The concentration-dependence is in general not so extensive that, at a concentration possibly different by several orders of magnitude, the picture of the applicability of some separation obtained at some other concentration is not valid. The concentration-dependence may be either positive or negative in direction.

The ionic strength of the solution does not generally have an appreciable effect on the extent of sorption, discounting the added ion of course. Primarily at high ionic strength ($>$1–2 M), there is some desorbent effect, presumably because of the influence on the hydration and the activity coefficients, and also the competing sorption.

The temperature of the solution does not affect the sorption greatly either in most cases. If there are different predominant processes in two different pH ranges, then the temperature-dependence too may be of a different nature. If the actual process is governed mainly by the OH^- concentration, the dependence of the sorption on the pOH is influenced to a negligible extent by the temperature [16].

Foreign ions. If some substance is present which forms a complex with the ion to be kept in solution, this will naturally have a desorbent effect. If a cation is present which is bound in an analogous complex, but precipitates better, this too will exert a desorbent effect to some extent. The effect can be exceptionally effective; an example of its use in analytical practice is the inhibition of sorption of UO_2^{2+} by the addition of Th^{4+} in carbonate medium [16].

The precipitation sequence. In general this does not have an essential influence on the sorption. In most cases nearly the same equilibrium partition can be attained by joint precipitation of the carrier and the trace element, or by addition of the prepared precipitate of the carrier to the solution.

Aging of the precipitate. Very many precipitates are known to "age" on standing. From the aspect of the chemical processes this phenomenon is manifested in increases in the degree of polymerization and the reticular structure, and macroscopically in an increase in the degree of ordering (a more crystalline structure). It is well known that this process decreases the solubility of the precipitate; obviously, therefore, the microcomponent incorporated into the precipitate is less easily dissolved out by a complex former or by washing of the precipitate, for example. The only way then to diminish the sorptional loss is dissolution of the precipitate and repeated precipitation.

If the precipitate is filtered off relatively quickly (within one hour), which is usual in analytical practice, the aging does not cause a change.

Table 3.2 presents some examples that can be applied well in analytical practice to keep trace elements in solution.

Ion and froth flotation

In the concentration processes discussed so far, the material was trans-

TABLE 3.2.

Some examples of holding trace elements in solution during precipitation

Composition of solution	Precipitate	Ions kept in solution (less than 3% loss)
≥ 2 M $NH_4OH +$ $\geq 2MNH_4^+$	Metal hydroxides	Zn^{2+}, Co^{2+} Co^{3+}, Cu^{2+}, Ni^{2+}, Cd^{2+}, Hg^{2+}, Ag^+
Alkali metal hydroxide pH > 13	Metal hydroxides	$Zn(OH)_4^{2-}$, $Al(OH)_4^-$, VO_3^-, MoO_4^{2-}, WO_4^{2-}, ReO_4^-, CrO_4^{2-}, SeO_3^{2-}, $Ga(OH)_4^-$
$NH_4OH + \geq 1$ M NH_4Cl pH ≤ 8	Metal hydroxides	Ca^{2+}, Mg^{2+}
0.2 M $H_2C_2O_4$ pH 0.7–2.0	CaC_2O_4	Fe^{3+}, Al^{3+}, Mg^{2+}
0.02–0.10 M $H_2C_2O_4$ pH 0.7–2.0	CaC_2O_4	Ln^{3+}
1 M $KIO_3 + HNO_3$	$Th(IO_3)_4$	Ln^{3+}
≥ 0.2 M $Na_2CO_3 + Th^{4+}$; pH 9–11	Metal hydroxides	UO_2^{2+}

ferred from one phase to another (crystallization), or possibly kept in the same phase (evaporation), but in the present case the concentration occurs at a liquid–gas interface [17–19].

The essence of the process is that a surface-active substance of opposite charge is added to the ion to be concentrated (which may even be a colloid), after which the resulting compound is passed to the surface of the solution by bubbling air (possibly nitrogen) through it. The monolayer can be created in a suitable apparatus and separated there too. In one variant of the method, the surfactant can be replaced with a trace-collecting precipitate, e.g. Fe(OH)$_3$ for the separation of tin, with paraffin dissolved in alcohol as the reagent [20].

The pH naturally plays an important role in the process. A possibility for the concentration of heavy metal traces (Cr, Mn, Co, Ni, Cu, Cd and Pb) from sea-water is provided by precipitation on indium hydroxide and flotation with sodium oleate-dodecylsulphate [21].

Before determination, the concentrate must generally be subjected to dry or wet ashing.

3.2. Ion-exchange. Chromatographic procedures

Ion-exchange and chromatographic separations have spread rapidly in analysis in the past 10–20 years. Since this theme is dealt with in many articles and monographs [22–33], it will be discussed only briefly here.

Ion-exchange resins

The functioning of ion-exchange resins is based on the exchange of the H$^+$ or OH$^-$ ions of their functional groups by the cations or anions of the solution, to an extent depending on the complex stability constants. The technical performance of the ion-exchange process may be static, when the resin is added to the solution, or it may be dynamic, when the solution to be examined is passed through an ion-exchange column packed with the resin. The latter procedure is much more widespread, for in this way the partition coefficient is multiplied considerably. This technique also provides a possibility for the chromatographic separation of ions with similar charge densities.

The matrix of an ion-exchange resin is a cross-linked organic macromolecule; these were earlier prepared by polycondensation, but are nowadays produced mainly by polymerization. The macromolecules bear functional groups containing the ions to be exchanged. The ion-exchange is a reversible process leading to equilibrium. The reaction between the mobile counter-ions of the ion-exchanger (e.g. Ca^{2+} and Cl$^-$) is stoichiometric. The quantity of counter-ions of the resin that are capable of exchange, referred to unit mass or unit volume of the resin, is termed the capacity.

Aims in the analytical use of ion-exchangers:

(a) Separation from ions disturbing the determination and from complex-forming ligands. In this case, either the interfering ions or the ion to be determined are bound.

(b) Concentration from dilute solution of the ion to be determined (seawater, liquid foodstuffs, urine, etc.). If the trace elements can be dissolved off the resin with a low volume of eluent, a concentration of even 100–1000-fold can be attained [31]. Myasoedova and Savvin [31a] have given a detailed review of the application of the new chelate sorbents. Drugov [31b] has made a critical evaluation of the concentration of atmospheric microelements that are harmful to the health.

(c) Reagent purification. In this case, naturally, the contaminants are bound. Anion-exchange is suitable, for instance, for the purification of concentrated HCl from Fe^{3+} and from other impurities forming anionic chloro complexes [32].

Classification of ion-exchangers. The properties of an ion-exchange resin are determined primarily by the nature of the active groups on the polymer matrix. The following classification may be made on the basis of their structure and mode of functioning:

1. Cation-exchangers. Weakly, moderately and strongly acidic products are distinguished. In the strongly acidic cation-exchangers the functional group is SO_3^-, in moderately acidic ones it is PO_3^{2-}, HPO_2^-, AsO_3^{2-} or SeO_3^-, and in weakly acidic ones it is COOH or phenolic OH.

The selectivity of the ion-exchange depends on the nature of the active groups of the resin. On a strongly acidic cation-exchange resin, the sequence for cations of the same valency coincides with the Hofmeister lyotropic series. As a consequence of the large dissociation constant of the SO_3H group, H^+ is bound least strongly (discounting Li^+). The H^+ ion displaces all other cations from weakly acidic cation-exchangers. The weakly acidic cation-exchangers are particularly selective for multivalent metal ions.

The selectivity sequence relating to the metal ions is modified considerably if complex-forming ligands are present in the solution. For example, the metal ion is converted to an anionic complex which is not bound at all on the cation-exchanger.

2. Anion-exchangers. Again, weakly, moderately and strongly basic products are distinguished. The exchanger group is generally an amino nitrogen, from secondary to quaternary, bearing a positive charge. On the

53

weakly basic ones HCO_3^- is bound least strongly, and OH^- most strongly; on the strongly basic resins OH^- is bound least strongly, in accordance with the large dissociation constants, and ClO_4^- most strongly.

3. Chelate resins. These are produced with a view to increasing the selectivity; a wide selection of complex-forming groups of use in analysis are available as their functional groups. Best known are the products containing N-polycarboxylic acid functional groups (e.g. iminodiacetic acid), for instance Dowex A-1. There is a constant increase in the number and importance of products containing the most varied analytical reagents, such as arsenazo derivatives, as functional groups [34]. Since the process here is not simple cation-exchange, but the formation of a chelate complex, the selectivities of products that can otherwise be regarded as weakly acidic on the basis of the acid strength of the functional group are governed by the complex-chemical equilibria involving the functional group in question in solution. The possibilities of application of chelate-forming resins have been surveyed by Blasius and Brozio [35], Vernon [36] and Myasoedova *et al.* [37].

4. Other ion-exchangers: Apart from the customary forms of ion-exchange resins, the following products are also used, particularly in chromatographic procedures:

(a) *paper impregnated with ion-exchange resin,* e.g. with Dowex A-1 [32];

(b) *cellulose products* containing various functional groups (e.g. PO_3H_2) [38], or cellulose filter paper containing, for example, 2,2'-diaminodiethylamine (DEN) functional groups [39] and iminodiacetic acid–ethyl cellulose [40];

(c) *extracting agents* [TBP, di(2-ethyl)-hexylphosphoric acid, etc.] and other complex formers adsorbed *on a solid support* [silica gel, poly(trifluoroethylene), etc.].

We may include here dithizone, diethyldithiocarbamate, etc. on foamed plastics [41–43], active carbon or some other support. These products can in fact be classified in the group of chelate resins too, though the functional group is not part of the molecule of the support. With an appropriately chosen system, however, the sorption is so strong that the capacity of the resulting charge does not decrease even after many cycles. A great advantage is the simple preparation; an example is the saturation of silica gel with di(2-ethyl)-hexylphosphoric acid by simple soaking for one day. Their importance lies predominantly in "reversed phase chromatography".

5. Inorganic ion-exchangers. These include natural and artificial zeolites, many metal hydroxides, phosphates or heteropoly salts (e.g. ammonium phosphomolybdate), and also $SiO_2 \cdot nH_2O$ or Al_2O_3 for use in thin-layer chromatography [44]. At present they are not of great importance in photometric trace analysis, but it is worthy of special mention that some of them display ion-specific properties, e.g. they may be utilized well to separate alkali metal ions from one another.

The technique is also known of applying the ion-exchange resin together with another compound; an example of this is the concentration of trace elements from sea-water with a mixture of ion-exchange resin and $Ti(OH)_4$ [45].

Requirements to be met by ion-exchangers used in trace analysis

Selectivity. The concept used in ion-exchange is in effect the same as the separation factor referring to two components (A and B) to be separated:

$$K_d = \frac{[A]_o[B]_r}{[B]_o[A]_r}$$

where $[A]_o$ is the concentration in the solution, and $[A]_r$ is the concentration in the resin phase.

The height of the column necessary for separation (HETP) can be determined via the theoretical plate number [32].

The height of one separation stage is otherwise 10–50 times the particle diameter, assuming the conditions necessary for the establishment of equilibrium, and above all the constancy of the flow rate. The selectivity sequence may be obtained if the water content of the resin prepared for the appropriate ionic form is known (a lower water content means a higher selectivity).

Exchange rate. A good kinetic property is one of the conditions of analytical application. From this aspect macroporous resins are advantageous. A high exchange rate permits a greater flow rate.

The exchange rate depends on the particle size of the resin and on the extent of reticulation, as well as on the radius of the hydrated ions and the solution concentration. The rate is controlled by the diffusion in the solution or in the interior of the resin particle, the correlations being relatively complicated [33].

Chemical resistance. Since a satisfactory separation is often possible only in strongly acidic or possibly strongly basic (pH>11) medium, resistance to acid or base is a requirement. There is no long-lasting demand in analytical application and thus, disregarding the weakly acidic or weakly basic resins, this requirement holds acceptably.

Temperature-resistance is a rarer requirement. An example of this, however, is the dissolution of uranium(VI) from rocks [1]. Our investigations show that the capacity of the Varion AP anion-exchange resin undergoes no change during stirring for 6 hours at 95°C in 0.2 M Na_2CO_3+0.1 M $NaHCO_3$ solution.

The least possible adsorption on the matrix. This demand is met mainly by highly reticulated products of high capacity. An unfavourable nature in this respect causes a deterioration in selectivity.

Chemical purity. The resin should not contain the component to be determined, or any interfering contaminant from which it will not be separated later [46]. In practice, the commercially available resin is purified by a series of appropriate elutions.

Performance of the ion-exchange

Static method. The resin is placed in the solution to be examined, and left in contact with it for the time necessary for approximate equilibrium to be attained.

Passage through an ion-exchange column. The ion to be determined or the interfering ion is bound by passage through the resin column. In the latter case the determination is carried out on the effluent solution. In the former case the ion to be determined is eluted out with a suitable solution. Solutions of high salt concentration (>1 M) and appropriate pH are generally used for elution. This salt may contain an ion which is otherwise bound less strongly than the ion to be eluted out, but which nevertheless displaces it because of the concentration conditions.

Metal ions bound in the form of anionic complexes may be eluted out with a solution of low ionic strength.

Elution is frequently carried out with complex formers which are able to elute out the bound metal ions with high selectivity, because of the even greater stability of the complexes.

Elution cannot be attained simply in all cases. A possible solution then is the ashing of the resin or its wet decomposition with an oxidizing acid mixture (e.g. H_2O_2+H_2SO_4).

Ion-exchange chromatography. This procedure is primarily suitable for the separation of ions of similar charge and similar properties; in such cases the simple elution technique is not satisfactory. Depending on the selectivity conditions, a frontal, displacement or elution method is selected. The basic principle in the chromatographic separation of metal ions on a cation-exchange column is that the small quantities of different metal ions bound in the uppermost part of the column are eluted out with electrolyte of appropriate composition. In the course of the elution the metal ions migrate in the column at different rates, depending on their partition coefficients, and then appear with a shift in time in the solution emerging from the column. The efficiency of separation is often improved with complex-forming anions. For quantitative analytical purposes it is indispensable to follow the elution curves in some way (e.g. via isotopic labelling). A frequent and advantageous solution for the separation of metal ions is to form anionic complexes (chloro, sulphato, fluoro, etc. complexes). On complex formation the lyotropic series naturally loses its validity, and practically any binding sequence can be achieved. The literature publishes diagrams on the correlation between the ligand (or acid) concentration and the partition coefficients for many systems [33]. The resin suitable for the given purpose may be selected from tables describing the properties [33]. It is good advice not necessarily to adhere rigidly to the products described in the individual publications, but to try out other resin products that can readily be obtained. (The Hungarian Varion resins are excellently applicable for many analytical tasks [47].)

Chromatographic procedures

The chromatographic procedures comprise one of the fundamental and most complex groups of partition methods. The basis of the method is partition between a stationary phase (generally an aqueous medium) bound on some support (or in its pores) and a mobile phase immiscible with this (usually an organic solvent, in accordance with the foregoing). The complex former may be present in both phases.

From the aspect of photometric trace analysis, the only procedures that can be considered are those in which the ion to be determined remains in solution (elution chromatography), or the appropriate section of the chromatogram is separated and the ion to be determined is dissolved off it.

Chromatographic systems contain the following fundamental components:

The support: a solid phase ($SiO_2 \cdot nH_2O$, cellulose, etc.) not soluble in the solvents used.

The stationary phase: this is a solvent bound in the pores of the support, or a complex former adsorbed on the support (TBP, D2EHPA, dithizone, etc.).

The mobile phase: a solvent not miscible with the stationary phase, containing the necessary complex formers.

The effectiveness of chromatographic procedures is determined by the R_f values; these indicate the relative positions occupied by the individual ions between the start and end lines:

$$R_f = \frac{a}{b}$$

where a is the distance of the component to be investigated from the start line, and b is the distance of the pure solvent from the start line.

This quotient naturally lies between 0.00 and 1.00; the larger the difference in the R_f values of two ions, the better they are separated from each other. The R_f values are primarily of use in paper and thin-layer chromatography; in column chromatography the designations discussed in connection with ion-exchange are employed.

By variation of the compositions of the stationary and mobile phases, and by suitable choice of the technique, even ions with comparatively similar properties, e.g. the lanthanide metal ions, can be separated from one another.

3.3. Extraction

Extraction is a method of separation in which some ion (molecule) undergoes partition between two mutually immiscible liquid phases. One of the phases is usually an aqueous solution, while the other is an organic solvent.

In the sense of the Gibbs phase rule, in the case of two phases and a single component (undergoing partition between the two phases) at constant pressure and temperature the system is a univariant one. This means that the partition coefficient (*D*), i.e. the quotient of the concentrations of the component in question in the two phases, is constant:

$$D = \frac{C_{\text{organic}}}{C_{\text{aqueous}}} = \text{constant}$$

Although this rule presumes that the same ionic (molecular) species is present in both phases, in practice this is disregarded and the sum of the concentrations of all the forms of the ion in question is taken as the basis. If the compositions of the solutions are otherwise the same, this value can be treated as an apparent (conditional) constant.

Extraction separation has spread generally in the past few decades, and particularly in photometric trace analysis. Accordingly, this topic has an abundant literature [27, 48–63].

Extraction may be applied with the following main aims:

(a) concentration;

(b) separation from interfering ions (cations or anions causing absorbance by reacting with the photometric reagent; complex formers with masking action; high salt concentration; etc.);

(c) extraction of a coloured complex with a view to its determination in the organic phase;

(d) separation from a reagent excess;

(e) increase of the sensitivity of the determination (suppression of dissociation; inhibition of decomposition of reagent; formation of a mixed complex of higher stability);

(f) reagent purification.

Naturally, several aims may be satisfied at the same time. In the extraction–thiocyanate determination of niobium, for example, the extraction serves simultaneously for separation from the main interfering ion, TiO^{2+}, decrease of the dissociation, concentration, prevention of decomposition of the reagent, and finally the determination [64].

3.3.1. MECHANISM OF EXTRACTION. THE MOST IMPORTANT EXTRACTION SYSTEMS

The following extraction systems are distinguished on the basis of the nature of the complex formed in the organic phase:

Extraction of a simple molecule (e.g. I_2).

Extraction of metal chelates (metal complexes of alkylphosphoric acids, dithizone, oximes, dithiocarbamates, β-diketones, oxine, etc.).

Extraction of solvated acids and salts with a reagent containing a donor group. The species to be extracted may be a halometal acid ($HFeCl_4$) or a neutral salt [$UO_2(NO_3)_2$], while the extracting agent may be an ether, ketone, ester, alcohol, neutral phosphate ester, trialkylphosphine oxide or sulphide, etc. This was earlier known as extraction of "oxonium" type.

Ion-pair formation. The most important examples are the salts of protonated trialkylamines (phosphines) and quaternary ammonium salts; these primarily excel in the extraction of polyvalent anions, including of course anionic complexes containing central metal ions. In effect, the extraction of salts (e.g. metal compounds of fatty acids) also involves ion-pair formation.

Synergistic extraction. Mixed complexes are generally formed, involving the participation of a chelate-forming anion and a neutral molecule containing a donor group. This process stabilizes the complex and increases the partition coefficient, possibly by several orders of magnitude.

Alkylphosphoric acid esters and β-diketones (thenoyltrifluoroacetone, acetylacetone), or trialkyl phosphates, phosphine oxides and trialkylamines take part in the best-known systems. A certain effect may even occur between two neutral ligands; indeed, by logical extension of the phenomenon, the partial replacement of the more polar "inert" diluent with an apolar one also induces synergism [65, 66].

3.3.2. FACTORS INFLUENCING EXTRACTION

The pH of the aqueous solution. The pH of the aqueous solution influences numerous equilibria affecting the extent of extraction. This is most marked in the case of reagents that are weakly acidic in character, since the concentration of H^+ also figures in the expression for the equilibrium of complex formation, by virtue of the dissociation constant:

$$HL \rightleftharpoons H^+ + L^-$$

If the reagent is an acid, such as dialkylphosphoric acid (HL_{org}), dissolved in the organic phase, then the hydrogen ion concentration of the aqueous phase varies directly in the process of extraction too:

$$M^{n+}_{aq} + nHL_{org} \rightleftharpoons ML_n + nH^+_{aq}$$

The pH additionally influences the equilibria of the concurrent reactions, that is the hydrolysis of the metal ion or the protonation of the anion of the aqueous phase, and the equilibrium of complex formation between the metal ion and the anion.

In the simplest case, the partition coefficient depends on the free reagent concentration and on the equilibrium acid concentration of the aqueous phase as follows:

$$D = K * \frac{[HL_{org}]^n}{[H^+_{aq}]^n}$$

where $K*$ is an apparent (conditional) equilibrium constant.

It can be seen from the equation that decrease of the acidity by one order of magnitude has the same effect on the partition coefficient as increase of the reagent concentration by one order of magnitude. At constant reagent concentration, D_M is a function only of $[H^+]$:

$$D_M = \frac{K**}{[H^+]^n}$$

A plot of log $[H^+]$ vs. log D_M yields a straight line of slope n. This generally coincides with the charge on the cation (Fig. 3.1a) [48].

The possibility of separating certain cations can in principle be seen from the conditional equilibrium constant. This is simply obtained with approximate accuracy as the intercept of the linear log D vs. pH function, on the basis of the correlation $\log D = \log K* + n \log [HL] - n \log [H^+]$. If $[HL] = [H^+]$ (this may be selected to be so), then $D = K*$.

The slopes of the straight lines are not the same for ions of different charge. In such cases the semilogarithmic plots are more favourable for comparison, i.e. pH vs. $E\%$ ($E\%$ is the relative extent of extraction), as shown in Fig. 3.1b.

The pH_{50} value, relating to 50% extraction, is frequently used to express the degree of separability. The difference between two pH_{50} values is in effect analogous to the separation factor of the two ions: if $\Delta pH_{50} = 4$, the separation factor $= 1 \times 10^4$.

The dependence of the partition on the acid concentration may differ from this in certain systems at high acidity. The phenomenon is well known for dialkylphosphoric acid esters that at high acidity the partition again increases with the acid concentration. It may be seen in Fig. 3.2, for example,

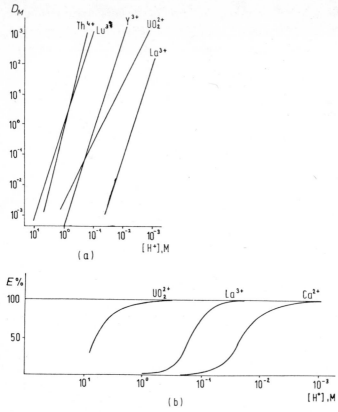

Fig. 3.1. Dependence of the partition of some metal ions on the acidity of the aqueous phase in extraction with HDBP.

(a) $\log [H^+]$ vs. $\log D_M$ ([HDBP]=0.005 M);
(b) $\log [H^+]$ vs. $E\%$ ([HDBP]=0.05 M)

that at high acidity the heavier Ln^{3+} cannot be dissolved by simple re-extraction from HDBP [67].

What has been said on the dependence of the partition on the acid concentration does not hold, of course, for systems in which there is no $H^+–M^{n+}$ exchange (extraction of solvate type; ion-pair formation). In these cases the pH of the solution varies primarily the equilibria in the aqueous phase. Thus, the formation of "metal acids" (e.g. $H[FeCl_4]$, $H[UO_2(NO_3)_3]$) becomes predominant at high acid concentration, permitting

62

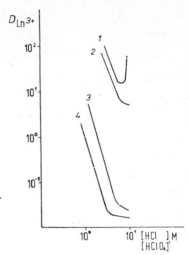

Fig. 3.2. Partition of some lanthanides as a function of the acidity in extraction with HDBP.

1—Yb^{3+}+$HClO_4$; 2—Yb^{3+}+HCl; 3—Gd^{3+}+HCl; 4—Tb^{3+}+HCl

solvation by extracting agents containing donor groups (ethyl ether, TBP, cyclohexanone, etc.).

The concentration of the reagent influences the partition in the manner described in the reaction equation. This is naturally valid for complex formation of all types; the slope of the linear plot of log $[HL]_{org}$ vs. log D_M provides the number of extracting agent molecules bound (solvated) in the complex molecule (if there is no stepwise complex formation and only a single complex is extracted).

The correlations containing all of the constants are essentially more complicated [48]; they are necessary only in the case of complex-chemical investigations. Naturally, calculations must then be performed with activities instead of concentrations. (The results of studies on equilibria in non-aqueous solutions [68, 69] can be utilized well to examine processes occurring in the organic phase.)

Certain reagents are present in a non-monomeric state in the organic phase. The dialkylphosphoric acids are dimerized in apolar diluents, and the monoalkylphosphoric acids may be polymers as well [70]; the alkylamines [71] and the basic dyestuffs too are generally polymers [72]. This fact does not alter the validity of the fundamental correlations: $\log D_M = K^* + n \log [HL] - n \log [H^+]$.

Concentration of the anion in the aqueous phase. The extraction of cations

63

is influenced by the anion in the aqueous phase if the anion takes place in complex formation in either the organic or the aqueous phase.

An example of the former case:

$$(UO_2^{2+})_{aq}+2(NO_3^-)_{aq}+2(TBP)_{org}\rightleftharpoons[UO_2(NO_3)_2.2\ TBP]_{org}$$

To a first approximation, the slope of the linear plot $\log[NO_3^-]$ vs. $\log D_{UO_2^{2+}}$ is then $+2$. This correlation holds only if $[H^+]$ is constant. Although it is true that $[H^+]$ does not feature directly in the complex-formation equilibrium, at high acidity it does affect the partition of UO_2^{2+} via the increase in the concurrent extraction of HNO_3, and it also influences the complex formation occurring in the aqueous phase.

In extractions of this type, the salting-out effect can be observed. Besides the earlier-mentioned influence of the anion concentration, the cation too plays a role. The mechanism of salting-out is that it decreases the hydrate sheath of the cation to be extracted, and facilitates the loss of H_2O from the environment of the cation, and hence the solvation by the extracting agent. Thus, the salting-out effect is proportional to the charge of the cation and the reciprocal ionic radius. Accordingly, it is popular to use $Al(NO_3)_3$ in the extraction of Th^{4+} or UO_2^{2+} with TBP from HNO_3 solution. A correctly chosen salting-out agent may cause a several-fold rise in the partition coefficient.

If the anion of the aqueous phase is the complex former of the cation to be extracted, this naturally reduces the partition, in accordance with the following correlation:

$$D=\frac{K_{extr}[HL]^n_{org}}{[H^+]^n(1+\beta_1[L']+\beta_2[L']^2+\ldots)}$$

where β is the stability constant and L' is the complex-forming ligand [51].

Masking is a very frequent solution to improve the separation. It is a condition, of course, that the separation factor should be improved. This means that the masking agent should not decrease the partition of the ion to be extracted, or only to a lower extent than that of the interfering ion.

Concentration of the ion to be extracted. The extent of extraction is generally independent of the concentration of the ion to be extracted. The causes of deviations from this are as follows:

If the equilibrium concentration of the extracting agent is decreased by a stoichiometrically not negligible quantity of the ion to be extracted.

In this case the decrease in the equilibrium concentration of the extracting agent must be taken into consideration.

If the ion to be extracted is not monomeric in some phase, or if the degree of polymerization is not the same in the two phases. For example, if the complex is binuclear in the organic phase, the general reaction equation is

$$2M_{aq}^+ + 2HL_{org} \rightleftharpoons M_2L_{2org} + 2H_{aq}^+$$

from which

$$K_{extr} = \frac{M_2L_{2org}[H^+]_{aq}^2}{[M^+]_{aq}^2[HL]_{org}^2} \text{ and } D_{M^+} = \frac{M_2L_{2org}}{[M^+]_{aq}^2}$$

The extent of the extraction in this case increases with the cation concentration (if $D > 1$). This is the situation, for instance, in the extraction of TiO^{2+} or $TiO(H_2O_2)^{2+}$ with HDBP [72].

The diluent (solvent). Even in current analytical practice, the diluent is often treated as an inert compound not participating in the process. However, it may have a considerable influence; it can cause a difference of even several orders of magnitude in the partition coefficient in some cases. The processes taking place in the organic phase (solvation, ionization, change in the dissociation of the complex, polymerization of the extracting agent or the complex, etc.) depend on the properties of the diluent, and above all on its polarity (relative permittivity). The correlations are to be found in more detail in the literature [65, 69].

From a practical aspect it may be stated that, in the event of a solvation mechanism, for example, an apolar diluent (e.g. petroleum ether) is more advantageous since it does not compete with the complex former (e.g. TBP). An active extracting agent may often be used even without a diluent (amyl acetate, methyl isobutyl ketone, etc.); naturally, the donor strength is important. This is of significance particularly in the case of complexes dissociating to a large extent in the aqueous phase (e.g. thiocyanates).

Basic dyes form a special case. With these, not even homologous solvents with very similar properties (e.g. benzene and toluene) can be interchanged, for the extraction of the dyes changes fundamentally, and thus the system becomes unsuitable for the determination (the spectra of the metal complexes are the same as those of the dyes) [72].

By a good choice of the solvent (diluent), the work can be accelerated too. If the extraction is followed by washing of the organic phase and then re-extraction, it is better if the specific weight of the solvent is less than that of the aqueous solution. If the extracting agent is added in several portions (e.g. dithizone), it is advantageous if the solvent has a higher density than that of the aqueous phase.

Ratio of volumes of organic and aqueous phases. The partition coefficient is naturally not affected by this ratio, but the extent of extraction is. The correlation is as follows: $E\% = 100 \ D/(D + V_{aq}/V_{org})$. In repeated extractions, the loss $(\%) = 100/(DV_{org}/V_{aq} + 1)^n$, where V_{org} and V_{aq} are the volumes of the organic and the aqueous phases, and n is the number of cycles.

From the latter formula it may be seen that it is more advantageous to add the given volume of extracting agent (or the aqueous solution suitable for re-extraction) in several portions. For instance, if $D = 9$, $V_{aq} = 10 \ cm^3$, $V_{org} = 9 \ cm^3$, 11% of the substance to be extracted remains in the aqueous phase. With $3 \times 3 \ cm^3$ of extracting agent, the loss falls to 2%.

If the partition between the two phases of the extracting agent or of some ligand (e.g. CNS^-) involved in the extracted complex is not sufficiently high (<50), the phase ratio affects the equilibrium concentration of the reagent in the organic phase, and thus the partition coefficient and absorbance of the ion to be extracted too. When the method is being elaborated in such a case, this effect must be taken into consideration as well.

The temperature influences the internal enthalpy of the system in the sense of the Arrhenius equation, and hence it affects the partition coefficients too via the change in the activation energy.

In analytical practice, however, the fluctuation in room temperature does not generally cause appreciable differences, and it is therefore usually not necessary to deal with the effect of temperature. Where this effect is a significant one (e.g. in the partition of basic dyes), this must naturally be borne in mind during preparation of the analytical procedure.

Duration of extraction. Although differences in order of magnitude may arise between the rates of the various processes, in analytical practice, i.e. when the two phases are shaken together in a separating-funnel, equilibrium is approximately established in 1–2 minutes. The required duration of shaking must naturally be determined experimentally.

In cases when an emulsion may form (e.g. in the presence of silica gel), the quantity of the emulsion increases with the duration of shaking; for this reason, among others, long or very intensive shaking is to be avoided. The duration of extraction is similarly important in the case of basic dyes.

3.3.3. WASHING OF ORGANIC PHASE. RE-EXTRACTION

Washing of organic phase. Whatever the aim of the extraction, there may be a need for washing of the organic phase. At any event, washing is advisable if the separation from some interfering substance is not satisfactory in one step, and if the partition coefficient of the ion to be determined is large enough for the washing not to cause an appreciable loss (and the partition of the interfering ion is sufficiently small ($D<0.1$) for the washing to be effective). The disturbing effect of Ti in the thiocyanate determination of Nb is eliminated in this way, for instance. Inclusion of the washing may be useful when the separation factor is satisfactory, but we wish to ensure that drops of the aqueous phase containing the interfering ion in high concentration should not pass into the re-extract.

A solution with the same composition as the initial aqueous phase is most generally used for washing. If the partition coefficient is large, the composition may also differ from this, but the partition coefficients and the possible consequences of modification (e.g. a change in colour due to stepwise complex formation or to hydrolysis) must be checked thoroughly during elaboration of the method.

Re-extraction. Unless the extracted complex is at the same time the absorbing compound on which the determination is based, the extracted ion must be returned to the aqueous phase or, more rarely, the coloured complex must be formed in the organic phase. For purposes of re-extraction, of course, a suitable solution must have a composition which ensures a low partition coefficient ($D<0.01$). The simplest (and most frequent case) is when this can be achieved merely by regulation of the acidity of the aqueous solution. In the case of chelate complexes, a solution more acidic by several orders of magnitude than the initial solution is generally suitable for this (e.g. re-extraction of the complexes formed between the lanthanides and 1-phenyl-3-methyl-4-benzoylpyrazolone-5 with hydrochloric acid [74]). From a solution of high acid concentration, complex metal acids may be re-extracted with a less acidic solution, and possibly with distilled water (re-extraction of UO_2^{2+} from TBP). Unusual techniques may also sometimes be resorted to; an example is the re-extraction of UO_2^{2+} from HDBP solution with NaOH solution [75].

Variation merely of the acidity does not always lead to a complete solution. There may then be a need for complex formers or, in the case of ions

67

of variable valency, for reducing agents. However, this is acceptable only if the complex former does not interfere with the further course of the analysis, or, in last resort, it can relatively easily be destroyed.

Partly because it is not very widespread, and partly because it is not generally applicable, re-extraction based on the antagonistic effect is not sufficiently well known, but it is a very convenient solution. Examples of its use are the re-extraction of lanthanides from HDBP solution by the addition of TBP [76], and the separation of the lanthanides from interfering elements by the combined application of D2EHPA and TBP [77].

In certain cases, the organic phase containing the extracted metal ion is shaken with the colour-forming reagent to form the coloured complex. Kletenik and Bikovskaya [78] determine TiO^{2+} extracted with diisoamyl-phosphoric acid in this way, with H_2O_2.

Finally, there is a possibility for an otherwise hydrophilic reagent to be added dissolved in a solvent miscible with the extract. This is recommended by Cerrai and Ghersini for the determination of Th^{4+} with arsenazo III; after the extraction of Th^{4+} with D2EHPA dissolved in cyclohexane, isopropanol and arsenazo III are added [79].

An obstacle to the spreading of this technique is the difficulty of ensuring optical purity.

In some cases there is simply no suitable means of re-extraction. The organic phase must then be decomposed. This is the procedure, for instance, in the determination of scandium [80], when the HDBP is destroyed with HBr after evaporation of the solvent. In other cases (e.g. with dithizones), long boiling with various acid mixtures and evaporation are necessary.

3.3.4. UNDESIRED PHENOMENA IN THE EXTRACTION

Co-extraction is a detrimental phenomenon; it has been studied in detail particularly by Zolotov *et al.* and Upor [16, 56, 57]. Co-extraction is known only in certain systems. The most important cases:

(a) The metal ion bound in an anionic complex forms an ion-pair with the cation of the aqueous phase (e.g. Ca^{2+}–$FeCl_4^-$).

(b) A poorly dissociating mixed complex is formed in the aqueous phase. An example is the extraction of vanadium as a heteropolyacid in the presence of phosphate and tungstate (which in another situation is utilized analytically for the determination of vanadium).

(c) The extraction of some microcomponent increases in the presence of a well extractable macrocomponent. The reason for this is that the latter alters the properties of the solvent, or enters the trace element undissociated ion-pair aggregate (e.g. $HFeCl_4$–In^{3+}).

It is important, therefore, to check the partitions in the presence of other ions too (e.g. the main components).

Another detrimental phenomenon may be the formation of insoluble copolymers in the organic phase. In the joint presence of Th^{4+} or Fe^{3+} and lanthanides, it has been observed that the Ln^{3+} ions are partially incorporated into the precipitate formed in the organic phase because of the low solubility of thorium dibutylphosphate, from which they cannot be dissolved out. In many other cases too, an emulsion or precipitate appearing at the interface between the two phases is an indication of loss. Such a situation arises particularly with organic dyes of poor quality (e.g. Ga^{3+}-malachite green).

3.3.5. NON-TRADITIONAL EXTRACTION SYSTEMS

Three-phase extraction. A condition for the development of a third phase in one case is the phenomenon of a solvent mixture the two components of which are not infinitely miscible with each other in the presence of the extracted ions, and thus two organic phases are formed. This is the situation, for example, in extractions with diantipyrylmethane [81], where the thiocyanate complexes of several ions can be concentrated in the lower organic phase of low volume. This method is preferably combined with other procedures, especially emission spectral analysis, but it also has a future in spectrophotometry.

Homogeneous and solid-phase extraction [82, 83]. The essence of the procedure is that one phase is formed transitionally from the aqueous solution and from the extracting agent which is solid at room temperature. This is achieved either by elevation above the critical dissolution temperature, or by the addition of an intermediary phase (promoting mixing). The latter may be a surfactant substance, or a volatile solvent that can later be removed with a stream of gas. The separated solid phase may be dissolved up in a small volume and subjected to photometry. An example is the determination of copper with PAN, extracted with benzophenone (in the presence of ethanol and a surfactant auxiliary reagent) [84].

3.4. Distillation methods

It follows from the essence of the method that its applicability is restricted to those compounds that can be converted to the gas or vapour phase. The methods can be classified as follows:

(a) isothermal distillation;
(b) distillation from solution, with heating;
(c) dry distillation.

Isothermal distillation. The two most important techniques here are microdiffusion and expulsion of a gaseous product from solution with a gas stream.

Microdiffusion (in effect gas diffusion on a micro scale) is a procedure developed by Conway [85]. Its essence is that a volatile compound of the element to be determined is produced in the sample to be analyzed, and the sample is then placed in a joint closed space together with an absorbing liquid above which the vapour pressure of the volatile compound is negligible. Thus, the component to be determined progressively diffuses into the absorbing liquid.

In the sense of Henry's law, the process is regulated primarily by the mole fraction of the dissolved volatile compound [86]. Accordingly, it is advantageous to prepare the sample in as concentrated a solution or in as thin a layer as possible. The duration of the diffusion is proportional to the square of the layer thickness and to the square of the distance travelled in the gas space.

It is mainly these factors which govern the technical design of the vessel used. The original Conway vessel was a thick-walled Petri dish, which was divided into two liquid compartments by a concentric (lower) glass wall of smaller diameter. Numerous other technical solutions may be employed. In the procedure of Teleshova [87], for instance, the CO_2 released from a rock sample by the addition of acid is absorbed in NaOH solution in a test-tube situated in an Erlenmeyer flask. Auffarht and Klockow [88] propose the addition of hexamethyldisiloxane (HMDSO) to quantify the diffusion, and the use of a rotating vessel, in the determination of the fluoride content of geological and aerosol samples.

The method is a very convenient one, permitting high-output separation and determination. Its limitation, of course, is mainly that it can be applied only in the case of a substance with high partial pressure, which dissolves to a low extent in the solution to be examined. In photometric trace analysis, the method is practically restricted to the separation of NH_3, F and SiO_2 (in the form of H_2SiF_6), as well as B (boric acid methyl ester) [89].

The method of driving out a compound in a gas stream under isothermal conditions (and naturally its absorption in a suitable solution) is performed primarily for the conversion of As, Sb and Ge to their hydrides. For arsenic and antimony the reduction is carried out with metallic zinc in acidic medium, and for germanium with sodium amalgam in NaOH solution. The gas stream is ensured by the H_2 evolved. AsH_3 is best absorbed in AgDDTC solution and determined photometrically [90]; SbH_3 is absorbed in mercury(II) chloride and determined by extraction–photometry [91].

Distillation by heating of a solution. Since the component to be determined is removed from solution by heating here, this is a more general possibility than isothermal distillation. The condition here is naturally that the boiling-point of the component to be separated should be relatively low. This is satisfied primarily by certain metal halides, but other possibilities too are known. The most widely-known examples from analytical practice are listed in Table 3.3.

TABLE 3.3

Separation of some elements by distillation

Element to be determined	Medium applied	Compound distilled
Ge	6 M HCl	$GeCl_4$
Cr	$H_2SO_4 + NaCl$	$CrOCl_2$
Sb, Sn, As, Os, Ge	$HBr + HClO_4$	MBr_n
F	$HClO_4 + H_2SO_4$	HF (H_2SiF_6)
Hg, Se	$HBr + H_2SO_4$	$HgBr_2$, $SeOBr_2$
Re	$HCl + H_2SO_4$	ReO_3

This procedure is suitable not only for the determination of a volatile element, but also for the elimination of its interfering effect.

Dry distillation (sublimation). In certain cases, compounds of low boiling-point can be separated from elements disturbing the determination, or from the element to be determined, by dry distillation, sublimation or pyro-hydrolysis.

The best-known example of the method is the determination of fluorine in rocks or other samples [92]. The sample is ignited in a stream of gas at 1200°C in a combustion furnace, in the presence of metal oxide (WO_3, Al_2O_3, UO_3), and the HF formed is absorbed in aqueous solution and determined. For determination of the iron oxide inclusion content of steels [93], the metallic iron is removed by vacuum-chlorination. A method is known for the determination of the germanium content of rocks, in which the $GeCl_4$ formed by ignition of the sample with CCl_4 is distilled off at 350°C (after reduction of the oxides with carbon).

3.5. Electrochemical methods

The theoretical basis of electrochemical methods of separation lies in the differences between the standard redox potentials of the elements, or the potential series for the metals.

If the standard electrode potentials of two metals in solution are sufficiently different, there is a possibility for their separation by electrolysis or cementation.

If some metal is connected as cathode, the ions of metals to the right of it in the series (i.e. more noble metals) will be deposited from the solution, while if this metal is connected as anode, it will dissolve up itself. Besides the fundamental validity of the Nernst equation, the process is modified by the polarization current, whereby the deposition of certain metals is shifted to a more positive potential [94]. The process is also influenced by the material and surface of the electrodes, and by the temperature. The method carried out at a controlled cathode potential is substantially better than electrophoresis at a constant terminal voltage. During the period of electrolysis, the terminal voltage is then continuously regulated.

The method can be utilized in two ways in metal analysis. In the more frequent, the component to be determined or the main component to be removed is deposited on the cathode (more rarely on the anode) from a solution of the sample. In the second procedure the sample is connected as anode (possibly in amalgam form); depending on the potential conditions, the component to be determined is deposited on the cathode or remains in the anode mud. Its quantitative collection may mean a difficulty in this case.

With a well-chosen complex former, the standard potentials can be modified so as to give a more favourable selectivity of separation.

In cases when the ion of a metal more negative than H_2 remains in solution, the otherwise widespread Hg-cathodic method is advantageous. This is used, for instance, for the determination of the aluminium content of an iron alloy or cast-iron [95], or for other elements which do not undergo amalgamation in the electrolysis.

In fact, paper-electrophoresis too can be classified here, which often gives a better separation than simple chromatography. The electro-ultrafiltration (EUF) method is suitable for the separation of the readily absorbed nutrient substances in soil analysis; this is a combination of electrodialysis and ultrafiltration [96, 97].

Cementation (or amalgam exchange) is the reduction of some metal ion by an amalgam, and its transfer into the amalgam without the introduction of current. This is analogous to the cathode process, while the dissolving out from the amalgam is analogous to the anode process. Here too the equilibria are regulated by the potential series; with the amalgam of a metal with a more negative electrode potential, a metal to the right of it in the series can be deposited from solution. As examples, Sn, Pb, Cu and Hg are deposited on zinc amalgam; Cd, Ni, Pb, Bi, Cu, etc. are deposited on magnesium amalgam.

In analytical practice, mainly sodium and zinc amalgams are used [98]. A very advantageous procedure is the selective separation of europium with sodium amalgam from a large quantity of other lanthanides [99]. By decomposition of the amalgam with HCl, the europium can be returned to the aqueous phase.

The selectivity of separations can be improved with complex formers. These are often necessary to keep certain ions in solution (e.g. Ln^{3+} with citrate).

Besides the works already referred to, good surveys of electrochemical separations are given by Schleicher [100], Lingane [101] and Charlot et al. [102].

3.6. Comparison of separation methods

The fields of application and the values of the individual separation and concentration methods are naturally not identical. To summarize, we shall present a brief survey of the advantages and disadvantages, as well as of the fields of application of the individual procedures.

1. Precipitation separation. This is suitable both for enrichment and for the separation of ions interfering in a determination. Precipitation on a carrier often permits separation from the disturbing substances at the same time. The use of an organic precipitating agent or a carrier not disturbing the further course of the analysis can in effect be regarded as "carrier-free" precipitation. The equilibria are not in general affected by the ionic strength of the solution. An excess of the precipitating agent does not usually disturb the further course of the analysis; after dissolution of the precipitate, the analysis can normally be performed immediately. The procedure is technically simple, and very advantageous for serial analyses. In many cases it is effectively not a separate operation, but equivalent to dissolving out following digestion.

A frequent drawback of precipitation separations is the unsatisfactory selectivity; this technique must therefore often be combined with another procedure, such as extraction. Coprecipitation and aging of the precipitate may cause deterioration of the selectivity or loss, and thus the investigation of these error sources is indispensable. Precipitation separation is suitable for practically any sample or element.

2. Extraction. This is a very simple procedure, and thus eminently suitable for serial analyses too. With variation of the composition of the aqueous phase, the complex formation permitting the desired extraction and the masking inhibiting the extraction of the interfering components can be solved in very many combinations. The selectivity can be improved by applying mixed ligand complexes. By washing of the organic phase, the separation factor can in most cases be increased to the desired extent even for the analysis of samples of extreme composition. The very wide choice of chelate formers permits the determination of almost every element in the organic phase immediately after extraction. Re-extraction can be performed simply.

The majority of the extracting agents and diluents are inflammable or poisonous; however, if the standard laboratory safety regulations are adhered to, this does not cause a problem.

In certain cases (mentioned above), coextraction may deteriorate the selectivity. With some high-stability complexes re-extraction cannot be achieved by simple means, and the organic phase must be decomposed in a difficult operation.

3. Ion-exchange and chromatographic methods. The most general possibility is the use of ion-exchange resins. This may be very advantageous for the concentration of the ion to be determined from a large volume. Similarly as in extraction, auxiliary complex formers provide a possibility for enhancement of the selectivity and the creation of a cationic or complex anionic form of the ion to be determined.

The method is more complicated than extraction or precipitation separation, but it is suitable for serial analyses. If the element to be determined is bound on a resin, in most cases it is obtained in a large volume after elution. Instead of elution, therefore, destruction or ashing of the resin is sometimes more favourable.

The possibility of chromatography may lead to great advantages in the separation of ions with very similar partition coefficients. Here, however, the reliable separation of the fractions is difficult.

In our view, ion-exchange is primarily of advantage for concentration from dilute solutions (e.g. water samples) and for binding of main constituents (analysis of metals and telecommunications materials). Anion-exchangers allow more variation, because of the varied complex-formation possibilities.

The methodology can be further developed via reversed-phase chromatography and the use of chelate resins containing the very many types of functional groups of organic reagents.

4. Distillation. The main restriction is the relatively small choice of components that are volatile (or can be made volatile) at low temperature; a further disadvantage is the technical complexity of the operations at elevated temperature.

However, microdiffusion is very advantageous and excellently applicable for serial analyses, for the determination of F, Si and N (NH_3), as is the distillation of AsH_3 for the determination of As, and the boiling of chloride-containing solutions for the removal of certain interfering elements (e.g.

Cr). Distillation at higher temperature (generally dry distillation) is predominantly of value in certain metallurgical analyses.

5. *Electrochemical methods.* These can mainly be applied in metal analyses. The very simple cementation (amalgam extraction) provides a satisfactory solution comparatively rarely.

A special apparatus is necessary for electrolysis, especially in the mercury-cathode variant common in metallurgical analysis. The method is primarily of great importance in the determination of the less noble metals (Ca, Al).

Detailed surveys of the methods of concentrating trace elements are to be found in the book by Zolotov and Kuzmin [102a] and in the review by Kuzmin [102b].

References

[1] Upor, E., Mohai, M.: *Ionenaustauscher und Ihre Anwendung.* Akadémiai Kiadó, Budapest 1965, p. 195.
[1a] Inczédy, J.: *Analitikai elválasztási módszerek korszerű irányai (Modern trends in analytical separation methods).* A kémia újabb eredményei, **10**, Akadémiai Kiadó, Budapest 1972.
[2] Lisenko, V. L., Kim, A. G.: *Trudy Komissii po anal. khim.* **15**, 200 (1965).
[3] Chuiko, V. T.: *Zh. Neorg. Khim.* **2**, 685 (1957).
[4] Chuiko, V. T.: *Trudy Komissii po anal. khim.* **15**, 236 (1965).
[5] Kuznetsov, V. I.: *Trudy Komissii po anal. khim.* **15**, 279 (1965).
[6] Skorokhod, O. R.: *Khimicheskii Analiz.* BGU, Minsk 1980.
[7] Kolarik, Z.: *Coll. Czech. Chem. Comm.* **27**, 938 (1962).
[8] Novikov, A. I.: *Vyssh. uchobn. Zav. Khim. i Khim. Techn.* **6**, 377 (1963).
[9] Novikov, A. I.: *Soosazhdenie s gidrotirovannymi okislyami.* Vol. 1, Dushanbe 1972.
[10] Kraus, K. A., Phillips, H. O., Carlson, T. A., Johnson, J. S.: *Proc. 2nd UN Internat. Conf. Peaceful Uses of Atomic Energy, Geneva, 1958,* Vol. 28, p. 3.
[11] Egorov, Yu. V.: *Statika sorptsii mikrokomponentov oksigidratami.* Atomizdat, Moscow 1975.
[12] Amphlett, C. B.: *Inorganic Ion Exchangers.* Elsevier, Amsterdam 1964.
[13] McGowan, I. C.: *Rec. Trav. Chim. Pays–Bas* **85**, 777 (1966).
[14] Upor, E., Nagy, Gy.: *Acta Chim. Acad. Sci. Hung.* **78**, 47 (1973).
[15] Upor, E., Nagy, Gy.: *Acta Chim. Acad. Sci. Hung.* **68**, 313 (1971).
[16] Upor, E.: Csapadékos és extrakciós elválasztási módszerek vizsgálata az urán és más elemek analitikájában (Study of precipitation and extraction methods of separation in the analysis of uranium and other elements). Thesis for C. Sc. Budapest 1966.

[17] Sebba, F.: *Ion Flotation*. Elsevier, Amsterdam 1962.

[18] Somasundaran, P.: *Separation and Purification Methods* **1**, 117 (1972).

[19] Kuzkin, S. F., Golman, A. M.: *Flotatsiya ionov i molekul*. Nedra, Leningrad 1974.

[20] Mizuike, A., Hiradije, M.: *Anal. Chim. Acta* **69**, 231 (1974).

[21] Hiradije, M., Ito, T., Baba, M., Kawaguchi, H., Mizuike, A.: *Anal. Chem.* **52**, 804 (1980).

[22] Delayette-Mills, M., Lya Karm, Janauer, G. E., Ping-Kwan Chan, Bernier, W. E.: *Anal. Chim. Acta* **124**, 365 (1981).

[23] Broekaert, J. A. C.: *Anal. Chim. Acta* **124**, 421 (1981).

[24] Rieman, W., Walton, H. F.: *Ion Exchange in Analytical Chemistry*. Pergamon, Oxford 1970.

[25] Shemyakin, F. M., Stepin, V. V.: *Ionoobmenyi khromatograficheskii analiz metallov*. Metallurgiya, Moscow 1965.

[26] Olshanova, K. M., Potapova, M. A., Kopilova, V. D., Morozova, N. M.: *Rukovodstvo po ionoobmennoi i raspredelitelnoi i osadochnoi khromatografii*. Khimiya, Moscow 1965.

[27] Marcus, I., Kertes, A. S.: *Ion Exchange and Solvent Extraction of Metal Complexes*. Wiley, New York 1969.

[28] Snyder, R. L., Kirkland, J. J.: *An Introduction to Modern Liquid Chromatography*. Wiley–Interscience, New York 1974.

[29] Kiselev, A. V., Dreving, V. P.: *Eksperimentalnye metody v adsorptsii i molekularnoi khromatografii*. Moscow University, 1973.

[30] Lederer, E., Lederer, M.: *Chromatography. A Review of Principles and Application*. Elsevier, Amsterdam 1957.

[31] Braun, T., Ghersini, G. (eds): *Extraction Chromatography*. Akadémiai Kiadó, Budapest 1975.

[31a] Myasoedova, G. V., Savvin, S. B.: *Zh. Anal. Khim.,* **37**, 499 (1982).

[31b] Drugov, Yu. S.: *Zav. Lab.* **48**, 1, 3 (1982).

[32] Inczédy, J.: *Analytical Application of Ion Exchangers*. Pergamon, Oxford 1966.

[33] Inczédy, J. (ed.): *Ioncserélők és alkalmazásuk (Ion exchangers and their application)*. Műszaki Könyvkiadó, Budapest 1980.

[34] Savvin, S. B.: *Organicheskie reagenty gruppy arsenazo*. III. Atomizdat, Moscow 1971.

[35] Blasius, E., Brozio, B.: In: *Chelating Ion Exchange Resins,* Flaschka, H. A., Barnard, A. J. (eds.). Marcel Dekker, New York 1967, p. 49.

[36] Vernon, F.: *Selected Ann. Rev. in Anal. Sci.* **2**, 83 (1972).

[37] Myasoedova, O. V., Eliseeva, O. P., Savvin, S. B.: *Zh. Anal. Khim.* **26**, 2172 (1971).

[38] Lásztity, A., Horváth, Zs.: *Acta Chim. Acad Sci. Hung.* **41**, 161 (1964).

[39] Smits, J., van Grieken, R.: *Anal. Chim. Acta* **123**, 9 (1981).

[40] Horváth, Zs., Falb, K., Fodor, K.: *Magy. Kém. Folyóirat* **83**, 254 (1977).

[41] Braun, T., Farag, B.: *Anal. Chim. Acta* **19**, 828 (1972).

[42] Braun, T., Abbas, M. N.: *Anal. Chim. Acta* **119**, 113 (1980).

[43] Braun, T., Farag, B.: *Anal. Chim. Acta* **99**, 1 (1978).

[44] Volinyets, M. P., Ermakov, A. N.: *Uspekhi Khimii* **39**, 934 (1970).

[45] Robakin, A. I.: *Radiokhimiya* **16**, 92 (1974).
[46] Mikhailov, G. I.: Ref. *Zh. Khim.* **11G**, 25 (1970).
[47] Varion Catalogue. NIKE, Fűzfő, Hungary, 1981.
[48] Morrison, G. H., Freiser, H.: *Solvent Extraction in Analytical Chemistry*. Wiley, New York 1957.
[49] Alimarin, I. P.: *Khimicheskie osnovy ekstraktsionnovo metoda razdeleniya elementov*. Nauka, Moscow 1966.
[50] Alimarin, I. P. (ed.): *Ekstraktsionnye metody v analiticheskoi khimii*. *Trudy Komissii po anal. khim.* XIV. AN SSSR, Moscow 1963.
[51] Stary, J.: *The Solvent Extraction of Metal Chelates*. Pergamon, Oxford 1964.
[52] Palei, I. P. (ed.): *Metody kontsentrirovaniya v analiticheskoi khimii*. *Trudy Komissii po anal. khim.* XV. Nauka, Moscow 1965.
[53] Korkisch, J.: *Modern Methods for the Separation of Rare Metal Ions*. Pergamon, Oxford 1969.
[54] Zolotov, Yu. A. (ed.): *Ekstraktsiya v analiticheskoi khimii i radiokhimii*. Inostr. Lit., Moscow 1961.
[55] De, A. K., Khopkar, S. M., Chalmers, R. A.: *Solvent Extraction of Metals*. Van Nostrand, New York 1970.
[56] Zolotov, Yu. A., Iofa, B. Z., Chuchalin, L. K.: *Ekstraktsiya galogenidnykh kompleksov metallov*. Nauka, Moscow 1973.
[57] Zolotov, Yu. A., Kuzmin, N. M.: *Ekstraktsionnoe kontsentrirovanie*. Khimiya, Moscow 1971.
[58] Nikolotova, E. I., Kartashova, N. A.: *Ekstraktsiya neutralnymi soedineniyami*. Vol. 1, Atomizdat, Moscow 1976.
[59] Mezhov, E.: *Ekstraktsiya solyami aminov i chetvertichnykh ammonievykh osnovanii*. Vol. 2, Atomizdat, Moscow 1977.
[60] Martinov, B. V.: *Ekstraktsiya organicheskimi kislotami i ikh solyami*. Vol. 3, Atomizdat, Moscow 1978.
[61] Vanifatova, N. G., Seryakova, I. V., Zolotov, Yu. A.: *Ekstraktsiya metallov neitralnymi serusoderzhashchimi soedineniyami*. Nauka, Moscow 1980.
[62] Smith, J., Nelissen, J., Van Grieken, R.: *Anal. Chim. Acta* **111**, 215 (1979).
[63] Schubiger, P. A., Bajo, S., Ferreira Marques, F., Wyttenbach, A., Jakob, K.: *Anal. Chim. Acta* **106**, 103 (1979).
[64] Mohai, M., Upor, E.: *Magy. Kém. Folyóirat* **72**, 394 (1966).
[65] Upor, E.: *Proc. 3rd Symposium on Coordination Chemistry*. Vol. 1. Akadémiai Kiadó, Budapest, 1970, p. 143.
[66] Dukov, I. L., Genov, L. Ch.: *Acta Chim. Acad. Sci. Hung.* **102**, 201 (1979).
[67] Navratil, O., Dubinyin, O.: *J. Inorg. Nucl. Chem.* **31**, 2927 (1969).
[68] Filakov, Yu. Ya., Zhitomirskii, A. N., Tarasenko, Yu. A.: *Fizicheskaya khimiya nevodnykh rastvorov*. Khimiya, Leningrad 1973.
[69] Gutman, V.: *Coordination Chemistry in Non-aqueous Solutions*. Springer, New York 1968.
[70] Ferraro, J. R., Peppard, D. F.: *Nucl. Sci. Eng.* **16**, 389 (1963).
[71] Shmidt, V. S.: *Ekstraktsiya aminami*. Atomizdat, Moscow 1970.

[72] Upor, E., Klesch, K.: *Magy. Kém. Folyóirat,* **73,** 270 (1967).

[73] Zadorozhnaya, E. M.: *Zh. Anal. Khim.* **29,** 993 (1974).

[74] Shizonenko, N. T., Gudzenko, L. V., Nikolenko, A. Ya.: *Zh. Anal. Khim.* **35,** 279 (1980).

[75] Mohai, M.: *Magy. Kém. Folyóirat* **81,** 164 (1975).

[76] Mohai, M., Upor, E.: *Magy. Kém. Folyóirat* **73,** 270 (1967).

[77] Bikovtsova, T. T., Cherkovnitskaya, I. A.: *Zh. Anal. Khim.* **35,** 1925 (1980).

[78] Kletenik, Yu. B., Bikovskaya, J. L.: *Zh. Anal. Khim.* **20,** 567 (1965).

[79] Cerrai, E., Ghersini, G.: *Anal. Chim. Acta* **37,** 925 (1967).

[80] Upor, E., Szalay, J., Klesch, K.: *Magy. Kém. Folyóirat* **74,** 438 (1968).

[81] Zhivopistsev, V. P., Petrov, B. I., Makhnev, J. A., Ponoshov, I. N.: Ref. *Zh. Khimiya* **21G,** 44 (1974).

[82] Flaschka, H. A., Barnas, R., Paschal, D.: *Anal. Letters* **5,** 523 (1972).

[83] Kuznetsov, V. I., Seryakova, I. V.: *Zh. Anal. Khim.* **14,** 161 (1959).

[84] Gorshkov, V. V., Orlova, L. P.: In: *Metody opredeleniya mikroelementov v pochvakh, rasteniyakh i vodakh.* Vazhenin, I. G. (ed.). Kolos, Moscow 1974.

[85] Conway, E. J.: *Microdiffusion Analysis and Volumetric Error.* Crosby–Lockwood, London 1950.

[86] Tolnay, P.: *A mikrodiffúziós technika alkalmazása a kémiai analízisben (Application of the microdiffusion technique in chemical analysis).* Felsőoktatási jegyzetellátó, Budapest 1955.

[87] Teleshova, R. L.: *Khimicheskii analiz mineralov i ikh khimicheskikh sostav.* Nauka, Moscow 1964, p. 140.

[88] Auffarht, J., Klockow, D.: *Anal. Chim. Acta,* **111,** 89 (1979).

[89] Novák, J., Tomásová, H.: *Chem. listy* **64,** 1039 (1970).

[90] Bode, H., Hackmann, K.: *Z. Anal. Chem.* **229,** 261 (1967).

[91] Schnepfe, M.: *Talanta* **20,** 175 (1973).

[92] Lorec, S.: *Chim. Anal.* **49,** 557 (1967).

[93] Klinger, P., Koch, W.: *Stahl und Eisen* **68,** 321 (1948).

[94] Kozlovskii, M. T.: *Trudy Komissii po anal. khim.* **15,** 132 (1965).

[95] Mika, J.: *Kohászati elemzések (Metallurgical analyses).* Műszaki Könyvkiadó, Budapest 1959.

[96] Németh, K.: *Applied Sciences and Development* **8,** 89 (1977).

[97] Németh, K., Varjú, M. (eds.): *First International Symposium on Electro-Ultrafiltration (EUF).* Budapest, 6–10 May 1980.

[98] Pugachevich, P. P.: *Rabota so rtutyu.* Khimiya, Moscow 1972.

[99] Poluektov, N. S., Kononenko, L. I.: *Spektrofotometricheskie metody opredeleniya individualnykh redkozemelnykh elementov.* Naukova Dumka, Kiev 1968.

[100] Schleicher, A.: *Electroanalytische Schnellmethoden.* F. Enke, Stuttgart 1947.

[101] Lingane, J.: *Electroanalytical Chemistry.* Interscience, New York 1958.

[102] Charlot, G., Badoz–Lambing, J., Tremillon, B.: *Electrochemical Reactions.* Elsevier, Amsterdam 1962.

[102a] Zolotov, Yu. A., Kuzmin, N. M.: *Kontsentrirovanie mikroelementov.* Khimiya, Moscow, 1982.

[102b] Kuzmin, N. M.: *Zav. Lab.,* **48,** 2, 11 (1982).

79

4. *Preparation of samples for analysis*

In this chapter we shall deal in detail with the operations whereby the component to be determined is transformed into the dissolved state necessary for the analysis, including the elimination of the circumstances inhibiting the determination or the required separations. As concerns the latter, we can include organic compounds causing absorbance or complexing the metal ions (biological materials, foodstuffs, natural waters, etc.).

Since the sampling, the processing of the sample or its storage cause difficulties, or must be carried out according to exact specifications, for certain types of samples, these too will be dealt with briefly here.

Excellent surveys of the application of sampling, storage and chemical pretreatment are to be found in the works of Sandell and Onishi [1], Reeves and Brooks [2], Dolezhal *et al.* [3], Gorsuch [4] and Bock [5].

4.1. Sampling

The sampling procedure must ensure that a sufficient amount of material is obtained for analysis, and that this is representative of the average composition of the substance in question.

Macroscopically, in the case of a substance in the liquid state that can be regarded as one phase (blood, milk, etc.), the quantity of sample is in effect not regulated by any restriction.

With non-homogeneous systems (e.g. sludge originating from the hydrometallurgical processing of ores or from sewage works), the sedimentation of the sample must be prevented by manual or mechanical shaking or by mixing during the sampling operation. To ensure that the sample has a composition representative of the substance, it is usually necessary to take a sample many times larger than that required for the analysis. Of course, care must be taken that in the following step, i.e. the measurement of the actual sample for analysis, the homogenization is performed equally scrupulously.

If only the solution above a sediment is to be analyzed, homogenization is not needed and is rather to be avoided; thus, mud should not be stirred up in the cas e ofa natural water. In most cases, however, it is necessary to separate the two phases as soon as possible, as otherwise a false result may be obtained. This is the situation, for instance, because of the further advance of the process in the work-up of ores, or because of the sorption of certain trace elements on the mud, or possibly their dissolution out from it, in the case of water samples. In the event of labile equilibria (such as in water samples where there is a relatively high OH^- concentration (10^{-6}–10^{-8} M) or breakdown of the CO_2–HCO_3^- equilibrium due to the release of CO_2), after filtration on a 0.6 μm filter it is advisable to acidify the sample, so that losses cannot occur as a result of sorption or even precipitation.

For sampling of an aerosol or of dust in air, the general solution is to bubble the material through an appropriate absorbing liquid or to suck it through a purpose-made filter (e.g. a Millipore filter). Sampling without loss or without fractionation is particularly difficult in the latter case. A more extensive treatment of this subject may be found in the works of Peregud and Garnet [6], Fuks [7], Jacobs [8] and Várkonyi and Cziczó [9].

The specifications for the sampling of solid material that is homogeneous, but consists of grains or lumps of various sizes, cover the quantity of the basic sample and the combination of comminution and aliquotting [10]. Detailed guides as to the sampling procedures, among others, are given by Laitinen and Harris [11, 12].

Correct sampling must ensure that the error arising from the dissimilarity is less than the analytical error. The control on the homogeneity cannot be carried out merely by evaluating the scatters in the analytical results of various origins, as these may be burdened by numerous other sources of error too. In general, the Fisher test is used to control the homogeneity. The essence of this is as follows: the compositions of some (e.g. 20) aliquot samples taken from the total mass of the sample are each determined once by the available method with the lowest scatter, and the composition of one of these aliquots is then determined in the same num .er of parallel measurements. The scatters of these two measurement series are calculated, as is the parameter F:

$$F = \frac{(S_{inter})^2}{(S_{intra})^2}$$

where S_{inter} is the empirical scatter calculated from the analytical results on the different aliquots, and S_{intra} is the analogous scatter for the one, given aliquot. In the event of ideal homogeneity, $F=1$; inhomogeneity increases the value of F. It is of value to carry out this examination for elements with different concentrations [13].

In practice, the sample-processing specifications are elaborated on the basis of approximate calculations and experiments. In these, consideration must naturally be paid to whether sampling is performed continuously or sporadically. For instance, Sporbeck [14] reports a continuous ore sample-processing system, in which four-stage comminution is used to decrease the grain size of the original sample from 100 mm to 0.15 mm, and its quantity from 10 tons to 0.36 kg.

For the preparation of ore, slag and coal samples, Dymov [15] proposes the correlation $Q=kd^2$ between the average grain diameter d and the necessary sample quantity Q, where k is a proportionality factor with a value between 22 and 0.25, depending on the composition.

The sampling of high-value ore shipments is also regulated by standards, because of the variable composition and quantity.

Depending on the circumstances and the requirements, the means of sampling may be very varied; they are well reviewed by Sporbeck [14].

In addition to the simple intermittent manual sampling in the case of liquid, solid, granular or powder samples, intermittent or continuous mechanical sampling is also possible; for important technological controls the latter is indispensable. In certain cases, continuous sampling forms a part of continuous or intermittent automated analysis.

Samples for the examination of solids of larger dimensions (e.g. metal sheet or plastic blocks) are produced by some kind of cutting.

It is important that the sample should not be contaminated with an element to be determined or with elements disturbing the analysis. In most cases this means the danger stemming from contamination by the material of the sampling equipment, or possibly from the insufficient cleaning of this. Attention must often be given to the possibility that the surface of the sample is contaminated. It is well known, for example, that plant samples may be contaminated with dust, fertilizer or some spray solution. The samples must in this case be washed with water (possibly with the addition of a wetting agent). Although this washing may give rise to losses in certain elements, experience nevertheless shows [16] that the error is lower this way. In the analysis of high-purity metals and semiconductors (e.g. silicon single-crystals), the sample is purified from possible surface contamination or an oxide layer by mild treatment with acid [17].

4.2. Storage of samples ▌

During the transporting and subsequent storage of a sample, care must be taken that it does not undergo any change which will falsify the analytical results. Primarily with solution samples, changes must be prevented which alter the original equilibria. On the storage of water samples or aqueous solutions, it is possible to prevent precipitate formation or hydrolysis (which promotes sorption on the walls of the vessel) by appropriate acidification of the solution, possibly by the addition of complex formers that do not disturb the subsequent course of the analysis, and by the use of suitable vessels [18–20].

With water samples from deep strata, where the liberation of CO_2 or the entry of O_2 would immediately lead to precipitate formation [$Fe(OH)_3$, $CaCO_3$, etc.], it is particularly important to prevent these processes. This can be achieved during the sampling, with a specially made "deep-water sampler".

Unless they are analyzed within a short time, biological materials (blood, tissues, urine) and perishable foodstuffs must be stored in a refrigerator. One possible solution is to measure a suitable aliquot of the sample and to add the reagents required in the first step (e.g. acids needed for decomposition).

In the case of non-labile powder-like samples, care must be taken that they do not undergo self-fractionation as a consequence of the inhomogeneous grain distribution; accordingly, before the sample is measured out it must be homogenized (preferably in a sealed vessel which is only partially occupied by the sample).

When samples sensitive to moisture and oxidation (e.g. sulphide minerals) are left to stand in the air, changes may occur in their composition. Such samples must be carefully stored in sealed bottles [21].

Samples may also be contaminated by the material in which they are packed or the vessels in which they are kept. Thought must be given to atmospheric dust too, particularly in laboratories situated in an area of industrial installations or in their vicinity. If metal powders are stored in plastic vessels for a prolonged period, their Cl and Sb contents can increase [22]; the same is true for the boron contents of plant samples stored in wrapping-paper.

4.3. Sample processing and pretreatment

Certain samples must be dried before the sampling. This occurs for sludgy samples, where the moisture must be removed prior to the dry-milling. It also holds for plant samples, since the initial samples with high moisture content cannot be comminuted. Drying is best performed for a prolonged period at lower temperature (80°C).

Plant samples are often crushed in a hammer mill or possibly a ball mill. Special care must be taken with regard to contamination here, because of the low concentrations of the trace elements. This can be solved partly by choosing a suitable structural material and partly by allowing only a short contact time.

As already mentioned in connection with the sampling, samples are prepared from metals by machining, filing or some other mechanical operation. Drillings or filings need no further comminution before the dissolution operation preceding the analysis. More recent examples include the use of spark machining and even a laser technique [23].

Before the analysis of ores, rocks, soils and coals, the sample must be crushed to a particle size smaller than 100 μm. Samples consisting of large-sized lumps can be brought into such a state by means of a system of various crushing and milling equipment (a crushing roller, a jaw crusher, a ring roll mill, a ball mill, etc. [14]). It is indispensable to check the efficiency of the operations with sieves of appropriate mesh size, and to crush the non-fine material further.

This is particularly important in the final stage: it may not be possible to attack the larger grains completely (especially since some of the hardest minerals are also among the most chemically resistant ones).

Crushing is often carried out in a mortar. This is generally made of porcelain, corundum, boron carbide or agate; the latter almost totally excludes the possibility of any contaminants entering the sample during the operation, and no losses occur via sorption on the walls of the vessel.

For plant analysis or in the determination of iron, Chapman and Pratt [24] recommend the use of a ball made of phosphor bronze instead of iron in the ball mill.

Besides the material of the equipment itself, another potential source of contamination is some other sample. Thus, in cases when the concentration of the element to be determined may differ by even several orders of magnitude in samples that are to be processed one after another (e.g. in drill core samples originating from geological research or from mine drilling), the equipment must always be cleaned very thoroughly. Among the methods for this is "flushing-out" with an inert sample.

It is generally not necessary to powder malleable or tough plastics, foodstuffs and biological materials. If this is nevertheless required, they can be brought into a brittle state by cooling to solid-CO_2 or liquid-nitrogen temperature, when they can readily be comminuted.

4.4. Pretreatment of samples of high organic matter content

Biological substances, plant samples, organic solvents, plastics, foodstuffs, carbonolites and many other industrial products or waste materials contain (or even predominantly consist of) a large proportion of organic compounds. The original medium is most often not suitable for analysis without pretreatment, for the following reasons:

(a) the element to be determined is kept in complex form by the organic acids or other ligands present in the sample;

(b) the basic colour of the sample disturbs the photometric determination (blood serum; humic acid in natural waters; etc.);

(c) frothing, emulsion formation or precipitation would occur in the course of the analysis;

(d) the element to be determined is not in the dissolved state.

Naturally, it does not hold generally that the organic compounds in the sample exert a disturbing effect. The extent to which they must be removed for a given analytical task to be solved depends on the nature (composition) of the sample, the component to be determined, and the quantity of sample to be taken for analysis.

The procedures ensuring the removal of interfering organic compounds and the dissolution of the element to be determined can be classified as ollows:

(a) there is no need for pretreatment; the analysis can be performed directly (on a solution) or after dissolution of the sample;

(b) evaporation (possibly in the presence of certain additives);

(c) boiling with acid or base in order to decompose complex compounds; partial decomposition of the organic matter;

(d) wet decomposition of the sample;

(e) digestion.

Determination without pretreatment

This is primarily applicable in non-photometric methods (neutron activation, atomic absorption, etc.), but in certain cases it is possible here, too.

The photometric determination of lead is known after the precipitation of Pb^{2+} on a CaC_2O_4 carrier [25], as is the determination of phosphate in the form of molybdenum blue following monochloroacetic acid precipitation from blood plasma [26]. The denaturing of proteins (and protein decomposition products) with mineral acids or with trichloroacetic acid is otherwise a frequent procedure in the case of biological materials; however, this must be classified not as sample pretreatment, but as separation from disturbing components.

In the analysis of red wine, the HCN traces remaining from the $K_4[Fe(CN)_6]$ used for clarification are driven out of the sample by boiling in a stream of CO_2 (distillation of the sample is prevented with a reflux condenser), and are determined after sorption in $AgNO_3$ solution [27].

Determination after evaporation

The components disturbing the determination can in certain cases be removed by evaporation of the sample. However, it is generally advisable to carry out this operation in the presence of some additive which prevents the removal of the volatile elements. For example, in the determination of the metal impurities in various organic solvents (chloroform, carbon tetrachloride, ethyl acetate, amyl alcohol, etc.), the solvent is evaporated off in the presence of oxalic acid. For measurement of the nitrite and nitrate contents of red wine, 10 cm^3 of sample is evaporated to dryness after the addition of active carbon, NaCl and glacial acetic acid, and the determination is then performed with diphenylamine [27]. The active carbon is added in this case not to prevent the loss, but to bind the residual colouring matter.

The evaporation method can naturally be used only for samples with a comparatively low dry-matter content (such as in the above cases). At times, the residue is further ignited, decomposed in the wet, or fused. For the determination of non-volatile elements in samples with high organic-matter contents, it can also happen that the evaporation is carried out without additives to remove volatile elements interfering in the determination.

Acidic or alkaline boiling to decompose complex compounds; partial destruction of organic matter. This can similarly lead to a satisfactory solution on occasions. Simple boiling with acid is sufficient, for example, in the determination of the copper or iron content of blood [4]. In the determination of the phosphate content of sewage-works sludge, the organic matter is oxidized with $K_2S_2O_8$ in alkaline medium [28]. In the determination of the uranium content of urine, boiling with nitric acid is sufficient to release the UO_2^{2+} from organic complexes and to allow its extraction with HDBP. This boiling also eliminates the emulsion formation which otherwise inhibits separation of the phases [29].

Wet decomposition of the sample

This is the most generally applied method of bringing samples containing organic compounds into a state suitable for analysis.

Depending on the nature of the sample, the mechanism of the reaction producing the decomposition can be primarily oxidation, reduction, esterification, hydrolysis, dehydration, nitration, or even a combination of some of these. Accordingly, the reagents are various mineral acids, oxidants and reductants, and their appropriate combinations. A very detailed survey of this topic is to be found in the work of Gorsuch [4].

The mechanisms of action of the most important reagents may be summarized as follows:

H_2SO_4: Its effect is dehydration and oxidation ($H_2SO_4 \rightarrow H_2O + SO_2 + O$). When used together with other acids, its main role is possibly the enhancement of their effects by means of its higher boiling-point. One mode of its application is continuation of the H_2SO_4 treatment after the addition of another reagent. It has the drawbacks that it is relatively difficult to remove the excess because of its high boiling-point, and that further treatment of the sample is necessary in the presence of ions forming insoluble sulphates (Pb^{2+}, Ba^{2+}, much Ca^{2+}).

HNO_3: This readily reacts with aromatic and even aliphatic compounds in oxidation, esterification or nitration processes. It is commonly used together with H_2SO_4 or $HClO_4$. In certain cases its nitrating ability is disadvantageous, for in this way stable compounds are formed; the sample in these cases is first treated with dilute HNO_3 or H_2SO_4.

$HClO_4$: This is used alone or with other acids (H_2SO_4, HNO_3). Its main advantage is its oxidative ability in the presence of organic compounds. Because of its tendency to decompose, the use of $HClO_4$ is permitted only with appropriate safety precautions. (The boiling-boint of the 72% azeotrope is 203°C; its vapour or perchlorates deposited on the overheated parts of the vessel are liable to explode. No explosion occurs if a comparatively small amount of sample is taken or if overheating and evaporation to dryness are prevented.)

H_2O_2: Its mechanism of action is an oxidative–hydrolytic process; it mainly converts the $C=C$ bond to an easily hydrolyzable tertiary alcohol group. The H_2SO_4 used as additive readily restores the $C=C$ bond by dehydration, and thus it is preferable to add the H_2O_2 after charring. The strong oxidative effect is a result of the formation of persulphuric acid (H_2SO_5). When it is used in combination with HNO_3, the electron-acceptor property of NO_2 is utilized; the incorporation of this facilitates the entrance of an OH group.

There are many other possibilities besides the most important reagents listed; some of these will be mentioned below.

The $HNO_3 + H_2SO_4 + HClO_4$ acid mixture can be used most generally, e.g. for the analysis of fat, flour, meat, jam and various liquid samples. Solid or syrupy samples are measured out onto filter paper and transferred together with the paper to the Kjeldahl flask used for decomposition.

The wet decomposition of plant samples is carried out with $H_2SO_4 +$

$+ HNO_3$ [30], $H_2SO_4 + HClO_4$ [31], or $HClO_4 + HNO_3$ [32], after preliminary drying and mincing (or chopping or grinding). In the latter case a 1 : 5 $HClO_4 : HNO_3$ mixture is used; the sample is first left to stand overnight in the acid mixture, and this is then heated on the following day. As a safety precaution the largest sample taken is 10–12 g.

A suitable method for the dissolution of the elements Fe, Mn, Zn and Cu in plant matter is to hydrolyze the sample with HCl in a closed space at 110°C. The dissolution is quantitative and can be performed simply [33].

$H_2SO_4 + H_2O_2$ is also of use in plant and foodstuffs analysis [34]. This method is also employed in the laboratory to destroy anion-exchange resin. The otherwise stable pyridine is likewise decomposed in this way.

For the analysis of biological samples, HNO_3, $HNO_3 + H_2O_2$, and $HClO_4$ (alone or with another acid) are equally useful. In the frequent situation when there is only a small quantity of sample, HNO_3 in particular permits a rapid and often low-temperature reaction.

Kovács and Orosz [34a] discuss the dry-ashing of mixed feedstuffs and the subsequent wet digestion. In the latter step they favour dissolution with $HNO_3 + H_2O_2$.

$KMnO_4$ too can be used to advantage in certain cases. For example, this is a standard method for the determination of the volatile mercury in the analysis of jam, urine or faeces, and dissolved-out humic substances are also decomposed by $KMnO_4$ [35].

A procedure similarly known in soil investigations is oxidation with $H_2O_2 + Fe(OH)_3$; here, the presence of OH exerts a catalytic effect [36].

The oxidizing action of nitrogen oxides is made use of to decompose organic matter in soil samples; nitrogen oxide gases developed from $H_2SO_4 + HNO_3$ are led into soil samples that have been converted into muds with an acid mixture in a flask [37].

To destroy the interfering yellow nitro compounds remaining after such decomposition, Pearson [38] recommends evaporation with saturated $(NH_4)_2C_2O_4$ solution up to the appearance of fumes of SO_3 formed from the H_2SO_4 added for the decomposition.

Pearson likewise proposes "cold" oxidation, primarily for the analysis of fats. In this procedure the sample is repeatedly evaporated to dryness with HNO_3 containing a little H_2SO_4, at 310–350°C.

More recently, a mixture of H_2O_2, H_2SO_4 and HNO_3 in suitable proportions has been suggested for the decomposition of organic substances (e.g. carboxymethyl cellulose) [39].

The book of Pomeranz and Meloan [40] contains a good compilation of the apparatus used for wet decomposition. The aim in the construction of this apparatus is to minimize losses. This refers to mechanical losses (sputtering, frothing) and to losses due to the volatilization of compounds with low boiling-points.

The main danger of frothing generally arises in the first stage of decomposition of organic compounds, while sputtering naturally occurs later, when the solutions have become concentrated. Steps must be taken to prevent such losses not only by the use of suitable apparatus (e.g. a Kjeldahl flask or a flask permitting reflux condensation and distillation, i.e. the free departure of the gases), but also by the regulation of temperature. As already mentioned, it is customary to commence the decomposition of the sample at room temperature [32] and to raise the temperature gradually; it is even possible that the heating must be stopped periodically, e.g. by removing the gas flame from under the Kjeldahl flask.

Decrease of the mechanical losses is also the aim if the samples can possibly be evaporated to low volume before the decomposition, e.g. for alcohol-containing fluids, or if heating is applied until the bulk of the material taken has been removed, e.g. fats, oils [38].

Anti-foam agents, e.g. silicone oil, alcohols are added to diminish the frothing of liquids containing high quantities of protein (blood, milk), and a mild current of air is blown onto the surface of the solution [5].

Wet decomposition has many advantages in the analysis of samples containing much organic matter. These advantages in most cases include speed, the simple apparatus, and the fact that the element to be determined is generally obtained in an appropriate solution. It is not necessary to fear that losses will be caused by incorporation into the material of the vessel. This procedure has the drawbacks that the sample taken cannot be too large, and the decomposition process requires constant supervision. When the method is used, attention must be paid to the possibility of losses due to volatility, and also to any impurities added with the relatively large amount of reagent. The former is usually less of a difficulty than in dry ashing, but the latter point (contamination) can give rise to serious problems.

There may be many causes for the losses due to volatility, depending on the nature of the sample being examined and the reagents used for the decomposition. In the determination of the Ni, Cu, V and Fe contents of petroleum derivatives, for instance, the loss may be caused by volatile porphyrin complexes [40]. As a consequence of the low boiling-points of many metal chlorides (Cr, Zn, Sb, Cd, Ga, Hg, Sn, V), losses may occur during the boiling of these metal ions with HCl or NH_4Cl. Many compounds of Hg, Se and As are volatile. The investigations by Bortlisz [41] revealed that, if the temperature is not controlled sufficiently, the use of a sand-bath to evaporate neutral or acidic solutions to dryness is accompanied by the loss of 7–27% of the Zn, Cu, Ni, Fe, Cr, Mn and Pb content.

Losses can be avoided by means of the correct choice of the digestion mixture and the technical conditions, including the use of a reflux condenser to prevent the departure of compounds with low boiling-points.

Heating in a closed space with nitric acid (or possibly with an acid mixture) has the aim of the definite exclusion of losses, and at the same time the rapid performance of the decomposition. A long-known variant of this is heating in a glass ampoule placed inside a protective cover (Carius method). Difficulties here involve the danger of injury when the ampoule is opened, because of the pressure of the gaseous decomposition products, and also the possibility of losses due to the spurting of the material.

Technically a very good solution is digestion in a teflon bomb with 65% or 100% HNO_3. By digestion in $HF + HNO_3$ in a teflon vessel simultaneously suitable for distillation, elements forming volatile fluorides (B, Si, Ge, etc.) can at the same time be separated from disturbing components.

Dry ashing

In principle, the course of this can be broken down into the following stages:

(a) removal of moisture;

(b) removal of volatile substances, including the products formed in thermal decomposition or oxidation (these may possibly be further oxidized in the gas phase);

(c) progressive oxidation of the non-volatile residue until the organic matter has been totally decomposed.

The individual phases of the process are naturally not differentiated and, depending on the character of the sample, they are not all of the same importance.

If these processes are separate in the technical solution, the temperatures of the first two steps are generally lower in the interest of inhibiting the loss of material. For instance, a rotary evaporator is recommended for the evaporation in the analysis of sewage samples [41], and heating at 30°C in a heated vacuum drier for the determination of heavy metals in blood serum [42]. A general technique is drying under an infrared lamp or under a bulb on a heating plate.

The sample is ashed in a closed space, in a gas stream, or in an open vessel.

Ashing in a closed space. A method that has long been known for the analysis of organic substances is the "oxygen-bottle method" (Hempel, Schöniger). In this the sample is burnt in oxygen; various technical procedures are used, but the combustion is generally begun with a platinum catalyst. For samples that are oxidized with difficulty (halogen compounds, certain biological materials, plastics), an additive promoting combustion (sugar, Na_2O_2, hydrocarbons) is also mixed in. The component to be determined is obtained in the absorbing solution (bases, acids, distilled water, etc.) situated in the flask.

The method has the advantages of technical simplicity, speed, low temperature and freedom from loss because of the closed space. However, it also has the great disadvantage that it is suitable only for small samples as the available amount of oxygen for stoichiometric combustion is slight. It is mainly of value, therefore, if the concentration of the element to be determined is not too low. Combustion in an oxygen bomb helps in this situation. The O_2 pressure at room temperature can then be increased to even 4 MPa, and hence the sample taken can be much greater (1–4 g). The sample is placed in a quartz or platinum vessel into the bomb. It may possibly be enclosed in a gelatin capsule [43]. The combustion is started electrically. The rate can be regulated by the addition of substances that slow down or accelerate the combustion process (e.g. SiO_2 or NH_4NO_3). This procedure is used to obtain aqueous solutions of the fluorine content of plastics (absorbed in 1 M NaOH), mercury in rice (1 M HNO_3) or lead in foodstuffs [44].

Ashing in a gas stream. This technique can be used primarily for the determination of volatile elements, or elements which form volatile compounds in an oxidizing, or more rarely a reducing atmosphere. The traditional procedure is heating at 900–1400°C in a current of oxygen, air or some other gas in a quartz combustion tube, the combustion product containing the element to be determined being led into a suitable absorber (usually a liquid).

The modern variant of combustion is "low-temperature ashing" with excited oxygen [45, 46]. In this, O_2 at 100–1000 Pa is led through a high-frequency (microwave) field, when atomic oxygen is formed with fairly high efficiency; this promotes the oxidation much better. The short lifetime of atomic oxygen (1 s) means that the sample must be situated directly in the field. The temperature of combustion is 100–200°C, which allows the loss-free determination of volatile elements (Se, As, Hg, etc.). The method is also a suitable one for the analysis of materials that burn with difficulty (plastics, mud, soot). One slight drawback is that the reaction is slower because of the low oxygen pressure, but even so the duration of ashing is not more than one hour.

Ashing in an open vessel. This is the oldest and to a certain extent the simplest means of removing organic compounds. In general this is of use in rock and ore analysis or with organic samples, to eliminate the humic compounds that disturb the analysis. In the knowledge of the preliminary treatment in the investigation of ore concentrates, for instance, the removal of the flotation oil residue requires heating of the sample at 500°C before the digestion [47].

When samples of mineral origin are heated, there are relatively few elements (Hg, Ge, Se) for which a slight loss must be feared, for the compounds comprising the rock or ash are generally natural "additives" which inhibit loss. Two large sources of error must be reckoned with in the ashing of mineral oil products or biological samples: one is the loss due to the volatility, while the other is the partial incorporation of the element to be determined into the material of the vessel ("retention loss") or contamination from the material of the vessel.

The procedure followed in ashing is generally that the sample is spread out in a thin layer in a suitable vessel (ashing bowl, crucible) and first dried (under an infrared lamp, on a heating plate, over a gas flame, in a vacuum-drying apparatus); it is then heated by gradual elevation of the temperature. This is best done in an ignition furnace, i.e. at a regulated temperature and at the same time slowly, because of the relatively closed space. If possible, it must be avoided that the sample bursts into flame, as this causes losses.

The material is first heated at the temperature of tarring or charring (300° C), and then at 450–550°C. The optimum conditions must naturally be decided in accordance with the natures of the sample and the element to be determined; this may possibly necessitate a higher temperature.

Ashing has the advantages that even large samples can be taken and there is no need for supervision during the otherwise relatively slow operation. Its drawback is the possibility of losses, mentioned in the introduction.

For a preliminary assessment of the losses due to volatility, the following possibilities must be considered:

(a) the possibility that the element to be determined is present in volatile form (lead tetraethyl, mercury in virtually all compounds, etc.);

(b) the possibility of some inorganic reaction in which a volatile compound is formed (reaction with the ash component, or perhaps with an inappropriately chosen additive);

(c) the possibility of formation of a volatile compound with the organic compounds present (the lead in PVC may be converted to $PbCl_2$; volatile metals (Cd) may be produced on the reducing action of carbon formed during the heating).

Detailed data on the losses occurring during combustion may be found in the works of Gorsuch [4], Bock [5] and Varjú [48, 49].

A general source of error in ashing is the retention loss, when the element to be determined undergoes partial incorporation into the vessel used for ashing.

The material of the vessel is usually porcelain, quartz or platinum. Rowan *et al.* [49a] use a nickel crucible in the dry-ashing of foodstuffs. Errors may be caused in porcelain by the metal oxides in the glaze, and in quartz by the silica, which forms complex silicates. For the latter reason, particularly the well polarizable PbO can be incorporated into the active sites in the crystal lattice of quartz, but metals (Cu, Ag) formed in the course of reduction processes during ashing may also be incorporated. These processes can occur with the ash components too, and they may mean a loss in the event of the customary dissolution of the ignition residue, or may necessitate digestion.

The NaCl frequently present in samples may increase the losses in an interesting way, over and above the losses due to the volatility of covalent metal chlorides, for Cl^- weakens the quartz crystal lattice and thus facilitates incorporation.

When platinum vessels are used, the danger exists that certain metals (primarily the more noble metals) can give rise to alloying.

Ashing with additives

It is frequent for ashing to be carried out in the presence of some additive. This addition may have the following aims:

(a) promotion of oxidation;
(b) prevention of losses due to volatility;
(c) decrease of losses due to incorporation.

There is a fairly wide choice of additives varying in chemical nature; the additive chosen depends on the above aims, the nature of the sample, the properties of the element to be determined, and the vessel used.

The most important of the additives promoting oxidation are HNO_3 and nitrates, e g. $Mg(NO_3)_2$. (Although we speak of "dry" ashing, the salts are added in aqueous solution, so as to soak into the entire sample during evaporation.)

The addition of HNO_3 usually takes place towards the end of the ashing, with a view to the oxidation of the charred residue, mainly in the analysis of plant and biological samples [50].

In fact, oxidizing fusion may be classified here too; for example, in the presence of Na_2O_2 (possibly $+ NaClO_4 + NaNO_3$) in a Parr bomb [4], the oxidation is complete within a few minutes at 300–400°C. This procedure can be used favourably for the analysis of polyethylene, animal blood, wheat flour and other samples, for instance. It has the advantage that the alkalinity of the melt means that there is no loss of volatile elements (Sb, As, etc.). A disadvantage is the danger of contamination from the large quantity (20-fold) of fusion mixture.

Another group of additives bind compounds of low boiling-point, but because of their chemical nature and diluting action they generally decrease the incorporation into the material of the vessel too.

A frequently employed additive is H_2SO_4; this decomposes volatile organic complexes and transforms chlorides into sulphates. Various basic compounds (alkali metal and alkaline earth metal hydroxides, carbonates and acetates) mainly bind anions. As an example, for the determination

95

of fluoride in a foodstuff sample, the material is heated in a mixture with $Ca(OH)_2$ [38]; in the case of plants, analysis follows impregnation with magnesium acetate solution and ashing [51].

Boron may be lost from certain plants even during the drying; in boron determinations, therefore, the wet sample must be dried after spraying with NaOH solution [23]. For the determination of Ge in coals, the sample must be mixed with CaO before the ashing [4].

Another additive is the Eschka mixture ($MgO + Na_2CO_3$), primarily used in the determination of boron [52].

As the above examples show, ashing in the presence of additives is very favourable in many cases, but care must be taken to examine the purity of the additive to be used.

4.5. Digestion of rocks and ores

In the foregoing subsections, many methods have already been considered which are suitable for the digestion of rocks and ores and for the dissolution of the component to be determined.

In the present section, however, for the sake of a concise treatment we shall deal separately with the digestion of rocks and ores, and samples to be treated in a similar way to these (ceramics, etc.).

The procedure here is governed by the nature of the sample, and also by the nature and concentration of the element to be determined. The procedures recommended for the digestion of certain minerals are not always suitable for the analysis of samples originating from geological research, or samples containing the mineral in question in an impure state. The dissolution should, if possible, be quantitative for all components. (An example of this is the determination of the lead content of rocks, in which the incorporation of Pb^{2+} into the silicate lattice means that decomposition with HF is unavoidable; this is in contradiction with some literature publications, which prescribe simple digestion with mineral acids.)

In the selection of the means of digestion, attention must be paid to the dangers of losses due to the formation of volatile compounds, and also to the possibility of formation of compounds that are difficult to digest (e.g. $BaSO_4$, β-stannic acid); a medium must be created which fits in well to the following operations. Thus, it is advisable to remove SiO_2 even when

96

the digestion could be carried out by simple alkali metal hydroxide fusion without the need for a platinum vessel, but when the presence of silica would disturb the subsequent determination (the determination of tungsten with dithiol); in general in those cases where a silica emulsion would be formed in the extraction following the digestion.

A special position is occupied by the phase-analysis methods. The essence of these is that it is desired to determine only one or other of the forms of an element present in different oxidation states or in different compounds. The key question in these methods is usually selective dissolution, as the subsequent determination is normally not able to distinguish between the originally different forms.

Below we shall discuss the various groups of digestion agents and the operations performed with them. Since the reagents used for digestion also dissolve the material of the vessel to some extent, attention will be paid to the materials to be recommended or avoided in the individual procedures, in an effort to reduce the damage to the vessel and the contamination of the sample.

4.5.1. ACIDIC TREATMENT OF SAMPLES

Acidic treatment does not generally lead to the quantitative dissolution of the samples. Minerals not containing silicates may be exceptions (carbonates, acid-soluble sulphides, etc.); most often, however, even these do not occur in pure form.

Thus, acidic treatment means selective dissolution to a certain extent; its choice is decided by the composition of the sample, the chemical properties and concentration of the component to be determined, and the form in which it is bound.

Digestion with non-oxidizing acids can be achieved without any essential change in the initial oxidation state.

The valency state of certain elements may be altered in this way since, as a complex-forming ligand, the anion of the acid used may change the standard redox potentials of ions of variable valency. For example, the oxidation state of the uranium in rocks can not be established by means of any type of acidic digestion, for it depends on the anion whether the $Fe(II)-Fe(III)$ system behaves as an oxidant or a reductant.

Digestion is rarely performed with a single reagent, but in general with acid mixtures or in the presence of other additives. The most important systems in rock analysis are the following:

HCl: A frequent digestion agent. Among its advantages is the fact that it is a strong acid, and thus its dissolving effect is relatively high; it forms anionic complexes with many metal ions, which may be useful in the later separation operations; with only a few ions it forms insoluble precipitates; its excess can readily be removed by evaporation. It is even able to dissolve natural and artificial silicates (orthite, cement, etc.) containing silica in a not very polymerized state.

It has the disadvantage that certain chlorides (Ga, As, Sb, Sn, Se, Hg) are volatile, and special steps must therefore be taken to retain these; there are very many materials (SiO_2, pyrite, many types of silicate, bauxite minerals, oxides of multivalent metals, etc.) which it does not attack, or to only a slight extent.

A frequent additive in digestion with HCl is H_2O_2, because of the oxidizing action of the atomic chlorine formed.

HNO_3: This is often applied as its 1:3 mixture with HCl (aqua regia), but it is also used in other proportions and with other acids (H_2SO_4, H_3PO_4, $HClO_4$). An advantage is the fact that the nitrogen oxides originating from the decomposition of HNO_3, and the NOCl formed from aqua regia, have oxidizing properties. It is primarily advantageous and effective for the dissolution of sulphides in rocks. It does not form volatile compounds with any elements.

H_3PO_4: After appropriate pretreatment, this may be a very effective digestion agent. The heating of H_3PO_4 at 300°C leads to the formation of condensed phosphoric acid, which is a very aggressive digestion agent [53, 54]. It can be employed to advantage in the digestion of iron ores and bauxites [55, 56].

HF: This is usually applied in combination with other acids, particularly H_2SO_4 or HNO_3. The aim of the addition of the latter is the oxidation of sulphides and metal ions of lower valency. The addition of H_2SO_4 or some other acid may also have the purpose of elevation of the boiling-point of the mixture, or the extraction of water, in order to quantify the reaction $SiO_2 + 6 HF \rightarrow H_2SiF_6 + 2 H_2O$. This procedure is otherwise a double-edged one, for the higher temperature promotes the removal of HF (with the low boiling-point of 106–112°C) before the effect of digestion has been

98

exerted, while in certain cases the digestion process is "frozen" in the presence of mineral acids. In the majority of cases, the most reasonable procedure is as follows: HF is poured onto the sample and the mixture is left to stand for a prolonged period (possibly overnight), heating being applied only after this; the heating may be carried out first in the presence of HF alone, the HF lost is replaced, and the other acid too is then added.

Digestion with HF is often an indispensable procedure for decomposition of the silicate crystal lattice, but it is accompanied by numerous difficulties and disadvantages. Whereas operations with other mineral acids can be carried out in glass vessels, the use of HF requires a platinum or an HF-resistant plastic (primarily teflon or polypropylene) vessel [57].

The fluorides of some elements (Ga, B, V, Se, etc.) are volatile, while those of others (Zr, Th, Ca, Ln) form stable complexes that dissolve with difficulty, especially in the presence of alkali metals. The addition of other acids, predominantly H_2SO_4 or $HClO_4$, may help here.

The residual fluoride traces may disturb the further course of the determination. A possible solution then is heating with H_3BO_3, with the addition of Al^{3+} or Be^{2+} [58].

A generally favourable procedure for minerals that are difficult to decompose is digestion with HF in a teflon bomb at 150–160°C. It must be pointed out that the digestion is not complete in some cases (tourmaline, zircon, cassiterite, titanium dioxide, and some polymorphous aluminium oxide variants) [59].

The losses observed during digestion in a teflon vessel have been examined by Van Eenbergen and Bruninx [60].

A digestion bomb with a vitreous carbon lining may be used successfully to digest organic and inorganic substances. This vessel tolerates even 200°C. The digestion acid is HF, HCl, HNO_3 or a mixture of these. The application of higher temperature increases the oxidation potential and hence shortens the digestion time [61]. In our own experience, complete dissolution is promoted if the vessel is shaken continuously during the heating period (the teflon bomb is built into a shaking-machine).

A good survey of digestion with HF can be found in the publications by Langmyhr and Paus [62–67].

Other combinations too are possible for the digestion of rocks with acids, but these will not be discussed separately here. Instead, examples are tabulated for the digestion of various ores (Table 4.1).

TABLE 4.1.

Reagents used for acidic digestion of various ores

Sample for analysis	Element to be determined	Digestion agent applied	Refs
Silicates	General trace analysis	$HF + H_2SO_4$	[62–67]
Chrome ore	General trace analysis	$H_2SO_4 + HCl + NaCl*$	[68, 73]
Zircon	General trace analysis	$HF + H_2SO_4$; subsequently fusion with KHF_2	[69]
Magnetite	Ge	$H_3PO_4 + KMnO_4$	[69]
Sphalerite	Tl, Co, Ni	$HCl + HNO_3$	[69]
Tetrahedrite	Fe	$H_2SO_4 + (NH_4)_2SO_4$	[69]
Fahl ore	As	HNO_3; subsequently H_2SO_4	[69]
Spinel	General trace analysis	$H_2SO_4 + (NH_4)_2SO_4$	[70]
Pyrochlore	General trace analysis	$H_2SO_4 + (NH_4)_2SO_4$	[70]
Pyrite	General trace analysis	$HCl + HNO_3$	[71]
Monazite	Ln	$H_2SO_4 + H_2O_2$	[72]
Manganese ores (pyrolusite, etc.)	General trace analysis	HCl; the residue must possibly be treated separately	[73]
Fluorite	General trace analysis	H_2SO_4; HCl (possible digestion of residue)	[74]
Mercury ores	General trace analysis	$HCl + HNO_3 + H_2SO_4**$	[74]
Bauxite	General trace analysis	$H_2SO_4 + HCl + HNO_3$	[74]
Sulphide ores	Cu	$H_3PO_4 + HCl$	[75]

* the NaCl serves to form and remove volatile $CrOCl_2$;
** the $HgCl_2$ evaporates.

100

In analyses requiring a lower accuracy, the demand for loss-free dissolution may be forced into the background by the need for simplicity or speed. This is the situation, for example, in the rapid, informatory on-site procedures in geological research [76].

4.5.2. DIGESTION BY FUSION

Digestion by fusion is primarily a suitable procedure for the decomposition of compounds not soluble in acids. The reactions occurring in the course of the digestion can be interpreted on the basis of the Lewis acid–base theory. It holds generally that a basic reagent must be used to digest acidic samples, and an acidic reagent to digest basic samples. The temperature of fusion (or sintering if milder heating is applied) is in the range 500–1000°C. Heterogeneous reactions are accelerated considerably by higher temperature. In a favourable case, a well-selected digestion may require some minutes; for minerals that are digested with difficulty, the process may last 1–2 hours.

The compounds used for digestion can be classified into the following groups, depending on their chemical action and oxidative nature:

(a) alkaline digestion (alkali metal carbonates, borax, alkali metal hydroxides);

(b) acidic digestion (alkali metal pyrosulphates or hydrosulphates, potassium hydrogen fluoride, boron trioxide);

(c) oxidizing digestion (sodium peroxide, alkali metal nitrates with alkali metal carbonate or hydroxide);

(d) reducing digestion (PbO, KCN with Na_2CO_3 or borax);

(e) other digestions ($K_2CO_3 + S$; $Na_2S + S$; etc.).

The glass vessels generally used for acidic dissolution cannot be employed here because of the higher temperature and the corrosive nature of the melts. Suitable materials for the vessels in which the most frequently used digestion agents are applied are listed in Table 4.2.

The temperature and duration of the fusion are governed not only by the melting-point of the digestion reagent, but also by the composition of the substance to be digested. Thus, the digestion of those oxides (iron, chromium and titanium oxides, corundum, refractory materials) or silicates

101

TABLE 4.2

Materials used in fusion digestions

Digestion mixture (compound)	Temperature applied, °C	Material of digestion vessel
Alkali metal carbonates	500–1000	Platinum, zircon, graphite (vitreous carbon)
Alkali metal hydroxides	400–500	Iron, nickel, silver, gold
Na_2O_2	500	Iron, corundum, silver, nickel, zircon
$Na_2CO_3 + Na_2B_4O_7$	800–1200	Platinum
B_2O_3	600	Platinum
$K_2S_2O_7$ (KHSO$_4$)	500	Porcelain, platinum (glass)
KHF_2	700–800	Platinum
Alkali metal carbonate + S	300–400 (sintering)	Porcelain
Alkali metal carbonate + CaO (MgO)	300–400 (sintering)	Porcelain, platinum

(zircon, tourmaline, etc.) that are particularly difficult to digest must be performed for a longer time at higher temperature. On the other hand, it is beneficial to work at the lowest possible temperature in order to keep the corrosion at a lower level and hence to minimize the contamination. This is achieved with an additive depressing the melting-point (the borax may play such a role in digestions with alkali metal carbonates), and by choosing from among similar compounds the reagent with the lowest melting-point. Some examples of melting-points: NaOH 328°C, KOH 360°C, Na_2CO_3 850°C, K_2CO_3 984°C, NaKCO$_3$ 500°C. Trofimov and Busev [76a] give a detailed review of the dissolution of the material of the digestion vessels.

A well-known procedure is heating at a temperature lower than the melting-point. This is sintering [e.g. $Na_2CO_3 + MgO$ (CaO or ZnO) or Na_2O_2]. The sample must be heated for a more prolonged period in this case, because of the much lower rate of the solid-phase heterogeneous reaction.

The digestion conditions depend among others on the sample weight. Especially with samples that are difficult to digest, it is better to take a small (a few tenths of a gram) sample, if the homogeneity of the sample and the sensitivity of the method permit this. In general it is not recommended that the sample should be in excess of 1–2 g, or the quantity of fusion mixture more than 10–15 g. The latter is limited by the size of the digestion vessel, and by the volume of the solutions formed in the later operations.

The material of the vessel used for the digestion may be selected on the basis of the chemical action and corrosive properties of the melt; the most important examples are given in Table 4.2. If the possibility arises, the composition of the sample too is taken into consideration. However, this is not always known exactly, and there is not always an opportunity to exchange the material of the vessel at will. It is known, for instance, that sulphides, arsenates and phosphates, and the metals partially reduced to the elemental state under the conditions of digestion, may form alloys with platinum, which thereby becomes brittle. Rock samples always contain a certain amount of sulphide (pyrite), and thus such damage cannot be prevented completely.

When a platinum vessel is used, it is always necessary to reckon with the digestion of the platinum itself to some extent. As an example, in a pyrosulphate fusion the quantity of platinum may attain even 1 mg; this can disturb certain determinations so much that platinum is not a suitable material for the fusion vessel in these cases (e.g. the determination of niobium with thiocyanate). The most commonly used additive for fusions with alkali metal hydroxides is the oxidant Na_2O_2. Nickel crucibles are particularly susceptible to attack by this. Since the vessel is generally made from the pure metal (but only from an alloy of the platinum metals in the case of a "platinum vessel"), the effect of the dissolved metal reaches appreciable proportions in disturbing the determination only in certain methods.

In recent years, zircon and graphite (vitreous carbon) crucibles have been recommended for certain ores that undergo fusion with difficulty [77, 78].

Some examples of fusion possibilities that have proved effective in practice are listed in Table 4.3.

The fusion is designed to produce a dissolved form of the sample. Following digestion, however, total dissolution is far from being general. In cases when the cationic and anionic components of the original compound to be digested are both to be found in the system after the digestion, an attempt to achieve total dissolution would result in the re-formation of the initial compound that has once been decomposed (e.g. $BaSO_4$, insoluble phosphates). In such cases, the following step in the analysis is aqueous leaching

103

TABLE 4.3.

The most important fusion agents for the digestion of rocks and ores

Sample to be examined	Fusion agent	Notes
Silicates	$NaOH + Na_2O_2$	
	$Na_2CO_3 + Na_2B_4O_7$	
	$LiBO_2$	
	$(H_3BO_3 + LiF)$ [79]	
Silicates digested with difficulty	KHF_2	
Oxide ores (Al, Fe, Ti, Cr)	$Na_2CO_3 + Na_2B_4O_7$ [80]	
	B_2O_3	
	$Na_2CO_3 + H_3BO_3$	
Zircon, tourmaline, beryl, chromite	$Na_2CO_3 + MgO$	Sintering
Oxides; Al and Fe ores	$KHSO_4$ (or $K_2S_2O_7$) [76]	
Zircon, monazite	$KHSO_4 + KHF_2$	
Oxide ores	$Na_2CO_3 + ZnO$	Sintering, e.g. for V determination
All types of ores	Na_2O_2 [81]	Determination of Ta
Molybdenite; silicates	$CaO + KMnO_4$ [81]	Determination of Re
Pt ore concentrate, Sb ore, etc.	Na_2O_2	
Pyrite, chromite	$Na_2CO_3 + NaNO_3$	
	$KOH + NaNO_3$	
Barytes	Na_2CO_3	
	$(NaPO_3)_n$	
Cassiterite	$NaCN$	
Sulphide ores; any sample	$Na_2CO_3 + Na_2B_4O_7 +$ $Pb_3O_4 +$ organic reductant (flour, etc.) [82]	Determination of Au, Ag; elements to be determined obtained in Pb alloy after fusion

(e.g. the sulphate after digestion with Na_2CO_3, the sulphate is removed from the formed $BaCO_3$ by aqueous leaching).

In other situations, although this danger would not threaten, we should give up the advantage provided by the applied digestion. There is a possibility of this primarily in alkaline fusion or sintering. Of the examples in the Table, mention may be made of the water-soluble VO_3^- and ReO_4^-, and of the polysulphides (sulpho salts) formed in the Freiberg digestion.

Fusion with alkali metal hydroxide and subsequent aqueous leaching is suitable for separation from certain disturbing components, if the ion to be determined remains in the hydroxide precipitate. Many examples of this are to be found in the present book, in connection with the determination of the individual elements.

In the determination of multivalent ions that are very susceptible to hydrolysis, and ions that form insoluble compounds in acidic medium (e.g. β-stannic acid), special care must be taken to prevent precipitate formation leading to losses; this is done by the addition of appropriate auxiliary complex formers when the melt is dissolved. In the determination of Ti or Zr, the most frequent reagent is SO_4^{2-}, or possibly $C_2O_4^{2-}$ or F^-; in the case of Nb^{5+}, Ta^{5+}, Sb^{5+} or Sn^{4+} the monomeric state is ensured with organic oxyacids (tartaric acid, citric acid). Care must naturally be taken that the auxiliary complex former does not interfere in the later course of the analysis.

4.5.3. OTHER MEANS OF DIGESTION

Besides the methods mentioned so far, there are some other procedures which are much less general, but which are simpler or more effective in certain instances. The most important of these are:

Digestion with ion-exchange resin. This is a procedure known for the analysis of compounds that dissolve with difficulty, e.g. apatite, barytes or gypsum in ore analysis [83, 84]. Its advantage is that it separates the cation and the anion of the insoluble compound, for the resin binds one of these while the counter-ion passes into solution.

Digestion in a gas stream. This leads to separation at the same time as digestion; at the temperature of the digestion (600–1300°C) the element to be determined is generally transformed into a volatile compound. By means of heating in a current of oxygen or air, for instance, it is possible to determine sulphide sulphur [85] and mercury [86], and even to separate the rhenium from the molybdenum in the analysis of molybdenite [87].

It is advantageous to heat oxide ores in a stream of hydrogen (SnO_2, TiO_2, Ta_2O_5, etc.). Metals with low boiling-points (Zn, Cd, Pb) may possibly be distilled off at the same time in a stream of nitrogen.

The most general procedure is chlorination, which may be carried out with gaseous Cl_2 (for the determination of S, As, Sb, Sn, Hg, Bi and Se), with gaseous HCl (for the digestion of $BaSO_4$, and for the determination of V and Mo, which form volatile chlorides) or with CCl_4 (to remove Fe and Al from the lanthanides).

Bromination too is of certain importance in the analysis of sulphide ores [88].

Pyrolysis and pyrohydrolysis. This is a method widely used primarily for the determination of fluoride [89]. The sample to be examined is mixed with an additive (WO_3, UO_3, Al_2O_3) promoting the decomposition of silicates, and is heated to 1200°C in the presence of water vapour or moist air; the HF formed is absorbed in water, and the fluoride is determined photometrically or by titration.

4.6. Dissolution of metals, alloys and certain metal compounds

Metallurgical substances, together with telecommunications materials and semiconductors, which have become of great importance in recent decades, comprise a very significant group of samples to be examined. In the treatment of these substances, therefore, special mention should be made of the means of dissolving up the element to be determined.

In contrast with the situation in rock analysis (discussed in the previous section), fusion digestion is of subordinate importance in metal analysis. A fundamental method here is dissolution with various acids and mainly with acid mixtures. Additionally, electrochemical methods in particular come into consideration, but in certain cases heating in a stream of gas is also of use.

In the analysis of metals, whether or not dissolution occurs in a non-oxidizing acid is decided by the electrochemical series. As stated by Bock [90], "oxidation with H^+" occurs in M/M^+ systems with normal potentials more negative than that of the H_2/H^+ system, i.e. from the alkali metals to lead, inclusively. In the case of alloys, the electrochemical conditions are more complicated; the more noble metal too may enter solution. For instance, in the analysis of magnesium alloys, dissolution with HCl is the appropriate method for the determination of copper too [91].

In dissolution, not only the electrochemical series is of importance, but also the anion of the acid, because of complex formation. This is important in the development of the equilibria, but additionally in the increase of the dissolution rate and also in the inhibition of hydrolysis and polymerization. For this reason HF is advantageous for the digestion of niobium and tantalum, which form fluorine complexes of high stability [92], while H_2SO_4 is a beneficial additive in the presence of HCl for determination of the titanium content of aluminium alloys [91].

For the dissolution of metals more positive than hydrogen and in the event of high concentrations of these in alloys, an oxidizing acid or an acid mixture is needed. This is similarly a useful procedure when, although the metal does dissolve in a non-oxidizing acid too, the further course of the determination demands that it be present in a higher oxidation state (this may refer either to the element to be determined or to some ion to be removed). Oxidizing dissolution is likewise necessary, of course, in the analysis of luminescent powders and semiconductors containing sulphide or selenide [93].

In the selection of the acid, attention should be paid to the phenomenon of passivity, which inhibits dissolution. For example, dissolution with HNO_3 is not suitable for the analysis of aluminium, titanium or chromium.

In certain cases the dissolution necessitates not oxidation, but reduction. An instance of this is the dissolution of cerium dioxide in H_2SO_4 in the presence of zinc filings [94]; Ce^{3+} can be obtained in much higher concentration and in a more stable solution than can Ce^{4+}.

With ferrites, reduction may be performed with a current of H_2; this is the procedure, for example, in the determination of silicon, as the operation preceding dissolution of the sample in H_2SO_4 [95].

An electrochemical method is frequently applied to achieve dissolution [96]. An example is the determination of the silicon content of aluminium, when the sample to be examined is connected as anode, and 2.5 M HNO_3 is used as electrolyte.

Depending on the electrochemical (and chemical) properties of the individual metals and on the composition of the electrolyte, in anodic dissolution the various metals accumulate in an undissolved state in the anode space, remain in solution, or are deposited on the cathode. This procedure is of particular importance in inclusion analysis.

Some of the best-known methods for the digestion of metals and their alloys are given in Table 4.4.

Even from this listing it may be seen that the procedure to be adopted is

TABLE 4.4.

Procedures used to dissolve metals and alloys

Digestion agent	Metal (alloy)	Element to be determined	Refs
$H_3BO_3 + HF + HNO_3$	Sn–Pb alloy	Al, Ag, As, Au, Bi, Cd	[97]
Na_2O_2, $NaKCO_3$	Ferrotitanium, ferrochromium, ferromanganese, ferrosilicon		[98]
$HCl + Br_2$ $(HBr + Br_2)$	Lead, antimony, tin, copper, magnesium; their alloys		[5]
NaOH solution	Aluminium, magnesium, silicon		[5] [91]
Aqua regia	Copper, nickel, steel, platinum alloy		[5, 5a]
$HCl + H_2O_2$	Tungsten lead alloys		[5]
	Aluminium	As	[91]
$HClO_4$	Chrome steel	V, Cr	[5]
HCl	Al alloy	Cu, Zr, Co, etc.	[91]
	Mg alloy	Al, Fe, Be, etc.	
$H_2SO_4 + HNO_3 + H_3PO_4$	Al alloy	Mn, Cr, B	[91]
$HF + HNO_3$	Niobium	P	[99]
HCl	Misch metal, ferrites	Co, Ni, Cr, Cu	[95]
$H_2SO_4 + HCl + HNO_3$	Molybdenum	Mn	[92]
HF	Nb, Ta, W	Mn	[92]
HNO_3	Silver		[96]
HNO_3 or $H_2SO_4 + Br_2$	CdS, CdSe		[93]

determined not only by the nature of the sample under examination, but also by whether this procedure fits in with the subsequent course of the analysis. This question, however, belongs rather in the sections dealing with the planning and elaboration of the individual methods.

4.7. Phase analysis

Because of its importance, we must deal briefly with the case when the determination of some element is not sufficient, but we are interested in that part of it which is present in a certain oxidation state or in a definite form of chemical bonding. In the stricter sense, we shall discuss the case when, in the study of a sample in the solid state, we wish to make use of selective dissolution to separate the various forms of the element to be determined. A rarer case, but similarly one belonging here, is when the dissolution itself is not selective, but the determination is. Examples of this are the determination of MnO_2 in the presence of Mn(II) in manganese ores, by dissolution in oxalic acid and measurement of the oxalate excess [74], and determination of Fe(II) in the presence of Fe(III) in rocks [100].

Phase-analysis methods are primarily of great importance in the following fields:

(a) soil examinations;
(b) analysis of ores and ore-processing products;
(c) metallurgical analyses.

4.7.1. SOIL EXAMINATIONS

In soils subject to agricultural cultivation, determination of the soluble concentrations of the individual ions is an important plant-physiological question, especially as the deficiency of certain elements is deleterious, but so is their presence in excessive concentrations (e.g. too much aluminium is poisonous for both plants and microorganisms) [101]. Depending on the aim of the analysis, the components to be determined can be classified as follows:

(a) Cations that can be exchanged in the soil. They can be determined after being shaken with a solution of NH_4Cl, ammonium acetate or KCl.

(b) The mobile trace elements in the soil, i.e. those that can be absorbed by the plants. One very important element is boron; this is determined after dissolution with boiling water. Mn, Cu, Co, Mo and Zn ions are dissolved out with dilute (<0.5 M) mineral acid, or with sodium or ammonium acetate or oxalate solution at pH 3.7–4.7.

(c) Of the elements added with the nutrients (fertilizer), determination of the readily soluble phosphate is of the greatest importance. Many conventional solvents are known for its dissolution: 0.2 M HCl; 0.5 M CH_3COOH; 0.001 M H_2SO_4; citric acid solution of pH 2.3; 1% $(NH_4)_2CO_3$ for study of carbonate containing soil.

109

(d) The nutrients that are readily absorbed into plants from the soil. A procedure suitable for the dissolving-out of these is the electro-ultra-filtration (EUF) method. With this it is possible to measure the rates of desorption of cations, and the rates of dissolution of phosphates, carbonates, etc. [102, 103].

4.7.2. ANALYSIS OF ORES AND ORE-PROCESSING PRODUCTS

In order to determine the genetics of rocks and ores, the value of ores to be exploited and the optimum processing technology, it is necessary to establish the oxidation states and binding modes of the individual elements. This is particularly important in the case of physical enrichment (flotation, gravity separation) or hydrometallurgical processing, e.g. bauxite digestion.

Thermal (derivatography) and X-ray structural examination (X-ray diffractometry) methods have long been known for this purpose. More recently, nuclear spectroscopic procedures (Mössbauer spectroscopy) have become available for the study of iron minerals [104]. Although these methods are indispensable for mineralogical investigations, in the present section we deal only with analytical procedures based on selective dissolution.

The theoretical basis of the phase-analysis method and the means of carrying it out are briefly discussed in the works of Filippova [105, 106], whose findings can be summarized as follows:

In selective dissolution, either one or all but one of the minerals to be differentiated is dissolved up. This operation may possibly be accompanied by the development of a new solid phase; the anion or cation of the mineral in question take part in this, e.g. the determination of $CaSO_4$ by dissolution with Na_2CO_3, via the dissolved SO_4^{2-}.

It is fundamentally the following processes that may make the separation possible:

(a) formation of a new insoluble phase (see the above example);
(b) formation of soluble anionic or cationic complexes;
(c) change in valency, e.g. oxidation of sulphides.

110

The applicability of the method with a given reagent is governed by the "separation factors" of the two or more minerals to be separated. In the event of complex formation, for instance, this is

$$\frac{K_{1_{equilib.}}}{K_{2_{equilib.}}} = \frac{K_{A_{solub.}} K_{B_{instab.}}}{K_{A_{instab.}} K_{B_{solub}}}$$

where $K_{A_{solub.}}$ and $K_{B_{solub.}}$ are the solubility products of the ions forming the minerals A and B, respectively, while $K_{A_{instab.}}$ and $K_{B_{instab.}}$ are the instability constants of the complexes formed between the reagent and the interacting ion from minerals A and B.

The foreign substances present may give rise to secondary processes which may falsify the result, e.g. the sorption of Zn^{2+} by $Fe(OH)_3$, etc.

The dissolution is affected by many additional factors: temperature, reagent concentration, grain size, solid/liquid phase ratio, duration of digestion, etc.

Since these procedures are conventional methods, the prescriptions must be adhered to strictly.

The accuracy of the methods does not generally attain that of the procedures used to determine the total quantities of the individual elements. It is not an easy task to check their accuracy; the addition method is not conclusive here, as the artificial compounds and the minerals do not generally behave absolutely identically.

Despite the listed sources of error and difficulties, the phase-analysis methods are indispensable, and many well-tried procedures are to be found among them. Some of these are mentioned in Table 4.5. In addition to the works of Filippova already referred to, the review by Steger [109] presents a survey of the application.

4.7.3. PHASE ANALYSIS IN METALLURGY

In metallurgy and the related processing industries, primarily inclusion analysis is of great importance, but the determination of the presence and quantities of metal oxides, for instance, is likewise very important. Some examples of these methods are presented in Table 4.6.

It is a special problem in metallurgical analysis that it is often necessary to know the composition corresponding to the molten state, but cooling is

TABLE 4.5.

Some examples of phase-analysis studies of minerals

Mineral dissolved	Mineral remaining undissolved	Reagent	Refs
Anglesite ($PbSO_4$)	Cerussite ($PbCO_3$)	Cl^-	[105]
Lead minerals with the exception of PbS	Galena (PbS)	EDTA	[105]
Galena (PbS)	Chalcopyrite ($CuFeS_2$)	Fe^{3+}	[105]
Uranium(VI) compounds	Uranium(IV) compounds	CO_3^{2-} + anion-exchange resin (inert atmosphere)	[107]
Colloidal Sn minerals	Cassiterite (SnO_2)	conc. H_2SO_4	[105]
Molybdenum minerals with the exception of MoS_2	Molybdenite (MoS_2)	conc. HCl	[105]
Silicates	Quartz (SiO_2)	$H_4P_2O_7$	[108]

TABLE 4.6.

Some examples of phase analysis used in metallurgical analysis

Component to be determined	Sample examined	Treatment	Refs
Al_2O_3	Metallic aluminium	$CaCl_2$ + NH_4Cl solution	[91]
C (graphite)	Cast iron	Anodic oxidation	[5]
Tl sulphide	Metallic tellurium	Boiling with $AgNO_3$ solution (TlS dissolves)	[110]
Tl	TlS	Dissolution of Tl in ethanol	[110]
Oxide, nitride and carbide inclusions	Steel, chromium, aluminium, non-ferrous metal alloys	Br_2 + CH_3OH (or acetate esters)	[5]
Mn	Slag from manganese ore processing	$CuSO_4$ (Mn dissolves)	[110]
Ti(II) + Ti(III)	Slag from titanium ore processing	Fe^{3+} after dissolution of metallic Fe with $CuSO_4$	[110]
CaF_2	Fluorine-containing slags	Dilute CH_3COOH (CaF_2 does not dissolve)	[110]

accompanied by the occurrence of further reactions, to extents depending on the conditions. Thus, the analytical results always relate only to the solidified sample; in the knowledge of the processes, we can at best extrapolate to the composition of the melt phase.

Another difficulty, which to some extent occurs in rock and ore analysis too, arises from the redox processes taking place during dissolution; these may change the original composition. This is an especially large problem in systems containing several metals and metal oxides. On the basis of the Nernst potentials for various metal pairs, Filippova reports data on what the ratio of the metal ions in solution may be for the redox process not to occur: for Fe–Cu this is 10^{-26}, for Fe–Cd 10^{-2}, for Sn–Cr 10^{-16}, and for Sn–Pb 10^{-3}.

If the appropriate ratio is ensured, the regulated complex formation elaborated on the basis of the stability constants can be utilized to advantage.

References

[1] Sandell, E. B., Onishi, H.: *Photometric Determination of Traces of Metals. General Aspects.* Wiley, New York 1978.

[2] Reeves, R. D., Brooks, R. R.: *Trace Element Analysis of Geological Materials.* Wiley, New York 1978.

[3] Dolezhal, J., Povondra, P., Shultsek, Z.: *Metody razlozhdeniya gornykh porod i mineralov.* Mir, Moscow 1968.

[4] Gorsuch, T. T.: *The Destruction of Organic Matter.* Pergamon, Oxford 1970.

[5] Bock, R.: *Aufschlußmethoden der anorganischen und organischen Chemie.* Verlag Chemie, Weinheim 1972.

[5a] Bukhtiarov, V. E.: *Ionoobmennye metody v analize metallov i splavov.* Metallurgiya, Moscow, 1982.

[6] Peregud, E. A., Garnet, E. V.: *Khimicheskii analiz vozdukha promyshlennykh predpriyatii.* Khimiya, Leningrad 1973.

[7] Fuks, N. A.: *Kolloidnyi Zhurnal* **37**, 427 (1975).

[8] Jacobs, M. B.: *The Analytical Toxicology of Industrial Inorganic Poisons.* Interscience, New York 1967.

[9] Várkonyi, T., Cziczó, T.: *A levegőminőség vizsgálata (Air-quality investigations).* Műszaki Könyvkiadó, Budapest 1980.

[10] Cochran, W. G.: *Sampling Techniques.* Wiley, New York 1953.

[11] Laitinen, H. A.: *Chemical Analysis. (An advanced text and reference.)* McGraw–Hill, New York 1960.

[12] Laitinen, H. A., Harris, W. E.: *Chemical Analysis.* McGraw–Hill, New York 1975.

[13] Zentay, P.: Paper presented at the 24th Hungarian Congress on Spectral Analysis. Miskolc, Hungary, 15 June 1981.

[14] Sporbeck, H.: *Z. für Anal. Chem.* **209**, 60 (1965).

[15] Dymov, A. M.: *Tekhnicheskii analiz rud i metallov.* Metallurgizdat, Moscow 1949.

[16] Paesh, K., Tracey, M. V.: *Moderne Methoden der Pflanzenanalyse.* Vol. 4. Springer, Heidelberg 1954.

[17] Sugawara, K. F., Yao-Szin, Su: *Anal. Chim. Acta* **80**, 143 (1975).
[18] Robertson, D. E.: *Anal. Chim. Acta* **42**, 523 (1968).
[19] Shendrikar, A. D., Dharmarajan, V., Walker–Merrick, H., West, P. W.: *Anal. Chim. Acta* **84**, 409 (1976).
[20] Subramanian, K. S., Chakrabarti, C. L., Sueiras, J. E., Maines, I. S.: *Anal. Chem.* **50**, 444 (1978).
[21] Steger, H. F.: *Talanta* **23**, 643 (1976).
[22] Tölg, G.: *Talanta* **19**, 1489 (1972).
[23] Ulmer, W., Tandler, W. S.: *Laser Angew. Strahlentechn.* **1**, 17 (1970).
[24] Chapman, H. D., Pratt, P. F.: *Methods of Analysis for Soils, Plants and Waters.* Univ. California Press, 1961.
[25] Lucas, C. C., Ross, J. R.: *J. Biol. Chem.* **111**, 285 (1935).
[26] King, E. J.: *Biochem. J.* **26**, 292 (1936).
[27] *Erjedésipari termékek minőségi követelményei és vizsgálati módszerei (Quality conditions and investigative methods for fermentation industry products).* Collection of Hungarian Standards. No. 36. Közgazdasági és Jogi Könyvkiadó, Budapest 1963.
[28] Fuchs, G. W.: *Int. J. Environ. Anal. Chem.* **1**, 123 (1971).
[29] Upor, E., Demkó, E., Fekete, L.: *Kísérletes Orvostudomány* **12**, 91 (1960).
[30] Győri, D.: *Agrokémia és talajtan* **10**, 425 (1961).
[31] Cresser, M. S., Parsons, J. W.: *Anal. Chim. Acta* **109**, 431 (1979).
[32] Tölgyesi, Gy.: *A növények mikroelemtartalma és ennek mezőgazdasági vonatkozásai (Microelement content of plants and its agricultural aspects).* Mezőgazdasági Kiadó, Budapest 1969.
[33] Varjú, E. M., Zsoldos, L.: *Agrokémia és talajtan* **23**, 149 (1974).
[34] Evans, W. H., Jackson, F. J., Dellaer, D.: *Analyst,* **104**, 16 (1979).
[34a] Kovács, E., Orosz, M.: *Magyar Kémikusok Lapja,* **35**, 1, 28 (1980).
[35] Keresztény, B., Marton, M.: *Agrokémia és talajtan* **8**, 265 (1959).
[36] Székely, Á.: *Gyors talajkémiai vizsgálatok (Rapid soil-chemistry investigations).* Mezőgazdasági Könyvkiadó, Budapest 1956.
[37] Rinkis, Ya. G.: *Pochvovedenie* **3**, 74 (1960).
[38] Pearson, D.: *The Chemical Analysis of Foods.* Churchill, London 1962.
[39] Cela Torrijos, R., Perez-Bustamante, J. A.: *Analyst* **103**, 1221 (1978).
[40] Pomeranz, Y., Meloan, C. E.: *Food Analysis. Theory and Practice.* AVI Publ. Co., Westport 1971.
[41] Bortlisz, J.: *Vom Wasser* **40**, 1 (1973).
[42] Leitner, S. S., Savary, I.: *Anal. Chim. Acta* **74**, 133 (1975).
[43] Conrad, A. L.: *Microchemia* **38**, 514 (1975).
[44] Buss, H., Kohlschütter, H. W., Preiss, M.: *Z. Anal. Chem.* **214**, 106 (1965).
[45] Gleit, C. E., Holland, W. D.: *Anal. Chem.* **34**, 1454 (1962).
[46] Kaiser, G., Tschöpfel, P., Tölg, G.: *Z. Anal. Chem.* **253**, 177 (1971).
[47] Simova, L., Petkova, L., Cekova, I.: *Sb. Trudy n–i i projekt in–t rudodob i obogat. Obogat.* (Ref. *Zh. Khim.* **8G**, 143 (1972).
[48] Varjú, E. M.: *Agrokémia és talajtan* **21**, 139 (1972).

[49] Varjú, E. M.: *Thermal Analysis. Proceedings Fourth ICTA, Budapest.* **2**, 245 (1974).
[49a] Rowan, C. A., Zajicek, O. T., Calabrese, E. J.: *Anal. Chem.*, **54**, 149 (1982).
[50] Heanes, D. L.: *Analyst* **106**, 172 (1981).
[51] Jacobs, M. B.: *The Chemical Analysis of Foods and Food Products.* Van Nostrand, New York 1959.
[52] Eschka, A.: *Öster. Z. Berg- u. Hüttenwes.* **22**, 111 (1874).
[53] Kiba, T., Takagi, T., Yoshimura, Y., Kishi, I.: *Bull. Chem. Soc. Japan* **28**, 641 (1955).
[54] Iordanov, N., Nikolova, B., Havezov, I.: *Talanta* **25**, 275 (1978).
[55] Mizoguchi, T., Ishi, H.: *Talanta* **25**, 311 (1978).
[56] Mizoguchi, T., Ishi, H.: *Talanta* **26**, 33 (1979).
[57] French, W. J.: *Anal. Chim. Acta* **66**, 324 (1966).
[58] Beyermann, K.: *Z. Anal. Chem.* **190**, 4 (1962).
[59] Barredo, F. B.: *Talanta* **23**, 859 (1976).
[60] Van Eenbergen, A., Bruninx, E.: *Anal. Chim. Acta* **98**, 405 (1978).
[61] Kotz, L., Henze, G., Pahlke, P., Veber, M.: *Talanta* **26**, 681 (1979).
[62] Langmyhr, F. J., Paus, P. E.: *Anal. Chim. Acta* **43**, 397 (1968).
[63] Langmyhr, F. J., Paus, P. E.: *Anal. Chim. Acta* **44**, 447 (1969).
[64] Langmyhr, F. J., Paus, P. E.: *Anal. Chim. Acta* **45**, 157 (1969).
[65] Langmyhr, F. J., Paus, P. E.: *Anal. Chim. Acta* **45**, 173 (1969).
[66] Langmyhr, F. J., Paus, P. E.: *Anal. Chim. Acta* **47**, 371 (1969).
[67] Langmyhr, F. J., Paus, P. E.: *Anal. Chim. Acta* **50**, 515 (1970).
[68] Ponomarev, A. I.: *Metody khimicheskovo analiza zheleznykh, titanomagnetitovykh i khromovykh rud.* Nauka, Moscow 1966.
[69] Berneman, I. O.: *Metody khimicheskovo analiza mineralov.* AN SSSR, Moscow 1961.
[70] Jeffery, P. G.: *Chemical Methods of Rock Analysis.* Pergamon, Oxford 1970.
[71] Viktorova, M. E., Isaeva, K. G.: *Zh. Anal. Khim.* **25**, 1140 (1970).
[72] Rodden, C. J.: *Analytical Chemistry of the Manhattan Project.* McGraw-Hill, New York 1950.
[73] Lyalikov, Yu. S., Tkachenko, N. S., Dobzhanskii, A. V., Sakonov, V. I.: *Analiz zheleznykh, margantsevykh rud i aglomeratov.* Metallurgiya, Moscow 1966.
[74] Knipovich, Yu. N., Morachevskii, Yu. V.: *Analiz mineralnovo syrya.* Khimiya, Gos. Nauchno tekhn. Leningrad 1959.
[75] Hoyle, W. C., Diehl, H.: *Talanta* **18**, 1072 (1971).
[76] Simó, B.: *Magy. Kémikusok Lapja* **26**, 408 (1971).
[76a] Trofimov, N. V., Busev, A. J.: *Zav. Lab.* **49**, 3, 5 (1983).
[77] Wall, G.: *Rep. natn. Inst. Metall. Johannesburg* No. 1798, 6 (1976).
[78] Bhargava, Om P., Gmitro, M., Grant Hines, W.: *Talanta* **27**, 263 (1980).
[79] Ingamells, C. O.: *Anal. Chim. Acta* **52**, 323 (1970).
[80] Bennett, H., Hawley, W. G.: *Methods of Silicate Analysis.* Academic Press, London 1965.
[81] Pakhomova, K. S., Pensionerova, V. M. (eds.): *Metody khimicheskovo analiza mineralnovo syrya.* And Ed. Gosgeoltekhizdat, Moscow 1963.

[82] Barishnikov, I. F. (ed.): *Probootbiranie i analiz blagorodnykh metallov*. Metallurgiya, Moscow 1968.

[83] Inczédy, J.: *Ioncserélők analitikai alkalmazása (Analytical application of ion-exchangers)*. Műszaki Könyvkiadó, Budapest 1962.

[84] Ronald, J.: *J. Chem. Educ.* **40**, 414 (1963).

[85] Faynberg, S. Yu., Filippova, N. A.: *Analiz rud tsvetnykh metallov*. Metallurgizdat, Moscow 1963.

[86] Onufrienok, I. P., Tereshkova, Z. N.: *Trudy Tomsk Politechn. Inst.* **102**, 151 (1959).

[87] Duca, A., Stanescu, D., Puscasu, M.: *Stud. Cercet. Chim. Acad. RPR Cluj* **13**, 197 (1962).

[88] Zverev, L. V., Petrova, N. V.: *Zav. Lab.* **23**, 1403 (1957).

[89] Nikolaev, N. S., Suvorova, S. N.: *Analiticheskaya khimiya ftora*. Nauka, Moscow 1970.

[90] Bock, R.: *Methoden der analytischen Chemie. Eine Einführung.* Vol. 1. *Trennungsmethoden*. Verlag Chemie, Weinheim 1974.

[91] Budanova, A. M., Volodarskaya, R. S., Kanayev, N. A.: *Analiz alyuminievykh i magnievykh splavov*. Metallurgiya, Moscow 1966.

[92] Donaldson, E. M., Inman, W. R.: *Talanta* **13**, 489 (1966).

[93] Alimarin, I. P.: *Analiz poluprovodnikovykh materialov*. Nauka, Moscow 1968.

[94] Mohai, M., Klesch, K.: Unpublished work.

[95] Funke, A., Lauhner, H. J.: *Z. Anal. Chem.* **249**, 26 (1970).

[96] Mika, J.: *Kohászati elemzések (Metallurgical analyses)*. Műszaki Könyvkiadó, Budapest 1959.

[97] Yae Y. Hwang, Sandonato, L. M.: *Anal. Chem.* **42**, 744 (1970).

[98] *Handbuch für das Eisenhütten Laboratorium*. Verlag Stahleisen, Düsseldorf, 1966.

[99] Chernikov, Yu. A., Dobkina, B. M.: *Nauchnye Trudy Giredmeta* **3**, 79 (1961).

[100] Meyrovitz, R.: *Anal. Chem.* **42**, 1110 (1970).

[101] Arinushkina, E. V.: *Rukovodstvo po khimicheskomu analizu pochv*. Moskow. University 1970.

[102] Németh, K.: *Applied Sciences and Development* **8**, 89 (1977).

[103] Németh, K., Varjú, E. M. (eds.): Paper presented at the *First International Symposium on Electro-Ultrafiltration (EUF)*. Budapest, 6–10 May 1980.

[104] De Coster, M., Pollak, H., Amelinak, S.: *Phys. Stat. Sol.* **3**, 283 (1963).

[105] Filippova, N. A.: *Fazovii analiz rud tsvetnykh metallov i produktov ikh pererabotki*. Metallurgizdat, Moscow 1963.

[106] Filippova, N. A.: *Fazovii analiz rud i produktov ikh pererabotki*. Khimiya, Moscow 1975.

[107] Upor, E., Mohai, M.: *Ionenaustauscher Symposium in Balatonszéplak*. Akadémiai Kiadó, Budapest 1965, p. 195.

[108] Goldman, I.: *Ind. Eng. Chem. Anal. Ed.* **13**, 798 (1941).

[109] Steger, H. F.: *Talanta* **23**, 81 (1976).

[110] Ponomarev, A. I. (ed.): *Khimicheskii i spektralnii analiz v metallurgii*. Nauka, Moscow 1965.

116

5. *The most important factors influencing the accuracy of determination. Error sources*

In earlier chapters we have already dealt with the errors in the photometric measurement, with the effects of interfering components, and in part with the means of eliminating these. In the present chapter the following topics will be discussed:

1. losses occurring during analysis;
2. contaminants;
3. errors in the solution equilibria, and establishment of these; problems of standard solutions and reagents;
4. control of the suitability of the final solution for determination;
5. control of the accuracy of the methods with standard samples.

5.1. Losses in the analysis

The losses may have various causes:

(a) inexact compilation of the analytical specifications;
(b) application of the method to a sample for which the possibility of its use has not been confirmed;
(c) incorrect performance of the method.

Losses may arise in any stage of the analysis: in the sampling, in the storage of the sample, in the processing of the sample, in the dissolution

117

of the component to be determined and the related operations, and in the separations.

The losses that can occur during the storage of the standard solutions will be dealt with separately.

Losses in sampling, and in storage and processing of sample. Since these were considered in the previous chapter, they will be referred to only briefly here.

In the sampling, losses can only arise when the material is not homogeneous. This applies both to the case when the sample does not consist of one phase (dust, aerosol) and to the case when the component to be determined is not distributed uniformly (various parts of plants, natural substances with various grain sizes, etc.). The loss may stem from the fact that, for example, the dust or aerosol is not trapped completely in the sampler, and thus the sample taken is incorrect. In the second case the sample aliquot taken may possibly not correspond to the average sample composition. This can occur particularly with sludgy samples (intermediates obtained in ore processing by precipitation from solution, semi-finished products, etc.), when the sedimentation, filtration or drying processes may be accompanied by self-fractionation according to the particle size.

A similar error can occur with air samples, if the sampling system is not suitable for trapping the smaller particles, or if the gaseous component to be examined is not absorbed quantitatively in the absorbing liquid.

Losses during the preparation and dissolution (digestion) of the sample and in the related operations. As in the previous point, these are error sources that are comparatively difficult to check. Also included here are those losses occurring during possible heating of the samples before analysis. It is well known that prior to the determination of mercury, for instance, it is not permitted to heat the sample because of the volatility of mercury compounds [1]; in the same way, in the determination of germanium, coal samples must not be heated unless an additive retaining germanium is added. For such reasons, great care must be taken when samples are subjected to dry ashing or ignition.

Losses can be caused not only by volatility, but also by the incorporation of the trace element into the material of the vessel, e.g. by the formation of an insoluble silicate in the case of lead. The additives recommended to prevent such losses are not always suitable. For instance, it is suggested that NaCl should be used as an additive in the ashing of organic materials,

118

but the chloride content loosens the silica lattice of porcelain, and hence increases the possibility of incorporation of certain metal ions. In apparatus made from metal, such losses must be reckoned with especially as a result of reduction and alloying.

Similar phenomena may be experienced during digestion too, both during the customary procedures performed in a platinum vessel and in the alkaline fusions carried out in nickel, silver or iron vessels.

In certain cases the digestion is accompanied by the formation of insoluble compounds which can be dissolved out of the vessel only with loss, and are also difficult to remove mechanically. For example, the polyphosphates formed from phosphates give very poorly soluble compounds with trivalent and tetravalent cations, while insoluble sulphates ($PbSO_4$, $BaSO_4$) may be obtained in the treatment of sulphide ores with an oxidizing acid mixture. The formation of some insoluble compound could well be the reason why, in contrast with our earlier publication [2], losses of thorium are observed in the fusion digestion of certain rock samples with alkali metal hydroxide.

In many cases of the analysis of rock samples, there is no need for the sample to be digested completely. Boiling with a solution of $HCl + H_2O_2$ is generally sufficient in the determination of uranium.

Even the boiling of solutions must be carried out with care in the case of certain elements. The loss is well known to be high when selenium solutions in HCl or H_2SO_4 are boiled [3]. Fuming with HF too is of danger from this respect in certain cases (Table 5.1). We have observed losses leading to false results in the determination of vanadium; the explanation of this is that the boiling-point of VOF_3 (or $VOCl_3$) is merely around 130°C. Detailed data on these losses are to be found in the works of Bock [4], Dolezhal *et al.* [5] and Gorsuch [1].

Because of the demands of serial analysis, it is widespread in practice for solutions to be evaporated to dryness not on a water-bath, but on a hot-plate or a sand-bath, the work being accelerated in this way. At such times, however, care must be taken that the sample is not overheated, for this can result in further losses, via the volatility or the formation of insoluble polymers in the case of multivalent metal ions, for instance.

Losses in the separation operations. In both precipitation and extraction separations, most errors are caused by the incorrect pH of the medium.

In precipitation separation, the phenomenon of sorption or coprecipitation on metal hydroxides is a function primarily of the pH (pOH). Some

119

TABLE 5.1.

Losses occurring because of volatility in the dissolution or digestion of samples

Operation performed	Element lost due to volatility (causing total loss or a large error)
Fusion with NaOH	Se, Hg
Fusion with Na_2O_2	Hg
Reducing digestion in acidic medium	As (AsH_3)
Boiling in HCl solution	As, Sb(III), Sn(IV), Ge, Se(VI), Os (OsO_4), Cr ($CrOCl_2$)
Boiling in $HClO_4$ solution	Se, Re
Boiling in $HF + HClO_4$ solution	As, Se, B
Boiling in $H_2SO_4 + HF$ solution	V, Se

Note: The extents of the losses depend strongly on the duration and temperature of the boiling.

ways in which the sorption can depend on the pH are illustrated in Fig. 5.1, without numerical values. It may be seen that the influence of the pH is manifested in both quantitative and qualitative differences. Since the course of the curve is affected not only by the equilibria in solution, but also by the properties of the macrocomponent metal hydroxide, variations

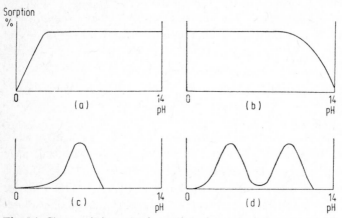

Fig. 5.1. Characteristic types of sorption of certain ions on a metal hydroxide.

(a) Th^{4+}, Ln^{3+}, Mg^{2+}, etc. (NH_4OH or alkali metal hydroxide);
(b) VO_3^-, WO_4^{2-}, MoO_4^{2-}, $Al(OH)_4^-$, etc. (alkali metal hydroxide);
(c) Cu^{2+}, Zn^{2+}, Ni^{2+}, etc. (NH_4OH solution);
(d) UO_2^{2+}, ZrO^{2+}, Sc^{3+}, etc. (alkali metal carbonate solution)

120

can occur in the positions and widths of the maxima or minima, depending on the composition of the sample.

Thus, it is of great importance to test the separations with actual samples, and to adhere strictly to the analytical specifications.

The equilibria existing in the period of precipitate formation may change as a consequence of aging of the precipitate, or the solution, during the time the solid and solution phases are standing together. In rare cases this has the result that certain ions initially separating out with the precipitate may partially dissolve up again; more frequently, however, ions from solution subsequently undergo sorption. In the event of such losses it is usually not enough to wash the precipitate; it must be redissolved and the precipitation repeated. Adherence to the analytical specifications must be emphasized here too. As an example, the separation of Ca^{2+} from $Fe(OH)_3$ by immediate filtration is not merely a possibility in the case of a fast analytical procedure; it is also compulsory if silicic acid too is present in the solution (due to the possible precipitation of calcium silicate).

It may happen that even before precipitation some component in the solution is not in the desired ionic form. Such a case arises from the presence of a complex former, from a non-monomeric state (e.g. for multivalent ions such as $Zr(IV)$, $Nb(V)$, $Sn(IV)$, etc. which tend to undergo hydrolysis and polymerization), and possibly from the existence of polynuclear hydroxo complexes (e.g. of $Ti+Hf$) already present. In this situation, of course, the values obtained with solutions containing the monomeric ions will not be valid for the partition between the precipitate and the solution phase. An error is frequently caused by an unsuitable oxidation state, for the analytical properties of ions of variable valency depend on the valency. For instance, even after fusion with $NaOH+Na_2O_2$, iron may remain partially in the $Fe(II)$ form, and in alkaline precipitation (either in approximately neutral, or in strongly basic medium) this causes the partial retention of the iron in solution. The errors stemming from this are increased further by the progressive precipitation of $Fe(OH)_3$ in the filtrate following oxidation. Similarly to Mg^{2+}, Mn^{2+} remains in solution in hydroxide precipitation at pH$<$9. However, if its oxidation is not inhibited, a loss occurs because of the precipitation of $Mn(OH)_3$.

In extraction separation, losses may be caused mainly by an incorrect pH and by possible complex formers. Mention must also be made of inappropriate reagents, however. Thus, the presence of the mono- or pyro-

121

phosphoric acid ester impurities in dialkylphosphoric acid esters can lead to errors, for they extract by totally different mechanisms and with different partition coefficients. As regards other reagents, the triphenylmethane derivatives are worthy of special mention; as a result of oxidative decomposition, these either give a precipitate at the interface between the two phases, e.g. gallium-malachite green, or give rise to a lower partition.

Errors due to coextraction have also been demonstrated in certain systems; this phenomenon is dealt with very comprehensively by Alimarin [6] and Zolotov [7].

Losses may occur in extraction if an emulsion is formed at the phase boundary, for part of the ion it is desired to extract may pass into the precipitate, e.g. in the extraction with HDBP of a thorium solution more concentrated than 10 $\mu g/cm^3$, thorium enters the precipitate. Losses may also be experienced if the phases are poorly separated.

In connection with other methods of separation, mention may be made of the errors arising in ion-exchange from the faulty preparation of the resin and the stretching-out of the elution curve.

5.2. Contamination during the analysis

In the course of analysis the sample can be contaminated either by the component to be determined or by any disturbing element, e.g. if vessels are insufficiently washed, residual acid traces may change the pH of the medium; a particle of rust may enter the sample after iron has been separated; etc.

The significance of the contamination dangers and the difficulty in excluding these depend basically on two things:

(a) The degree of difficulty of the task, i.e. the expected concentration of the component to be determined. It is more difficult to exclude contamination in the analysis of high-purity compounds, e.g. telecommunication materials, or in a sample containing only very low concentrations of the component to be determined, e.g. power-plant feedwater, sea-water, etc., than in the general case.

(b) The other factor is the "concentration" (occurrence) in the laboratory of the component to be determined. In the case of rhenium, for instance, no precautions at all are necessary from this aspect whereas it is extremely difficult to avoid contamination in the determination of iron traces.

The possibility of contamination is not restricted only to the period of the analysis, but also holds for the operations preceding this, including the stages of production and technological experiments.

Our considerations will be confined to two stages:

(a) sampling, processing and storage;
(b) the analysis itself.

5.2.1. CONTAMINATION DURING SAMPLING, PROCESSING AND STORAGE

In the course of the sampling and processing, predominantly those samples may undergo contamination which must be homogenized or comminuted. The contamination of plant samples is due primarily to dust and dried-on sprays; washing is recommended to decrease this, but this is not a perfect solution either [8].

In comminution or powdering, the greatest source of contamination is naturally the material of the equipment used. Plant samples are often prepared with a hammer crusher, rock samples are powdered in a disc mill or a ball mill, and metal samples are usually obtained in the form of turnings. In these operations, Cr, Mn, Ni or Fe contaminants may enter the sample. In the case of rock and ore samples, for instance, a further error may be caused by the fact that samples with very different concentrations are processed with the same equipment, but this may not have been cleaned sufficiently carefully between the two samples. It is therefore recommended that there should be an intermediate operation of the equipment with "empty" samples. High-purity materials are powdered in agate or possibly porcelain apparatus. It is useful to check the "background" value of the contamination with "empty" samples from time to time.

Solid samples may also undergo contamination during storage. An interesting case of this is the contamination of the Apollo-11 moon-rock samples with In and Ag from the wrapping material, but naturally there are many more everyday examples. For instance, when plant samples were stored for 15 weeks in brown wrapping paper the boron content

123

increased from 3.5 ppm to 31 ppm because of a volatile boron compound in the paper. Certain plastics may contain various quantities of metals as a result of the catalyst used for their production, the plasticizer in them, or some other reason. The Zn content of polystyrene preparations and rubber stoppers is often a source of danger.

Since preliminary evidence about such contamination is not always available, the possibility must be examined before use. Solution samples may be stored in glass or plastic vessels. In both cases attention should be paid to the risk of contamination, the extent of which can depend on the material of the vessel, the composition of the solution, the nature of the component to be determined and the duration of storage.

The view is widespread that more contaminants dissolve out of a glass vessel than a plastic one. The conclusions of authors who have examined this question in detail are not so clear-cut. As mentioned above, certain plastics may contain large amounts of metal compounds. Scott and Ure [9] report that high-density polyethylene contains unacceptably large quantities of Zn and Cd; adhesive tape used to secure stoppers in vessels may, depending on its nature, give rise to Cr, Mo, Pb or Cu contamination. The bakelite caps on screw-topped vessels contain many elements of importance in the investigation of biological materials (Cu, Mn, Fe, Zn). Attention should also be drawn to the Cu and Zn contents of the inks used to denote the contents of vessels.

The degree of contamination can depend strongly on the pH of the solution too. At least in the event of glass vessels, solution of pH 7 is the most favourable; increase of either the acidity or the basicity leads to an increase in the dissolution from the vessel. However, since the loss from the solution is greatest in the pH range where hydrolysis of the metal ions occurs (and it is advisable to acidify the solution to avoid this), it rarely happens that these two opposing conditions can be satisfied simultaneously. In the case of biological fluids or foodstuffs, the dissolving-out of contaminants from the vessel may be promoted by organic acids or other compounds which generally behave as complex formers.

The extent of the error that may be expected as a consequence of the contamination can be established by carrying out blank tests, involving the storage and analysis of "empty" solutions, and by performing experiments with actual samples.

Contamination during the period of the analysis may have the following reasons: the vessels used for the analytical operations; the reagents; the environment (the atmosphere, the working equipment, etc.).

The vessels used for the analytical operations. This may be a source of serious contamination in some cases, depending on the compositions of the reactants and the physical conditions of the processes (temperature, duration, etc.). In the ignition of samples with a high organic matter content (plants, coals, plastics, etc.), the combustion furnace or the apparatus may be a source of contamination. Metal vessels and porcelain vessels that are unglazed, or where the glaze is not intact, must be avoided. Porcelain or platinum vessels are generally suitable for the ignition of precipitates, but in the examination of high-purity materials frequently only quartz is satisfactory.

The danger of contamination is even higher in the digestion (dissolution) of samples, for the vessel then comes into contact with solution or melt at elevated temperature. Platinum is attacked by the pyrosulphate melts used in rock analysis; in contrast with certain literature reports, the determination of niobium with thiocyanate cannot be carried out by this route because of the presence of interfering platinum.

Glass is to be avoided completely in the determination of boron traces; in our experience, an error is caused even if a solution is stirred with a glass rod only once.

Robertson [10] has published a detailed compilation of the contaminants to be found in laboratory vessels. From this it may be learned that the Millipore filter, which is very advantageous for fast filtration, contains much Cr and Zn. The occurrence of contamination from the vessels, and the possibilities of eliminating this, are similarly discussed by Kosta [10a] and by Gretzinger *et al.* [10b].

Quartz, siliconized quartz, teflon and polyethylene vessels are to be recommended for the examination of high-purity materials. With traditional glassware, the contamination can be diminished by "aging", i.e. by carrying out a series of alternate treatments with acid and base (this is in contrast with the general conception that unused vessels are preferable for sensitive analyses).

Perhaps the greatest problem in trace analysis is caused by the impurities in the reagents. This refers particularly to the chemicals used in much greater quantities than the sample (acids and bases used for digestion, masking agents, buffer solutions, etc.). According to the certified lists of impurities and to the standards, an appreciable proportion of compounds of "analytical purity" cannot be regarded as pure from the aspect of trace analysis. As examples, the maximum heavy metal contamination permitted in NH_4Cl by the Hungarian standards is 10 ppm, while the maximum permitted iron content of $HClO_4$ is 2 ppm [11]; the corresponding values for these compounds in the Merck catalogue are 6 and 2 ppm, respectively [12], so that the differences are not too great.

Unfortunately, it is not rare in reality that certain products do not meet even such specifications.

Before reagents are used, therefore, it is necessary to examine their impurities. This can be begun most simply by performing a blank test with the addition of all of the reagents. If contamination in excess of the permitted error of the analysis is found, then the reagents can be examined individually, starting in the final step and then adding the preceding operation, this being continued until the compound responsible for the contamination is identified.

The following procedures may be considered if the impurities are present in disturbing amounts:

(a) compensation of the error by means of calculation or via the comparative blank test;

(b) increase of the quantity of sample taken;

(c) the use of purer reagents;

(d) purification of the reagents.

Compensation of the error. This can be conceived of as an indirect determination, as the result is calculated from the difference between two values. One requirement, therefore, is that the light-absorbance caused by the contaminants should not be comparable in magnitude with the "useful" absorbance. The permissible extent of contamination is governed overall by the size of the permitted error and by the scatter in the value caused by the contamination. As a guiding principle it may be accepted that at most half of the permitted error should originate from this.

In certain cases, the error caused by the contaminants is decreased by

126

increasing the amount of sample taken. A condition of this is that the quantity of the reagents should not be increased at all, or to only a lesser extent. This method is limited by the fact that the technical performance of some operations, e.g. ashing, precipitation separation, etc., is cumbersome or impossible above a certain amount.

Compounds with purities higher than the designation *"of highest analytical purity"* are available commercially, e.g. the "Suprapure" products of Merck, the "Osobo chisty" ("special-purity") reagents of Soyuzhimexport, etc. However, the impurities in these are generally not so much lower as to satisfy the high purity requirements, and in particular they are not prepared with a view to some definite aim. Accordingly, such reagents should be bought only if an assessment of the attainable increase in accuracy shows this to be worthwhile.

Purification of the reagents is often necessary, but this cannot always be carried out in an analysis laboratory. For instance, the KOH to be applied in the solid state for digestions can clearly not be purified in solution, and acids to be applied in concentrated form cannot be purified in a diluted state.

The following are the most commonly used procedures, depending on the nature and state of matter of the compounds: *Recrystallization,* or possibly sublimation. This can be employed both for inorganic and for organic compounds. If the impurity is incorporated homogeneously into the crystal lattice, some intermediate purification (extraction, sorption on active carbon, etc.) must be inserted. *Sorption* on active carbon, silica gel, a molecular sieve, a trace-collecting precipitate, etc. *Ion-exchange.* This may be suitable for the purification of acids, for example, and in the case of metal impurities which form anionic complexes on an anion-exchanger too ($[FeCl_4]^-$, $[ZnCl_4]^{2-}$, etc.). *Extraction.* Apart from the starting compounds, this is very advantageous for the purification of buffer and masking solutions (in methods involving dithizone, etc.). *Distillation.* This may be used for solvents and some inorganic acids. A very advantageous variant is isothermal distillation, due to the very slight danger of contamination. It has the drawback that the reagent obtained is more dilute than initially. It is widely used to purify HCl and NH_4OH [13].

Mitchell [13a] and Moody [13b] provide detailed descriptions of the purification of analytical reagents.

Contamination caused by the working environment.

This may have the following sources:

(a) the air in the laboratory;
(b) the furnishings and equipment;
(c) the analyst (hands, clothing, etc.).

The dust and gas impurities in *the laboratory air* depend basically on the location of the laboratory. It is clear that, in a laboratory situated within an ore-concentrating plant, it is difficult to carry out reliable analyses of certain metal contaminants; similarly, in the vicinity of heavy vehicle traffic it is difficult to determine lead without a positive error. Siting of the laboratory with consideration to the prevailing wind direction, the construction of a relatively sealed building, or filtration of the air may all help, but in some cases only the relocation of the laboratory is effective.

In the customary trace analyses, contamination originating from the air is generally not significant. Even then, however it is advisable to maintain certain precautions, such as storing clean vessels under cover or filled with water.

Dust contamination can be successfully avoided if the longer operations, e.g. evaporation to dryness, are performed in a covered vessel. In the event of necessity, all of the analytical operations must be carried out in a closed chamber (box).

Contamination may be caused indirectly by the laboratory air, the ceiling, corroded metal objects in the laboratory, the metal components of the lights, iron stands, and the fume cupboards. In practice, a particularly great danger arises in the case of systematically performed operations with acidic solutions (acidic digestions in rock and metal analyses; wet decomposition of biological materials; etc.), and not only in the examination of high-purity substances. Insofar as possible, defensive steps must be taken against this with painting and with substitution of the metal parts, if this can be achieved.

In high-purity laboratories, e.g. in the case of the examination of luminescent powders, metal equipment should be avoided (water taps, stands, clamps, etc.); if this is not completely practicable, such equipment should be covered with rubber or plastic (not containing metal impurities).

128

Contamination may also result from *the analyst's clothing* or other objects. Mention may be made, for example, of metal rings, watches (chromium), zinc-containing ointments, cosmetic preparations and even the ink of ball-point pens.

A detailed review of the possibilities of contamination occurring in the course of trace analysis is to be found in the monograph of Sandell and Onishi [14] and in the review of Tchöppel *et al.* [14a].

5.3. Errors in the solution equilibria. Standard solution and reagents

In this section we shall survey all the errors that can arise from an incorrect knowledge of concentration, from incorrect compositions of reagents, and from non-optimum parameters (pH, redox potential, concentration, etc.).

5.3.1. ERRORS ARISING FROM AN INCORRECT KNOWLEDGE OF CONCENTRATION

This is naturally of great importance primarily in connection with the standard solutions necessary to prepare the calibration curve, but the reagent concentrations too are important. When certain standard solutions are being prepared, in many cases the composition of a compound that is given on the label of the bottle should not be accepted without checking. Differences occur mainly in the following cases:

(a) crystal hydrates, or compounds which absorb moisture, when the water content may differ substantially from the theoretical value;

(b) basic compounds which may bind CO_2 from the air;

(c) organic dyes, the active ingredient content of which may vary because of the salt content, the latter otherwise not disturbing the determination;

(d) mineral acids, and primarily HCl (even if the difference here is relatively not very large, it can cause an error when the acid concentration must be maintained at an exact value).

5.3.2. ERRORS ARISING FROM INCORRECT COMPOSITIONS OF REAGENTS

We referred earlier to errors caused by reagents with inappropriate compositions. Disregarding crude errors, this occurs primarily with organic dyes, where in an extreme case the bulk of a preparation may be impurity. If the latter is indifferent, it influences only the concentration and can be taken into account, via recording of the calibration plot and other examinations. At times, however, a reagent is not fit to be used as a consequence of its poor quality or its decomposition during standing, even in the solid phase.

It is well known that the products of oxidative decomposition of dithizone (formazan disulphates) give rise to errors in analysis [15], and they must be removed from the reagent solution by washing with NH_4OH.

In our analytical practice we have encountered several other reagents, e.g. 3,3-diaminobenzidine, malachite green, arsenazo III, containing impurities which made the reagent unusable (precipitate formation in the organic phase, variable spectrum, deviation from Beer's law, lower selectivity, etc.). One di-n-butylphosphoric acid (HDBP) product was found to contain 50% monobutylphosphoric acid.

Contamination can be established in some cases by means of chromatography, paper-electrophoresis or other methods used in organic analysis. (Examples are the control of the purity of chromotropic acid bis-azo derivatives by thin-layer chromatography [15a], or the demonstration of impurities of di-2-ethylhexylphosphoric acid by means of a hydrolysis method [15b]). However, this generally demands special expertise, well-tried methods of examination, and additional working time. In general, therefore, we must be satisfied with the assessment of the suitability of a reagent on the basis of the behaviour with the interfering elements to be determined, via spectrum recording and absorbance values. Particularly in extraction–photometric determinations, great attention must be paid to the observation of the appearance of a third phase (precipitate, emulsion, incrustation).

In the event of a reagent of unsuitable quality, its purification may be attempted (recrystallization, extraction, distillation), but this does not always prove effective.

Much useful advice on the purification of laboratory chemicals is given in the book by Perrin *et al.* [16].

Differences of the equilibria in solution from the desired state can primarily be attributed to the following factors: incorrect pH; incorrect redox potential; undesired complex formation; deviation from the monomeric state.

These are naturally interdependent, but for the sake of clarity we shall treat them separately.

Incorrect pH can be the main cause of all errors, including losses occurring in precipitation separation, the undesired precipitation, or dissolution, of interfering components, a deterioration in the efficiency and selectivity of extraction and ion-exchange, and a decrease in the extent of the complex formation desired in the determination. By this is to be understood the phenomena of hydrolysis and polymerization, and at times the change of the redox equilibria too.

Buffer solutions are generally prescribed for attainment of the desired pH. Detailed compilations of such buffer solutions are to be found in monographs and handbooks [17, 18].

However, in our experience, the buffer mixtures described in the various publications are not always satisfactory. For instance, the $HCl + KCl$ solution suggested for pH 2.0 has an insignificant buffer capacity, while the $CH_3COOH + CH_3COONH_4$ solution recommended for maintenance of pH 3.0 in the analysis of lanthanides is not suitable either. The reason for such an insufficient capacity is often the low concentration of the buffer. However, an increase in the concentration may be impeded by the fact that the anion of the buffer is at the same time a complex-forming ligand. Thus, in the above case (determination of the lanthanides with arsenazo III) an increase in the concentration cannot be permitted because of the resulting masking effect. Similar difficulties are encountered with phosphate buffers too.

Whether or not the buffer can be dispensed with is governed above all by the curve of the pH-dependence of the actual process. For instance, in the determination of UO_2^{2+} with arsenazo III the absorbance does not change in the pH interval 1.8–3.0, and thus a good solution is neutralization and then addition of the calculated quantity of HCl. The situation is similar in the determination of the lanthanides, where neutralization is effected in the presence of phenolphthalein and the quantity of acid necessary to establish a pH of 2.5 is subsequently added. Our measurements have demonstrated that this value can be maintained within a pH deviation of 0.05 even under the conditions of serial analysis.

Universal indicator paper graduated in 0.2 pH units is useful for control of the desired pH value, but in a more demanding case pH measurement is essential.

A pH level higher than 3.0 can otherwise be maintained reliably without a buffer only if the desired value can be detected by the change in colour of some indicator, or if the requirements are not strict.

An incorrect redox potential can cause errors by giving rise to an unsuitable oxidation state of ions of variable valency, or to decomposition of the reagent.

In alkaline medium the metal ions have low standard redox potentials; thus, the reactions $Mn(II) \rightarrow Mn(III)$ and $Fe(II) \rightarrow Fe(III)$ begin even in response to atmospheric oxygen. Maintenance of the Mn^{2+} state is possible only in the presence of a reductant, e.g. NH_2OH. Shifts in the equilibrium may be promoted not only by OH^-, but by other ligands too. Thus, the potential of the Fe^{3+}/Fe^{2+} system decreases in the presence of F^- or $H_2PO_4^-$, and UO_2^{2+} is reduced. If NO_3^- is added to the manganese system, the Mn^{2+} is partly oxidized to Mn^{3+}.

It may occur that the solvents and extractants contain reducing impurities; alcohols, for example, may contain aldehydes. These may reduce the metal ions in the higher oxidation states or, for instance, the H_2O_2 present as reagent, and hence a false result emerges from the determination.

Thus, maintenance of the desired oxidation state is the more difficult, the lower the concentration and the higher the potential difference in the redox system in question, including not only the ion to be determined, but the other ions present too. The only solution in many cases is the application of a reductant or oxidant excess. If this is not permissible, the procedure used to eliminate the excess must clearly not affect the ion now in the desired oxidation state, e.g. the excess of MnO_4^- applied for the oxidation of V^{4+} can be eliminated with $H_2C_2O_4$, but the excess of Ti^{3+} added for the reduction of UO_2^{2+} can be oxidized only in a relatively low concentration with NH_2OH [19].

Na_2O_2 and H_2O_2 are frequently used in digestions. The H_2O_2 may

132

disturb the determination, e.g. by bleaching azo dyes, but its removal is not certain even if the acidic solution is evaporated to dryness.

Among the interfering components in *undesired complex formation,* mention has already been made above of the anions of certain buffer solutions and also anions which influence the redox potential. Masking effects may be exhibited by certain reagent impurities, e.g. the mono- and pyrophosphoric acid esters in HDBP of unsatisfactory quality. Disturbance can similarly be caused by complex formers initially present in the sample (phosphate, fluoride, organic acids, etc.).

These effects can be established with standard solutions, and more accurately with internal standard methods, and in part they can even be included in the calculations.

Some of the disturbing effects result from the formation of polynuclear or mixed ligand complexes.

Examples of the former are the formation of Mo^{5+} and Al^{3+}-containing thiocyanate (in the presence of much Al^{3+}) [20], or polynuclear complexes causing coextraction with halide complexes [6]. Predominantly recording of the spectrum is suitable for recognition of the error. It is not possible to compensate for the resulting error with an internal standard method, as a consequence of the differing ionic states. An attempt must be made to learn as much as possible about the composition of the solution to be examined and to exclude such errors by appropriate elaboration of the method.

The formation of mixed complexes increases the stability compared to that of the original parent complex, and in this way disturbance may be caused by ions which otherwise do not yield an appreciable absorbance with the reagent. This has been observed, for example, in connection with the formation of the mixed ligand complex Fe^{3+}–morin–humic acid when morin is used to determine UO_2^{2+} in natural water [21], and the mixed ligand complex Ca^{2+}–arsenazo III–HPO_4^{2-} in another case. Mixed ligand complex formation also influences the absorbance of the ion to be determined.

Among the interfering effects caused by undesired complex formation, special mention must be made of *hydrolysis and polymerization.* This heading also covers polynuclear hydroxo (oxo) complexes produced by copolymerization.

The phenomenon has important consequences, which can cause errors

133

during the storage of the samples, in the analysis, and in the preparation and keeping of standard solutions.

The OH^- ion is known to be one of the ligands that give rise to complexes of very high stability. The tendency to complex formation is, to a first approximation, proportional to the valency of the cation, or to the degree of oxidation. Whereas the alkali metal ions are present in the form M^+. aq in solution even at pH 13, Zr(IV) occurs as ZrO^{2+} even in strongly acidic medium, and may even not be monomeric. At higher degrees of oxidation oxyanions too are formed, which are already soluble, e.g. MnO_4^-. Deviations from the general rule may be caused by the ionic radius, the coordination number and other factors. For instance, Th and Zr are situated in the same column of the periodic system, but even in mildly acidic medium Th is to be found in the form Th^{4+}.

Hydrolysis is in effect stepwise complex formation; its products range from cationic complexes to polymeric hydroxide (oxide hydrate) precipitates. In the intermediate steps, polynuclear complexes may exist in solution too.

In the knowledge of the hydrolysis constants of the individual metal ions, it is possible to calculate how a metal is distributed in the various forms at given OH^- concentration (pH) and metal ion concentration. The distributions for certain ions are to be found very illustratively in the literature, as plots of pH vs. mole fraction; this is exemplified for Fe(III) in Fig. 5.2 [22].

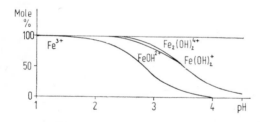

Fig. 5.2. Distribution of iron(III) compounds as a function of pH ($[Fe(III)] = 10^{-4}$ M) [22].

If more than one multivalent metal ion is present, then heteropolymers may be formed, depending on the individual characteristics; these are hydroxo complexes containing several metal ions in the inner sphere. Heteropolyacid formation may cause interference in acidic medium, but for only a few ions (VO_3^-, WO_4^{2-}, MnO_3^{2-}, HPO_4^{2-}).

The most frequent of the errors caused by hydrolysis are as follows: Loss due to *sorption on the wall of the vessel.* Sorption on colloidal particles in very dilute solutions (pseudocolloidal state) can also be included here, e.g. sorption on clay particles in water samples.

The decrease in the actual concentration of the metal ion to be determined because of the "aging" of the solution (or as a consequence of an incorrect pH) is in effect the masking of the metal ion by the OH^- ion. Babko [23] reports that the polymerization, occurring when a Zr(IV) solution in 0.5 M HCl is allowed to stand for 10 days, gives a 20% lower absorbance with alizarin than originally.

"Loss of the individual character" [23] because of the formation of a co-polymer containing another metal ion too. For instance, the determination of Mg^{2+} with titan yellow is not disturbed by the tartrate ion, but in the presence of Fe^{3+} a colour is not formed. The following experience and general notes relating to such phenomena are worthy of mention:

Many authors have dealt with the losses occurring during the storage of standard solutions. Their experiments lead to the conclusion that the cause of the losses is primarily an incorrect medium, and above all a pH promoting hydrolysis. Other factors, such as the material of the vessel, have only a secondary influence on the results. Hydroxo complexes are sorbed on virtually all surfaces.

This is the main explanation for the losses observed when water samples, urine samples, hydrometallurgical solutions, etc. are stored. Processes leading to a pH increase (loss of CO_2, decomposition of urea) and the progressive hydrolysis steps may take place during a particularly long time. Such processes may be countered by acidification of the samples, and in part by storing them in a refrigerator.

The monomeric ionic state can be ensured in dilute solution. Although it is true that the amount sorbed on the walls of the vessel may be relatively higher in dilute than in concentrated solution, it is easier to maintain the desired state in the bulk of a dilute solution.

Hydrolysis and polymerization that has already taken place can naturally not be eliminated by boiling, since this in fact promotes hydrolysis.

There are three possibilities to inhibit hydrolysis and in part to restore the ionic state: acidification of the solution to the appropriate extent; addition of an auxiliary complex former; addition of another metal ion which gives rise to a mixed nucleus hydroxo complex.

The solution must originally be acidified to the necessary extent, on the basis of the equilibrium constants or the diagram illustrating the distribution

135

in the individual forms (as examples, Fe^{3+} can be kept in a stock solution at 0.05 M HCl, but Nb^{5+} only at 10 M HCl). If acidification is necessary at a later time, depolymerization, which usually proceeds slowly, can be accelerated by heating in this medium.

In the preparation of standard solutions, the degree of acidity can in principle be optional, but the subsequent operations with too acidic solutions are frequently inconvenient, e.g. on the direct recording of the calibration plot. Independently of this, constant solutions of certain metal ions (Ta^{5+}, Sn^{4+}, Sb^{5+}, etc.) can not be prepared at high acidity.

Thus, there is another widely applicable possibility for maintenance of the reactivity of the ion to be determined: addition of an auxiliary complex former. A condition for the use of this is that it prevents hydrolysis, but at the same time impedes the "useful" complex formation to at most only a slight extent.

The procedure to be applied must be selected in accordance with the tendencies of the metal ions in question to undergo complex formation. For example, polymerization is prevented in the case of Sn^{4+} by the addition of Cl^- ($SnCl_6^{2-}$), in the case of Sb^{5+}, Ti^{4+} and ZrO^{2+} by the addition of SO_4^{2-} (sulphato complexes), and in the case of Nb^{5+} and Ta^{5+} by the addition of tartaric acid. Here boiling is of use for the transformation of hydrolyzed solutions, and possibly precipitates.

A procedure that is not used widely enough, but is often the best answer, is the formation of a copolymer with another metal ion. In contrast with earlier-mentioned examples of the undesired "loss of individual character", this situation is created in a controlled manner here with a cation which does not cause interference during the determination. In the preparation of a TiO^{2+} stock solution, such a relatively "innocent" cation is Al^{3+}; when the copolymer is later decomposed (possibly by the photometric reagent itself), monomeric TiO^{2+} is obtained [23]. ZrO^{2+} can similarly be recommended to stabilize a Ta^{5+} stock solution, and As^{5+} to stabilize an Sb^{5+} solution [24]. Of the procedures described here, the simplest must be chosen experimentally, conforming with the requirements of the analytical method.

The monomeric state, or more exactly the same state in the standard solution and in the solution to be examined, is one of the fundamental criteria of the accuracy of the method. A deviation from this situation cannot be compensated for with an internal standard method, and its recognition is not always easy either. It is very important, therefore, that the specifications be elaborated carefully, and strictly adhered to.

Finally, mention must be made of the error which may arise from dilution of the solutions taken for photometric measurement.

Even today, it is still possible at times to encounter in the literature the conception that stoichiometric ratios can be taken as basis (e.g. "a five-fold quantity of reagent is sufficient") instead of the concentrations governing the equilibria. This is clearly not logical; at best, it can be said that a satisfactory sensitivity can be ensured in this way too in the event of complexes of exceptionally high stability.

According to Babko and Pilipenko [25], a change in the reagent concentration, i.e. dilution without reagent, causes a change in the absorbance, this corresponding to the relation

$$\Delta A = \frac{K}{xc}(n-1)$$

where c is the concentration of the metal ion, K is the dissociation constant of the complex, x is a measure of the reagent excess, and n is the extent of its dilution.

5.4. Control of the suitability of the final solution for determination

Examination of the suitability of the solution taken for analysis comprises a very important part of the elaboration and control of the method. In effect, answers are expected here to two questions:

(a) Is there any difference in the absorbance of the ion to be determined in comparison to the values featuring in the calibration plot?

(b) Is there any foreign colour present which will disturb the determination?

A clear-cut answer can be obtained to the first question with an internal standard method. If a known quantity of the component to be determined is added to an aliquot of the solution to be examined and the determination is carried out on the mixture, the difference in the two values should give the value calculated via the calibration plot. A different, usually smaller, value may be obtained if:

(a) a complex former acting as a masking agent is present (from reagents added to the sample or in the pretreatment operations);

(b) the pH is inappropriate;

(c) the redox potential of the solution is not suitable (naturally, only in the case of ions of variable valency, or reagents sensitive in this respect);

(d) a polynuclear complex or possibly a copolymer containing another metal ion too is formed in the solution, or the reagent and some other anion form a mixed ligand complex with the ion to be determined (with a different specific absorbance).

The measurement can naturally not give an answer as to which of the listed possibilities has caused the difference, but it does indicate the fact that the difference exists and allows steps to be taken to eliminate the error.

If this difference is not too large, but it does not seem possible to exclude it in the course of the analysis, the calibration plot is then inapplicable and the internal standard method must be used instead.

A condition of application of the internal standard method is naturally that the added ion should be in the same state (oxidation state, monomer or the same degree of polymerization, etc.) as the ion to be determined, and its addition should not alter the character of the solution (pH, redox potential).

Of course, the internal standard procedure is not suitable for demonstration of contamination which increases the absorbance additively. An answer to this may be given primarily by recording the absorption spectrum of the solution, though when the interference is a large-scale one the error may possibly be visible to the naked eye.

If a recording spectrophotometer is available, recording of the spectrum may become a systematic control method. A good example of the necessity of spectrum recording is the recognition of the disturbing effect of Fe^{3+} in the determination of UO_2^{2+} with arsenazo III [26]. The determination is disturbed strongly by even a small quantity of Fe^{3+}. Since Fe^{2+} does not absorb even in high concentration, the Fe^{3+} is reduced with ascorbic acid. In the case of a large iron content, however, a certain absorbance increase is observed, this varying with time. As can be seen in Fig. 5.3, the cause of

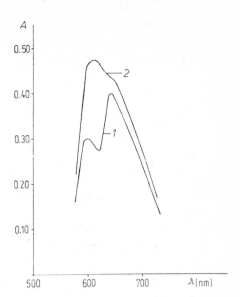

Fig. 5.3. Absorption spectrum of UO_2^{2+}–Fe^{2+}–arsenazo III–ascorbic acid system [26].

1—immediately after mixing; *2*—1 hour later

this is the re-formation of Fe^{3+} after the reduction. The explanation is that the redox potential of the Fe^{2+}–Fe^{3+} system is decreased by the ascorbic acid–dehydroascorbic acid and the arsenazo III, as complex formers with Fe^{3+}. Thus, the Fe^{2+} reduces the azo bond in the arsenazo III, with the formation of Fe^{3+}, which reacts with arsenazo III. Without spectrum recording, this error would hardly have been detected.

As this example reveals, certain disturbing effects may cause a change in the colour as a function of time. Since a rather long time may elapse under the conditions of serial analysis, it is advisable to extend the control to the effect of the time passing until photometry is performed.

Light may be shed on the occurring errors if the analysis is carried out with different solution aliquots. Particularly in the case of a masking effect, there is a high probability that the two absorbance values will not be proportional; a higher result is obtained from the smaller sample. If the contaminant giving the foreign colour is a complex of low stability which does not obey Beer's law, it similarly gives rise to less interference in a smaller sample. If the values obtained from the two samples coincide, however, there is only a small probability that a disturbing effect is exerted which remains undetected during the elaboration of the method.

To summarize, it may be stated that the above simple methods establish

139

with high probability whether the solution taken for the determination is suitable, and they demonstrate the existence of disturbing effects. It is important that they should be used in analytical practice.

5.5. Control of the accuracy of the methods with standard samples

In research with the aim of the elaboration of an analytical method (no matter how carefully this is planned and carried out), there may remain uncertainties which can be eliminated only with difficulty, or not at all, by the use of an artificial mixture or the control methods described above (internal standard, spectrum recording, etc.). For instance, the applicability of a digestion method in which the sample is not dissolved up completely can be established best with a sample of analogous composition in which the concentration of the component to be determined is known. When artificial mixtures are prepared, it is frequently not possible to take into consideration components which influence the accuracy of the method, e.g. organic compounds that may not be known exactly in the case of food products or biological materials.

In recent decades, standard samples have been prepared for a great range of substances to be examined, in all branches of industry. Some of these are carefully selected and processed and subsequently analyzed naturally-occurring substances (rock and ore samples) or industrial products, e.g. telecommunication materials of high purity. Another group of samples are in fact artificial mixtures, made from natural material, e.g. urine, blood, plant ash, etc., by the addition of trace elements in known concentrations.

The best-known early standards were the G–1 and W–1 granite and diabase samples of the United States Geological Service [27]. Since then standards have been made available by very many international and national organizations (International Atomic Energy Agency, United States Bureau of Standards, ASTM, Geological Standards Committee of Comecon, etc.), research institutes and companies. Such activity is likewise continued in Hungary (Hungarian State Geological Institute, Iron-Industry Research Institute, Metal-Industry Research Institute, Mining Research Institute, etc.).

Publications dealing with standard samples give a detailed account of the experience acquired in the analyses of the samples [28–34]. The most probable values for natural samples are the resultants of the analyses of these samples in many laboratories, and the results of the different analyses do not always display good agreement. It is quite common for years to pass before the officially accepted ("recommended", "suggested", "most probable", "mean") values emerge for certain elements that are especially difficult to analyze.

This process is slowed by the fact that there is not always a guarantee concerning the work of the laboratories performing the analyses. Abbey [28] has made an attempt to attain an objective assessment: the laboratories are classified on the basis of their achievements in the analyses of many standard samples, and subsequently the recommended values are calculated only from the data from those laboratories which consistently produce the best results, e.g. only 8 out of 25 in a given case.

In spite of all these difficulties, standard samples that have already been certified are naturally well applicable for the control of trace-analysis methods. However, we wanted to point out here that the performance of trace analyses of high accuracy is not always guaranteed even in internationally acknowledged laboratories.

As a consequence of the difficulties in acquiring standard samples, and their relatively high prices, locally produced standard samples are generally used for the control of serial analyses. If these are analyzed many times, and if possible in other laboratories too, individual samples may be of assistance in the control over a long period.

Depending on the requirements, the control can extend from the periodic analysis of a code-labelled sample to the inclusion of such a sample in every series. This latter procedure is justified only very rarely, however, not only for economic reasons, but also because it leads to the wooden performance of the work rather than the enhancement of the concentration.

References

[1] Gorsuch, T. T.: *The Destruction of Organic Matter.* Pergamon, Oxford 1970.
[2] Upor, E., Jurcsik, I., Mohai, M.: *Acta Chim. Acad. Sci. Hung.* **37**, 1 (1963).
[3] Bock, R., Jacob, D.: *Z. Anal. Chem.* **81**, 200 (1964).
[4] Bock, R.: *Aufschlußmethoden der anorganischen und organischen Chemie.* Verlag Chemie, Weinheim 1972.

[5] Dolezhal, J., Povondra, P., Shultsek, Z.: *Metody razlozheniya gornykh porod i mineralov*. Mir, Moscow 1968.

[6] Alimarin, I. P. (ed.): *Ekstraktsionnye metody v analiticheskoi khimii*. *(Trudy Komissii po anal. khim. XIV.)* AN SSSR, Moscow 1963.

[7] Zolotov, Yu. A.: *Zh. Anal. Khim.* **27**, 1118 (1972).

[8] Thomas, A. D., Smythe, L. E.: *Talanta* **20**, 469 (1973).

[9] Scott, R. O., Ure, A. M.: *Proc. Soc. Anal. Chem.* **9**, 288 (1972).

[10] Robertson, D. E.: *Anal. Chem.* **40**, 1067 (1968).

[10a] Kosta, L. *Talanta,* **29**, 985 (1982).

[10b] Gretzinger, K., Kotz, L., Tchöppel, P., Tölg, G.: *Talanta* **29**, 1011 (1982).

[11] Kolos, E. (ed.): *Vegyszervizsgálat (Examination of chemicals)*. Műszaki Kiadó, Budapest 1969.

[12] Merck Chemikalien–Reagenzien. Darmstadt 1974.

[13] Marecek, J., Ditz, J.: *Chem. Listy* **59**, 972 (1965).

[13a] Mitchell, J. W.: *Talanta* **29**, 933 (1982).

[13b] Moody, J. R., Beary, E. S.: *Talanta* **29**, 1003 (1982).

[14] Sandell, E. B., Onishi, H.: *Photometric Determination of Traces of Metals. General Aspects*. Wiley, New York 1978.

[14a] Tchöppel, P., Kotz, L., Schultz, W., Weber, M., Tölg, G.: *Z. f. Anal. Chem.* **302**, 1 (1980).

[15] Irving, H. M. N. H.: *Anal. Chim. Acta* **56**, 205 (1971).

[15a] Savvin, S. B., Akimova, T. G., Syanova, E. M.: *Zh. Ahal. Khim.* **37**, 640 (1979).

[15b] Ignatenko, A. B., Pastoukhova, I. V., Petrov, K. A., Smelov, V. S., Gauzov, B. A., Shmidt, V. S.: *Zh. Anal. Khim* **37**, 744 (1982).

[16] Perrin, D. D., Armarego, W. L. F., Perrin, D. R.: *Purification of Laboratory Chemicals*. Pergamon, Oxford 1966.

[17] Perrin, D. D., Dempsey, B.: *Buffers for pH and Metal Ion Control*. Chapman Hall, London 1974.

[18] Schwabe, K.: *Fortschritte der pH-Meßtechnik*. Steinkopff, Dresden 1963.

[19] Anselmi, D.: *Chim. ind.* **54**, 1081 (1972).

[20] Babko, A. K., Grudeshina, G. I.: *Zav. Lab.* **30**, 773 (1974).

[21] Upor, E.: *Hidrológiai Közlöny* **4**, 299 (1958).

[22] Sillen, L. G., Martell, A. E.: *Stability Constants of Metal-Ion Complexes*. Special Publication No. 17. The Chemical Society, London 1964.

[23] Babko, A. K.: *Talanta* **15**, 721 (1968).

[24] Lemerle, J.: Étude de la polimérisation de l'acide antimonique et de stabilisation en solution par l'acide arsénieux. Thesis, 1971.

[25] Babko, A. K., Pilipenko, A. T.: *Photometric Analysis*. Mir, Moscow 1971.

[26] Mohai, M.: Uranium determination methods based on spectrophotometry of the UO_2^{2+}–arsenazo III complex. Paper presented at the Second European Conference on Analytical Chemistry, Budapest, 1975.

[27] Fairbairn, H. W.: *U.S. Geol. Surv. Bull.* No. 980 (1951).

[28] Abbey, S.: *Canadian Spectroscopy* **15**, 2 (1970).

[29] Abbey, S.: *Geol. Surv. Canad. Paper* 72–30 (1972).

[30] Abbey, S.: *Geol. Surv. Canad. Paper* 73–36 (1973).
[31] Abbey, S.: *Geol. Surv. Canad. Paper* 74–41 (1975).
[32] Lontsykh, S. V., Parshin, A. K.: *Zh. Anal. Khim.* **34**, 2446 (1979).
[33] Lontsykh, S. V., Sidorovskii, A. J., Zakovirin, O. M.: *Zh. Anal. Khim.* **34**, 2453 (1979).
[34] Ostroumov, G. V. (ed.): *Metodicheskie osnovy issledovaniya khimicheskovo sostava gornykh porod, rud i mineralov.* Nedra, Moscow 1979.

6. Determination of the individual elements

Introduction

In this section of the book we shall deal with the determination of the individual elements. The scope available does not permit us to provide the reader with a full account of all the possibilities for determination of a given element; we shall merely give examples of the application of the reagents which we consider to be of the greatest importance.

In the short sections relating to the individual elements, the following procedure is generally employed:

1. A brief account is given of the analytical (complex-chemical) properties of the ions of the element in question.

2. When a fair number of well-proved reagents are known for the determination of an element, the fundamental properties of these reagents are tabulated.

3. If one reagent is outstandingly applicable for the determination in our experience, procedures based on the use of other reagents are not given.

4. For the individual elements an attempt is made to mention the variants suitable for the analysis of the most important types of samples.

5. A detailed description is given primarily of those determinations that have been elaborated in our laboratory or in the application of which we have experience.

6. If the element in question can be determined advantageously by some other method too, this is referred to.

6.1. Alkali metals

The alkali metals belong in column I/1 of the periodic system. The ability of their ions to form complexes is negligible. Thus, there are few possibilities for their direct photometric determination. The lithium ion, which has the greatest polarizing power among the alkali metal ions, is to a certain extent an exception. Thus, the lithium content of natural waters and mineral waters can be determined with nitroanthranilazo [1, 2].

A method was earlier recommended for the determination of 1 mg/1 K^+ in natural waters by precipitation with dipicrylamine, with photometric determination following dissolution of the precipitate [3].

After the separation of interfering cations, there is a possibility for the determination of sodium by the precipitation of acetates containing three cations, e.g. sodium zinc uranyl acetate [4]; the precipitate is then dissolved up, and the sodium is determined indirectly with an appropriate reagent by photometry of the uranium or the zinc. However, this can be performed with sufficient accuracy only for higher concentrations of sodium, and hence this cannot strictly be considered for the analysis of small traces.

Alkali metals may be determined simply and with adequate accuracy mainly by flame-emission (flame–photometric) methods.

*Determination of lithium in natural waters and mineral waters with
nitroanthranilazo* [1, 2]

Nitroanthranilazo (see Appendix) forms an orange-yellow complex with the lithium ion in alkaline aqueous solvent media. The following ions do not interfere: 100-fold Mg^{2+} and Pb^{2+}, 50-fold Ca^{2+}, 25-fold Zn^{2+}, Ba^{2+}, Na^+ and Sr^{2+}, 10-fold Al^{3+} and 2.5-fold Fe^{3+}. SO_3^{2-}, Co^{2+}, Ni^{2+} and Cu^{2+} can be permitted only in the same concentration as the lithium. For separation from interfering elements, LiCl is extracted with acetone.

Analytical procedure: A 25–50 cm^3 water is evaporated to dryness, the Li^+ is extracted 3 times with acetone, the acetone is evaporated off, and the residue is made up to 25 cm^3 with distilled water. An aliquot of at most 5.0 cm^3 is taken, 0.83 cm^3 0.002 M nitroanthranilazo (dissolved in dimethylformamide) and 0.83 cm^3 0.2 M KOH are added, and the volume is made up to 25 cm^3 with acetone. The absorbance is measured not at the maximum

(460 nm), but at 530 nm, where the difference between the absorbances of the reagent and the complex is the greatest. 50 μg/l Li^+ can be determined.

References

[1] Dziomko, V.M., Zelichenok, S.L., Markovich, I.S.: *Zh. Anal. Khim.* **23**, 170 (1968).
[2] Morgen, E. A., Vlasov, N. A.: *Zh. Prikl. Khim.* **44**, 2752 (1971).
[3] Lure, Yu. Yu.: *Unifitsirovannye metody analiza vod.* Khimiya, Moscow 1973.
[4] Lundell, G.E.F., Bright, H.A., Hoffman, J.I.: *Applied Inorganic Analysis.* National Bureau of Standards, Washington 1955.

6.2. Aluminium

Aluminium belongs in column III/1 of the periodic system. In its aqueous solutions it is trivalent. It is amphoteric in nature; the $Al(OH)_3$ precipitating out at pH 5 dissolves up as the complex $Al(OH)_4^-$ at pH >10. This fact offers two possibilities for the separation of Al^{3+} via precipitation: from Ca^{2+} and Mg^{2+}, for example, at pH 5, and from $Fe(OH)_3$ and other insoluble hydroxides at pH >12.

Of its complexes, the high-stability AlF_6^{3-} is primarily of importance in the determination of other elements. Also characteristic are its complexes with O and N-containing ligands. These include the photometric reagents for Al^{3+}, a detailed account of which is to be found in the publications by Tikhonov [1, 2].

The photometric reagents for Al^{3+} are anthraquinone derivatives (alizarin), triphenylmethane derivatives (aluminon, eriochrome cyanine R, chrome azurol S, pyrocatechol violet) or N and O-containing compounds (oxine). A brief survey of their properties is given in Table 6.1.

With the exception of oxine, these reagents do not yield true solutions but lakes, and thus a reproducible absorbance is possible only if the optimum conditions are carefully adhered to.

The extraction of Al^{3+} with oxine is of use not only for its determination, but also for its separation.

Aluminium may be masked with tartrate, citrate and fluoride. The masking effect may be eliminated by evaporation to dryness with conc. H_2SO_4.

146

TABLE 6.1.

More important photometric reagents for aluminium

Reagent	pH of determination	λ_{max} nm	$\varepsilon \times 10^{-3}$	Interferences	Refs
Aluminon	5.3	530	1.1	In presence of $(NH_4)_2CO_3$ and thioglycollic acid, only Be and Cr	[1, 2]
Eriochrome cyanine R	5.8–6.0	535	7.4	Many ions	[1, 2]
Chrome azurol S	5.7	545	5.0	Many ions	[1, 2, 7, 8]
Oxine	8	395	4.8	Many ions	[1, 2, 9]
Pyrocatechol violet + zephyramine	5.5–10	587	89	F^-, I^-, SCN^-, ClO_4^-	[6]
Nitrosulphophenol S	2.5	685	64	Interfering ions can be eliminated by addition of complex-formers	[10]

DETERMINATIONS WITH OXINE

Photometric determination of aluminium in digestion solution from a uranium plant (authors' procedure)

The digestion solution contains iron, magnesium, and uranium in addition to 100 mg/dm³ aluminium. In the first step the Al^{3+} is separated by alkaline precipitation, and the uranium remaining in solution is removed by extraction.

Analytical procedure: 20–25 cm³ solution is taken, the Fe^{2+} in it is oxidized with a few drops of saturated $KMnO_4$ solution, and the volume is made up to approximately 100 cm³. Trivalent metal ions are precipitated in NH_4Cl medium at the boiling-point with 1:1 NH_4OH. The precipitate is filtered off and dissolved in hot 1:1 HCl, and the solution is evaporated down to low volume. The volume is then made up to about 100 cm³ with hot distilled water. The resulting solution is poured into 100 cm³ hot 20% NaOH, and 3 cm³ 30% H_2O_2 is added in small portions. The solution is stirred, covered and boiled gently for 30 min. The precipitate is filtered off and washed, and after cooling the filtrate is made up to 500 cm³. In this

147

medium, at pH > 12, the uranyl hydroxide precipitate undergoes peptization and passes into solution [3], and thus the uranium must be removed from the solution by extraction.

The acidity of a 10–25 cm^3 aliquot from the stock solution is adjusted to 6 M with conc. HCl. The uranium is extracted from this medium with 2×10 cm^3 30 vol.% TBP in petroleum ether. The aqueous phase is evaporated to 2–3 cm^3.

2 cm^3 10% tartaric acid is added to keep the aluminium in solution, the mixture is diluted with 20 cm^3 distilled water, and the pH is increased to about 8 with NH$_4$OH solution in the presence of 1 drop phenolphthalein. Masking solutions are added to bind the metal traces (Fe, Ti, U, V, Sn, Co, Ni, Mn, Cu, Cr, Zn, Cd, Mo, Zr): 1 cm^3 3% H$_2$O$_2$, 5 min later 5 cm^3 20% Na$_2$SO$_3$, and 3 min later 5 cm^3 10% KCN. The solution is heated to 80°C to promote formation of the iron complex Fe(CN)$_6^{4-}$, and then cooled. The aluminium is extracted from the solution with $10+5+5$ cm^3 oxine solution in CHCl$_3$ (2 g oxine/100 cm^3 CHCl$_3$). The combined organic phase is washed with 10 cm^3 distilled water, and its volume is then made up to 50 cm^3 with CHCl$_3$. The absorbance of the solution is measured against that of pure CHCl$_3$ at 395 nm. The calibration curve is produced between the limits 20 and 200 μg Al^{3+}/50 cm^3.

Determination of aluminium in plant materials [4]

Analytical procedure: The finely-powdered leaf matter (0.4–0.8 g) is ashed for 2 hours at 450°C in a porcelain crucible. After cooling, the residue is heated with 6 cm^3 1:1 HCl, 10 cm^3 water is added, and the mixture is left to stand overnight. The precipitated silica is filtered off, and the volume is made up to 50 cm^3. A 2 cm^3 aliquot is pipetted into a shaking-funnel and made alkaline with 0.5 cm^3 1:1 NH$_4$OH in the presence of phenolphthalein indicator. In this step (if required), the iron can be extracted with 1,10-phenanthroline and measured photometrically. A further 10 cm^3 1:1 NH$_4$OH is added to the aqueous phase, 5 min later 4 cm^3 10% EDTA is added, and the mixture is left to stand for another 5 min. 2 cm^3 oxine reagent (5% in alcohol) is added, the complex is allowed to develop during 1 hour, and it is then extracted with 20 cm^3 CCl$_4$. The phases are left to separate for 15 min, the organic solution is filtered, and its absorbance is measured against pure CCl$_4$ at 395 nm.

Determination of aluminium in tantalum-niobium minerals [5]

Analytical procedure: 5–10 mg of the material to be analyzed is fused with $K_2S_2O_7$ in a platinum vessel. The melt is dissolved out with 10 cm³ 0.5 M H_2SO_4, then 1 cm³ 15% tartaric acid and 20–30 mg ascorbic acid are added, and the mixture is neutralized to pH 5 with NH_4OH. 5 cm³ 0.25% bipyridyl is added, the mixture is left for 1 hour, and the Al, Tl, U and Sn are extracted jointly with 5 cm³ 2% oxine in benzene solution. The organic phase is washed with 10 cm³ distilled water, and then re-extracted 3 times with 1:2 HCl. 1 cm³ 15% tartaric acid and 2 drops 30% H_2O_2 are added to the aqueous solution, and the pH is increased to 9 with NH_4OH. Extraction of the aluminium is repeated with 10+5+5 cm³ oxine, the volume is made up to 25 cm³ and the absorbance is measured at 395 nm.

DETERMINATIONS WITH OTHER REAGENTS

Determination of aluminium content of river-water with pyrocatechol violet + zephyramine [6]

Aluminium traces in water are determined by photometry on a $CHCl_3$ solution of the triple complex formed between aluminium, pyrocatechol violet (PV) and zephyramine (tetradecyldimethylbenzylammonium chloride; ZCl).

Analytical procedure: The water is acidified (0.5 cm³ conc. H_2SO_4/dm^3) and filtered (0.45 μm membrane filter), and a 1–5 cm³ aliquot is taken. 0.5 cm³ 0.05 M hydroxylammonium sulphate, 1 cm³ 1 M thiourea, 1 cm³ 0.01 M 1,10-phenanthroline and 1.5 cm³ 0.2 M acetate buffer (pH 5.6) are added, and the mixture is left to stand for 1 hour. Subsequently, 1 cm³ 5×10^{-4} M PV is added, the solution is diluted to 10 cm³, and after 5 min it is extracted with 5 cm³ 10^{-3} M ZCl in $CHCl_3$ (for 10 min). The organic phase is shaken for 10 min with 5 cm³ washing liquid [1 M NH_4OH– $(NH_4)_2SO_4$ buffer in 0.5 M NaCl, pH 9.5]. After 30 min, photometry is carried out in 1 cm cells at 587 nm. The limit of detection is 3 μg Al/dm^3, and the accuracy is 4%.

149

Determination of a small quantity of aluminium in zirconium ceramics with chrome azurol S [7]

Zirconium ceramics contain lanthanides, yttrium and calcium, all of which interfere in the determination of aluminium. An ion-exchange separation is therefore inserted before development of the colour.

Analytical procedure: 50 mg finely-powdered material is weighed into a Pt crucible (a Pt alloy containing 5% Au) and 500 mg digestion mixture (45 parts borax+16 parts boric acid+7 parts lithium hydroxide) is added. The mixture is fused for 10 min at 900°C. The melt is dissolved out with 5 cm³ 6.5 M HCl, heated up to about 60°C. A further 20 cm³ 6.5 M HCl and 20 cm³ ethanol are added. The solution is passed through an ion-exchange column (Dowex 50 W-X8, 100–200 mesh, 15×1 cm charge) at a flow rate of 0.5 cm³/min. (Pretreatment of column: passage of a mixture of 60 cm³ 6 M HCl+40 cm³ water, followed by 50 cm³ 3 M HCl in 50% ethanol.) The column is washed with 150 cm³ 3 M HCl in 50% ethanol, the washings are combined with the original effluent, and the volume is made up to 250 cm³.

A few drops of conc. H_2SO_4 are added to a 25 cm³ aliquot of this stock solution, and the mixture is evaporated to dryness. The residue is dissolved in 5 cm³ water; 2 cm³ ascorbic acid (50 mg/cm³), 10 cm³ buffer (27.5 g ammonium acetate+0.5 cm³ glacial acetic acid per 100 cm³) and 2 cm³ 0.2% chrome azurol S in 25% ethanol are added, and the volume is made up to 50 cm³. After 15 min, the absorbance is measured at 540 nm, in 4 cm cells.

The calibration curve is produced for the concentration range 2–10 mg Al^{3+}/50 cm³.

References

[1] Tikhonov, V. N.: *Zh. Anal. Khim.* **21**, 829 (1966).
[2] Tikhonov, V. N.: *Analiticheskaya khimiya alyuminiya*. Nauka, Moscow 1971.
[3] Upor, E.: *Kémiai Közlemények* **34**, 45 (1970).
[4] Paul, J.: *Microchim. Acta* **6**, 1075 (1966).
[5] Lyubomilova, G. V., Miller, A. D.: *Novye metody issl. po analizu redkomet. min., rud i gorn. porod*. Nauka, Moscow 1970.
[6] Korenaga, T., Motomizu, S., Toei, K.: *Talanta* **27**, 33 (1980).

[7] Kruidhof, H.: *Anal. Chim. Acta* **99**, 193 (1978).
[8] Nemodruk, A. A., Supastashvili, G. D., Arebadze, N. G., Kikapidze, T. A.: *Zh. Anal. Khim.* **37**, 1028 (1982).
[9] Nazarenko, V. A., Shitareva, G. G., Veshchikova, N. A.: *Zav. Lab.* **47**, 6, 19 (1981).
[10] Savvin, S. B., Petrova, T. V., Mongush, K. D.: *Zh. Anal. Khim.* **35**, 54 (1980).

6.3. Antimony

This element belongs in column V/1 of the periodic system. In its properties it displays many similarities to the elements above and below it in this column, i.e. arsenic and bismuth, but also to the neighbouring elements in the same period, particularly to tin. It is either trivalent or pentavalent in its compounds.

There is no selective photometric reagent for antimony, and its separation from interfering ions is therefore very essential.

In acidic medium, the standard redox potential of the $Sb^{5+}-Sb^{3+}$ system is $+0.75$ V, i.e. Sb^{5+} is a mild oxidant. NO_2^-, Ce^{4+} or MnO_4^- are of use for the oxidation of Sb^{3+}. Metallic antimony or SbH_3 (b.p. $-17°C$) can be obtained by reduction, e.g. with cementation on copper, or with zinc amalgam; this provides a means for separation.

$Sb(OH)_3$ (or H_3SbO_3) is markedly amphoteric in nature, while H_3SbO_4 ($H[Sb(OH)_6]$) yields a precipitate or at least a colloid at acidic pH. Precipitation of antimonic acid, e.g. on a stannic acid or $MnO_2.nH_2O$ carrier, is likewise a separation possibility.

The complex ions formed with O-containing ligands are of analytical importance partly to keep Sb^{5+} in solution (tartaric acid, etc.), and partly as photometric reagents. Both valency states of antimony form fairly stable complexes with halides, but whereas the chloro complexes of Sb(V) can be well extracted, of the Sb(III) compounds only SbI_3 (SbI_4^-) is suitable for quantitative separation.

Also of importance are the reactions of ion-pair formation with basic dyes, certain S-containing compounds (dithizone, dithiocarbamates) and N-containing ligands too (antipyrine derivatives).

Some of the most important reagents and determination possibilities are listed in Table 6.2. The reader is referred to the literature for a more detailed compilation of the photometric reagents for antimony [1] and its determination with basic dyes [2,3].

TABLE 6.2.

Main characteristics of some reagents suitable for photometric determination of antimony

Reagent (absorbing complex)	Conditions of determination	λ_{max} nm	Interferences	Refs
I^- (SbI_4^-)	1.2–1.9 M H_2SO_4	420	Bi^{3+}	[1]
Diantipyrylmethane [(DAM)SbI_4]	Extraction with $CHCl_3$ from 0.2–3.0 M HCl; shaking with KI	340	Pt, Pd, Os, Tl	[12]
Rhodamine B	Extraction with benzene from 6 M HCl	565	$Cr_2O_7^{2-}$, Fe^{3+} Ga^{3+}, Tl^{3+}	[13]
Crystal violet	Extraction with benzene from 2.5 M HCl	610	—	[14]
4,4'-Bis(N-methyl-N-benzylaminophenyl)-antipyrylcarbinol	Extraction with benzene from 1–3 M HCl	590	Tl^{3+}, Au^{3+}, Fe^{3+}, Sn^{4+}	[15]
Brilliant green	Extraction with benzene or toluene from 1–2 M HCl	620	Fe	[6]
Thionine	Extraction with nitro-benzene from 1–2 M HCl	600	As, Cd	[8]

Photometric determination of antimony in zinc and lead ores with thiourea and potassium iodide [4]

1.0 g sample is treated with conc. H_2SO_4 first in the cold, and then with heating for 15 min, after which the mixture is cooled and 100–150 cm³ distilled water is added. After the addition of 5 cm³ 10% tartaric acid, the solution is boiled in order to dissolve the antimony, the insoluble residue is filtered off, and the volume of the filtrate is made up to 250 cm³. A 25–30 cm³ aliquot of this solution is taken, and 15 cm³ 1 : 1 H_2SO_4 and 20 cm³ reagent (350 g KI+25 g ascorbic acid in 1000 cm³) are added. Following a waiting period of 2–3 min, 5 cm³ 10% thiourea is added, and the volume is made up to 100 cm³. The absorbance is measured at 425 nm, or with a colour filter corresponding to this, with a layer thickness of 5 cm. The calibration plot is obtained for the range 0.1–2.0 mg Sb^{5+}. The thiourea stabilizes the SbI_4^- ion.

Donaldson [5] uses sodium metabisulphite in 6 M HCl medium to reduce antimony oxidized during the dissolution of ores. The antimony is separated from the interfering elements by precipitation on iron hydroxide–lanthanum oxide. The precipitate is dissolved in 5 M HCl containing thiourea and tin(II) chloride. HF is added to bind the excess tin, and the antimony is then extracted as the xanthate. Washing with HCl is inserted to remove bismuth. The antimony is determined in the form of its iodide. The sensitivity of the method is 1 ppm Sb.

Determination of antimony traces in rocks with brilliant green [6,7]

An extractable ion-pair is formed between $SbCl_6^-$ and brilliant green. The oxidation of Sb^{3+} is achieved with $NaNO_2$.

Analytical procedure: 0.1–1 g sample is evaporated with a mixture of $HF + H_2SO_4$, the residue is ignited and digested with $K_2S_2O_7$, and the melt is dissolved in 6 M HCl. A 1–5 cm³ aliquot is measured into a shaking-funnel, and the volume is made up to 20 cm³ with 6 M HCl. 0.5 cm³ 5% $NaNO_2$ solution is added, followed 2 min later by the addition of 7 cm³ distilled water, 25 cm³ 5% hexametaphosphate and 5 cm³ 0.005% brilliant green, and extraction is performed for 1 min with 15 cm³ benzene or toluene. The absorbance of the organic phase is measured in a 1–4 cm cell at 620 nm. The extract of the reagents taken through the process is used as reference solution. The calibration curve is recorded for 1–10 μg Sb. The sensitivity of the method is 10 ppm Sb.

Note:

The analysis must be continued immediately after the oxidation; the phases must be separated without delay after the extraction; and the reagent solutions employed must be freshly prepared daily.

Determination of antimony content of lead, tin and bismuth-based alloys with thionine [8]

The method based on the extraction and photometric determination of the ion-association formed between Sb(V) and protonated thionine (diamino-phenothiazine).

Analytical procedure: 0.5 g of the material to be examined is dissolved by boiling in 10–15 cm³ of a 3 : 1 mixture of $HCl + HNO_3$. For the removal of HNO_3, the solution is evaporated to dryness 3 times with conc. HCl. It is then washed into a 50 cm^3 volumetric flask with conc. HCl, and the volume is made up to the mark with conc. HCl.

An aliquot (at most 2.5 cm^3) is measured into a shaking-funnel. If an aliquot smaller than 2.5 cm^3 is taken, the volume is made up to 2.5 cm^3 with conc. HCl. 20 cm^3 water and 2 cm^3 1.6×10^{-4} M thionine solution are added, and extraction is performed for 30 s with 25 cm^3 of a 4:1 mixture of chloroform+nitrobenzene. The extract is centrifuged, and the absorbance of the solution is measured in a 5 cm cell at 600 nm.

The method is disturbed by As and Cd only in more than a 20-fold excess.

For lead, tin and bismuth-based alloys, the sensitivity of the method is 50 ppm Sb; its error is $< \pm 10$ rel.%.

Some other possibilities for the photometric determination of antimony

Extraction with methylene blue is recommended for the determination of the antimony content of steels. Interference is not caused by 5,000-fold quantities of Cu, Sn, Bi, Ti, Mo, Co, Ni, Mn and Cr [9].

On analogy with the determination of As, a procedure is known for the absorption of SbH$_3$ in Ag-DDTC and its photometric measurement [10] Pyrrolidinedithiocarbamic acid is suitable for the determination of Sb in the presence of a large amount of Sn [11].

A small amount of antimony in sea-water is determined with diantipyryl-methane [12], Sb^{3+} being extracted into CHCl$_3$ from an iodide medium.

References

[1] Marczenko, Z.: *Spectrophotometric Determination of Elements*. Ellis–Horwood, London 1976.
[2] Blyum, I. I. A.: *Ekstraktsionnye-fotometricheske metody analiza*. Nauka, Moscow 1970.
[3] Kish, P. P., Onishchenko, Yu. K.: *Zh. Anal. Khim.* **29**, 102 (1974).
[4] Jankowski, J.: *Hutn. Listy* **20**, 205 (1965).
[5] Donaldson, E. M.: *Talanta* **26**, 999 (1979).
[6] Stanton, R. F.: *Analyst* **87**, 299 (1962).
[7] Schreck, M.: Unpublished work.

[8] Abrutis, A. A., Kazlauskas, R. M., Narushkyavichyus, L. R.: *Zh. Anal. Khim.* **34**, 1997 (1979).

[9] Yakovlev, P. Ya., Ribina, P. Ya., Petrova, L. G.: *Ref. Zh. Khim.* **11G**, 158 (1972).

[10] Dal Cortivo, L. A., Cefola, M., Umberger, C. J.: *Anal. Biochem.* **1**, 491 (1960).

[11] Kovács, E., Guyer, H.: *Z. Anal. Chem.* **186**, 267 (1962).

[12] Afanasev, Yu. A.: *Zh. Anal. Khim.* **30**, 1830 (1975).

[13] De Souza, T. L. C., Kerbysen, J. D.: *Anal. Chem.* **40**, 1146 (1968).

[14] Vasilev, P. I., Kuskova, N. K., Pakhomova, K. S.: *Metody khim. analiza mineralnovo syrya.* Vol. 8. Nedra, Moscow 1965, p. 191.

[15] Busev, A. I., Bogdanova, E. S., Tiptsova, V. G.: *Zh. Anal. Khim.* **20**, 585 (1965).

6.4. Arsenic

Arsenic belongs in column V/1 of the periodic system. In aqueous solution it is stable in the form of trivalent and pentavalent oxidation states. It is amphoteric, but its hydroxides are weak acids.

In certain of its properties, such as its tendency to form heteropolyacids, it closely resembles phosphorus. The insolubility of its sulphides and a certain tendency to undergo complex formation with S-containing ligands, however, also make it similar to antimony. This shows up in the fact that each of these two elements usually disturbs the determination of the other.

For separation, good use may be made of the fact that its ions are fairly easily reduced to elemental arsenic ($E_{AsO_3^{3-}/As°} = +0.24$ V), while more vigorous reduction leads to formation of the volatile AsH_3 ($E_{As°/AsH_3} = -0.54$ V).

The halides of As(III) can be extracted, and they can also be separated by distillation, preferably in an inert gas stream. When extracted from 8–12 M HCl with inert solvents, they are accompanied only by Ge^{4+} [1].

Reduction to AsH_3 is of particular importance in the analysis of arsenic; in the very simple Gutzeit method [2] the AsH_3 produced with metallic zinc in acidic medium is trapped on filter paper soaked with $HgBr_2$. In the event of a small quantity of arsenic, the yellow spot due to the compound $H(HgBr)_2As$ is evaluated with a comparative series. This separation can naturally be combined with other methods of measurement.

Primarily the following methods are used for photometric determination:

1. Heteropoly acids, and above all arsenomolybdic acid. The yellow complex $H_3[As(Mo_3O_{10})_4]$ can be extracted with alcohols and measured

155

by photometry ($\lambda_{max} = 330$ nm). A much more frequently applied procedure is to convert it to molybdenum blue ($\lambda_{max} = 840$ nm) by reduction. In more recent procedures the arsenomolybdic acid is determined after extraction with basic dyes [3]. The investigations by Lisitsina *et al.* [4] indicated that, of these methods, the best is molybdenum blue formation by reduction with ascorbic acid. However, the accuracy depends greatly on the reducing agent used, the acidity, the temperature and other factors. Since side-processes also occur [5], the given analytical specifications must be followed strictly. The determination is disturbed by SiO_3^{2-}, PO_4^{3-} and GeO_3^{2-}, which similarly form heteropolyacids, and by Sn^{2+}, Pb^{2+}, Bi^{3+}, Ba^{2+} and NO_3^-. Thus, these ions must be separated.

2. Reaction of AsH_3 with Ag-DDTC [6,7]. The sensitivity of the determination is increased considerably by the presence of organic bases (pyridine, brucine, etc.).

3. Measurement of the absorbance of colloidal arsenic is also of use; for instance, it is recommended for the determination of the arsenic content of metallic tin [8].

Compared to these three methods, other photometric procedures are rare.

A valuable survey of the determination of arsenic by physical and chemical methods is to be found in the book by Nemodruk [9].

Determination of the arsenic content of rocks with Ag–DDTC (according to [7] and the authors' experience)

Analytical procedure: A rock sample of at most 1.0 g is weighed, and fused with 10 g NaOH and a little Na_2O_2 in a nickel crucible above a gas flame. The melt is leached out with hot water and filtered, the precipitate is washed several times with 1 M NaOH, and the volume of the filtrate is made up to 250 cm³. The arsenic is in the filtrate, but sorption on the hydroxide precipitate may lead to a loss of a few per cent; in important analyses, therefore, the precipitation step is repeated. An aliquot of the filtrate, containing at most 40 µg As, is taken in an AsH_3-evolving apparatus [10], and acidified with HCl. The volume of the solution is made up to about 50 cm³ in such a way that the final HCl concentration corresponds to a dilution

156

of 1 : 4. 0.5 cm^3 25% SnCl$_2$ is added, and a small vessel containing 3–4 g zinc filings is stood on the bottom of the flask. The flask is fitted with the distillation head, the exit-tube of which leads into a test-tube containing 7–8 cm^3 Ag–DDTC solution. (Preparation of Ag–DDTC solution: 0.225 g NaDDTC is dissolved in a little water, and 11.0 cm^3 0.1 M AgNO$_3$ is added. The precipitate formed is filtered off, dissolved in 100 cm^3 CHCl$_3$, the solution is freed from water with CaCl$_2$, and 0.020 g brucine is dissolved up in it.) The zinc in the flask is allowed to come into contact with the acidic solution, and the evolution of AsH$_3$ begins. This process lasts for about 1 hour. The absorbent solution is then made up to 10 cm^3 with CHCl$_3$, and the absorbance is measured at 525 nm. The calibration plot is prepared with 5–10 μg arsenic. 1×10^{-3}% arsenic can be determined in this way ($\varepsilon = 1.0 \times 10^4$).

Only antimony and germanium interfere (SbH$_3$ and GeH$_4$), but their quantities in rock samples of average composition are much less than that of arsenic. Since sulphides undergo oxidation during the digestion, there is no need for a preliminary trap to sorb H$_2$S.

Determination of arsenic contamination in air [11]

Analytical procedure: 20–100 dm^3 of air is bubbled through two sampling vessels connected in series, each containing 10 cm^3 of a 1 : 1 oxidizing mixture of 0.01 M KMnO$_4$ + 0.05 M H$_2$SO$_4$. The contents of the vessels are separately evaporated to dryness in the presence of a few drops of 3% H$_2$O$_2$ solution. The residue is dissolved up in 10 cm^3 water. A 5 cm^3 aliquot is taken, 0.1 cm^3 ammonium molybdate solution is added (18.8 g ammonium molybdate + 80 cm^3 5 M H$_2$SO$_4$ in a volume of 250 cm^3), and 10 min later 0.1 cm^3 1% ascorbic acid solution is added. Photometry (or colour comparison) is performed after 40–50 min. The absorbance maximum is at 720 nm. The attainable sensitivity is 0.5 μg As.

Determination of arsenic content of aluminium alloys with dithiopyrylmethane (DTPM) [12]

In 1–3.5 M H$_2$SO$_4$ medium, As(III) forms a yellow complex with DTPM. As(V) is reduced by DTPM to As(III). The optimum absorbance is at

330 nm, with $\varepsilon = 2.6 \times 10^4$. Interference in the determination is caused by Bi, Sb, Cd, Pd, Pt and Sn.

Analytical procedure: 2 g of the material to be examined is dissolved in 40 cm³ 1 : 1 HCl, with the intermediate addition of 10 cm³ 3% H_2O_2. Air-cooling is applied to avoid the loss of As. After dissolution, the H_2O_2 is decomposed by boiling, the solution is cooled, and the volume is made up to 100 cm³ (the medium is then roughly 0.3 M in HCl). A 10–15 cm³ aliquot is passed through a cation-exchanger in H^+ form (KU-1, KU-2), the column is washed with 30 cm³ 0.1 M HCl, the effluents are combined, and the volume is made up to 50 cm³. An aliquot containing 10–90 μg As is pipetted into a 25–50 cm³ volumetric flask, 4 cm³ 1×10^{-2} M DTPM (dissolved in 1 : 1 acetic acid) is added, and the solution is made up to the mark with H_2SO_4 so as to give a final acidity of 2.85 M H_2SO_4. The absorbance of the solution is measured at 330 nm after 20 min. The As-free reagent solutions are used as reference. The calibration curve is recorded with 10–90 μg As. The method is suitable for the determination of 4 ppm As.

References

[1] Babko, A. K., Pilipenko, A. T.: *Fotometricheskii analiz. Metody opredeleniya nemetallov.* Izd. Khimiya, Moscow 1974.
[2] Satterle, H. S., Blodgett, G.: *Ind. Eng. Chem. Anal. Ed.* **16**, 400 (1944).
[3] Babko, A. K., Ivashkovich, E. M.: *Zh. Anal. Khim.* **27**, 120 (1972).
[4] Lisitsina, D. N., Shcherbov, D. P., Talatynova, N. A.: *Issledovaniya v oblasti khimicheskikh i fizicheskikh metodov analiza mineralnovo syrya.* ONTI Kaz. IMS, Alma-Ata 1973.
[5] Duval, L.: *Chim. Anal.* **51**, 415 (1969).
[6] Vasah, V., Sedivec, V.: *Chem. Listy* **46**, 341 (1952).
[7] Bode, H., Hackmann, K.: *Z. Anal. Chem.* **229**, 261 (1967).
[8] Dragomirecky, A., Mayer, V., Michal, J., Rericha, K.: *Photometrische Analyse anorganischer Roh- und Werkstoffe.* VEB D. V. f. Grundstoffindustrie, Leipzig 1968.
[9] Nemodruk, A. A.: *Analiticheskaya khimiya myshyaka.* Nauka, Moscow 1976.
[10] Fiest, J., Getrost, H.: *Organische Reagenzien für die Spurenanalyse.* Merck, Darmstadt, 1975.
[11] Várkonyi, T., Cziczó, T.: *A levegőminőség vizsgálata (Air-quality investigations).* Műszaki Könyvkiadó, Budapest 1980.
[12] Akimov, L. V., Efremova, L. V., Rudzit, G. P.: *Zh. Anal. Khim.* **33**, 934 (1978).

6.5. Barium and strontium

These elements belong in column II/1 of the periodic system. Their properties have been reviewed in detail by Frumina et al. [1].

The ability of the alkaline earth metal ions to form complexes decreases in the direction $Ca^{2+} \rightarrow Ra^{2+}$, in accordance with the decrease in the charge density. Their reactions based on complex formation are always lower in selectivity than those of the other metal ions, naturally with the exception of the alkali metal ions. A certain possibility for their separation is provided by the precipitation of $BaSO_4$ ($SrSO_4$) or $BaCrO_4$.

Their photometric determination is not advantageous, and it is advisable to choose some other method (atomic absorption, emission spectrum analysis) for their analysis. This applies particularly to barium and strontium; the comparatively more stable complexes of calcium offer a more general possibility for its separation (extraction with azoazoxy-BN and with di-alkylphosphoric acids) and determination (glyoxal-bis[2-hydroxyanil]) than in the case of barium or strontium.

Several chromotropic acid bis-azo derivatives are recommended for the photometric determination of barium and strontium [2,3]. Of these, mention may be made of orthanil S, chlorphosphonazo III or nitchromeazo (see Appendix). These form 1 : 1 complexes with Ba^{2+} and Sr^{2+} in weakly acidic medium, at pH 2–6 depending on the substituents. The formation is promoted by water-miscible solvents (alcohol, acetone). These reagents can be used successfully for the indirect determination of SO_4^{2-} too, by measurement of the excess of Ba^{2+} added to a solution freed from other cations.

Determination of barium with sulphonazo III [4]

An aliquot corresponding to 2–3 μg Ba^{2+} is taken from a solution not containing disturbing ions, and transferred to a 25 cm^3 volumetric flask. The pH of the solution is adjusted to 2 by the addition of HCl; 0.5 cm^3 sulphonazo III solution (1×10^{-3} M) and 12.0 cm^3 acetone or alcohol are added, and the solution is made up to the mark. The absorbance of the solution is measured at 640 nm.

The determination is not disturbed by a 30-fold amount of Ca^{2+}, Mg^{2+} and UO_2^{2+}, a 10-fold amount of Sr^{2+}, a 2-fold amount of Th^{4+} and Pb^{2+}, and an equivalent amount of La^{3+}. Of the anions, a 350-fold amount of CH_3COO^-, a 250-fold amount of F^- and a 1.5-fold amount of PO_4^{3-} are permissible. Small quantities of interfering metal ions can be masked with EDTA; however, a higher EDTA concentration ($>10^{-4}$ M) also decreases the colour of the Ba^{2+} complex.

Kemp and Williams [5] report that in the presence of EGTA (ethylene glycol-bis[aminoethyl ether]-N,N,N',N'-tetraacetic acid) and DCTA (*trans*-1,2-diaminocyclohexane-N,N,N',N'-tetraacetic acid) only Ca^{2+} and Cu^{2+} interfere in the determination of Sr^{2+} and Ba^{2+}. However, since these masking agents may be present only in a concentration of 10^{-3} M, the sensitivity of the determination is limited. By the separate addition of DCTA and EGTA, the Sr^{2+} and the Ba^{2+} can be determined individually, from two measurements.

Determination of strontium content of barium salts with nitchromeazo [6]

Up to a $Ba^{2+} : Sr^{2+}$ ratio of 200 : 1, the Ba^{2+} can be masked with SO_4^{2-}, while the presence of a 5-fold amount of Ca^{2+} is permissible.

Analytical procedure: 0.2 g of the sample to be examined is dissolved in water and the volume is made up to 250 cm³. A 1.0 cm³ aliquot is taken in a 10 cm³ volumetric flask, and 0.5 cm³ 0.06% nitchromeazo solution, 3.5 cm³ acetate buffer of pH 3.7 (10 cm³ 1 M sodium acetate +3.0 cm³ glacial acetic acid in 1,000 cm³) and 1.0 cm³ 0.08% Na_2SO_4 solution are added, followed 20 min later by 4 cm³ acetone. The absorbance is measured at 650 nm.

The method is suitable for the determination of 0.5% Sr in Ba, and hence for the control down to this level of the purification processes in barium salt production.

References

[1] Frumina, N. S., Goryunova, N. N., Eremenko, S. N.: *Analiticheskaya khimiya bariya*. Nauka, Moscow 1977.
[2] Savvin, S. B., Akimova, T. G., Dedkova, V. P.: *Organicheskie reagenty dlya opredeleniya Ba^{2+} i SO_4^{2-}*. Nauka, Moscow 1971.
[3] Lukin, A. M.: *Assortiment reaktivov na strontsii i barii*. NITEKHIM, Moscow 1972.
[4] Savvin, S. B., Dedkov, Yu. M., Makarova, V. P.: *Zh. Anal. Khim.* **17**, 43 (1962).
[5] Kemp, J., Williams, M. B.: *Anal. Chem.* **45**, 124 (1973).
[6] Kreshkov, A. P., Kuznetsov, V. V.: *Zav. Lab.* **34**, 134 (1968).

6.6. Beryllium

Beryllium belongs in column II/1 of the periodic system. Only complex ions of it are known, for the Be^{2+} ion has an electronic structure which allows it readily to accept 4 electron-pairs. Thus, complex ions such as $[Be(H_2O)_4]^{2+}$ and $[BeF_4]^{2-}$ are formed. As a consequence of its small ionic radius, and hence its high charge density, however, Be^{2+} does not resemble Mg^{2+}, the element below it in this column, so much as Al^{3+} and Zn^{2+} in its behaviour. Accordingly, it is amphoteric. $Be(OH)_2$ dissolves up at higher alkalinity. Because of the small ionic radius, Be tends to form covalent compounds. Besides the fluoride, stable complexes are primarily obtained with ligands containing O and N donor atoms. Such ligands of analytical importance are hydroxyacids, β-diketones (acetylacetone), chromotropic acid azo derivatives (beryllon II, III and IV), anthraquinone derivatives (quinalizarin) and triphenylmethane derivatives. Data on some reagents are listed in Table 6.3.

The main possibilities that may be considered for separation are precipitation with NH_4OH or $NaOH$, extraction with β-diketones and fatty acids, and ion-exchange.

TABLE 6.3.

More important photometric reagents for beryllium

Reagent	Reagent λ_{max} nm	Complex λ_{max} nm	$\varepsilon \times 10^{-3}$	Medium pH	Interferences	Notes	Refs
Acetylacetone	275	300	31	7–8	Many elements	EDTA addition, ion-exchange separation	[2, 9]
Beryllon II	550	620	12	12–13	Ca, Mg, Al, Cu, Co, Ni, Mn, Mo, Cr	EDTA, TEA addition	[2, 3, 7]
Beryllon IV	485	530	—	6–8	Al, Mg, Fe	EDTA addition	[2]
Sulphochrome	520	570	14	4.9–5.3	Ti, W, Mo, Cr	EDTA addition	[8]

Accounts of the up-to-date spectrophotometric methods for beryllium are given by Kuznetsov *et al.* [10], Khovoselova and Batsanova [2] and Gladinovich and Stolyarov [1].

Determination of beryllium with beryllon II in rock samples containing manganese dioxide (procedure of the authors)

Beryllon II (see Appendix) is soluble in water; in acidic solution it is red, and in alkaline solution violet. In the pH range 11.5–14 it forms blue complexes with Be, Ca, Mg, Al, Cu, Co, Ni, Mn, Mo and Cr [3]. Much of the disturbance in Be determination can be eliminated by EDTA addition [4]; if this is not sufficient, Be can be separated by extraction with acetylacetone [5].

The procedure given here is a modification of a method [6] reported for rock analysis. $H_2C_2O_4$ must be added in the digestion step when the MnO_2 is dissolved, and extraction of the Be^{2+} with acetylacetone must be included. Without this, the triethanolamine (TEA) applied for masking [7] gives a green colour with manganese, which disturbs the determination.

Analytical procedure: 0.3–0.5 g sample is weighed into a platinum dish, 1 cm³ conc. H_2SO_4 is poured over it, and 1 g oxalic acid is mixed in. After the effervescence has ceased, the mixture is evaporated with 15 cm³ HF; this process is then repeated with 1 cm³ 1 : 1 H_2SO_4 and 15 cm³ distilled water. If necessary, the residue is fused with KHF_2, the melt is dissolved out with H_2SO_4 solution, and the volume is made up to 50 cm³. 10 cm³ 15% EDTA is added to an aliquot containing 1–10 μg Be^{2+}, and the pH of the solution is adjusted to 6–7 with 1 : 1 NH_4OH. 0.25 cm³ acetylacetone and 10 cm³ CCl_4 are added to this solution in a shaking-funnel, and the mixture is shaken for 30 s and then separated. A further 0.1 cm³ acetylacetone and 10 cm³ CCl_4 are added to the aqueous phase, and the extraction is repeated. 3 cm³ conc. HNO_3 and 3 cm³ $HClO_4$ are added to the combined organic phase for destruction of the acetylacetone, and the solution is covered and evaporated, first on a water-bath, and then to dryness on a sand-bath. The residue is dissolved in 15 cm³ dilute HNO_3 (10 cm³ HNO_3/dm^3), the solution is transferred to a 50 cm³ volumetric flask, and 0.1 cm³ Fe^{3+} solution (25 g $Fe_2[SO_4]_3/dm^3$), 3 cm³ complex former (140 g EDTA + 14 cm³ TEA/dm³) and 2 drops 0.1% aqueous auramine solution are added. The solution is titrated with 40% KOH until the yellow colour of the indicator disappears, and 2 cm^3 beryllon II solution (0.025%, containing 1 cm^3 conc. HCl/dm^3) is added. The volume is made up to the mark with EDTA solution (14 g/dm^3) of pH 12–13. A solution containing the reagents is used as reference. After 20 min, the absorbance is measured at 625 nm with a spectrophotometer. The calibration plot (slightly curved) is recorded up to a concentration of 15 µg Be^{2+}/50 cm^3.

Determination of beryllium in alloys and ores with sulphochrome [8]

Sulphochrome (see Appendix) forms a complex with beryllium in the pH range 4.9–5.3. In the presence of EDTA, the determination is not disturbed by a 400-fold amount of Fe, a 50-fold amount of Ni, Co, Cu and Al, a 6-fold amount of Ti, W and Mo, a 3-fold amount of Cr, and appreciable quantities of Mg, Ca, Zn and Mn.

Analytical procedure for alloys: 0.5 g sample is dissolved in 1 : 1 HNO$_3$, the solution is evaporated almost to dryness, and the residue is dissolved in 100–200 cm^3 water. An aliquot corresponding to 1–15 µg Be is transferred to a 50 cm^3 volumetric flask, the pH is adjusted to 2–3 with dilute NH$_4$OH, then 15–20 cm^3 ammonium acetate buffer (pH 5), 2.5 cm^3 0.1 M EDTA and 5 cm^3 5×10^{-4} M sulphochrome are added, and the volume is made up to the mark with buffer solution. The absorbance is measured at 570 nm in 1 cm cells. The reference is the corresponding mixture of the reagents. The calibration plot is recorded in an analogous way.

Analytical procedure for ores: 0.1–0.5 g ore is digested with 5–6 g KF in a platinum crucible, 5 cm^3 conc. H$_2$SO$_4$ is added, and the mixture is evaporated until SO$_3$ fumes appear. Leaching is performed with 50–75 cm^3 hot distilled water, and the volume is made up to 100–200 cm^3. The subsequent steps are as above.

References

[1] Gladinovich, D. B., Stolyarov, K. P.: *Zav. Lab.* **47**, No. 5, 3 (1981).
[2] Khovoselova, A. V., Batsanova, L. R.: *Analiticheskaya khimiya berilliya.* Nauka, Moscow 1966.
[3] Lukin, A. M., Zavarakhina, G. V.: *Zh. Anal. Khim.* **11**, 393 (1956).
[4] Karanovich, G. G.: *Zh. Anal. Khim.* **11**, 400 (1956).

163

[5] Plotnikova, R. N., Ashaeva, R. P., Shcherbov, D. P.: *Zav. Lab.* **32,** 1063 (1966).
[6] Vasilev, P. I., Zemtsova, L. I., Pakhomova, K. S.: *Metody khimicheskovo analiza mineralnovo syrya.* Vol. 11. Nedra, Moscow 1968.
[7] Bichkov, L. A., Nevzorov, A. N.: *Zav. Lab.* **38,** 927 (1972).
[8] Arkhangelskaya, A. S., Molot, L. A.: *Zav. Lab.* **46,** 883 (1980).
[9] Merill, J. R., Honda, M., Arnold, J. R.: *Anal. Chem.* **32,** 1420 (1960).
[10] Kuznetsov, V. I., Bolshakova, L. I., Fan Min-e: *Zh. Anal. Khim.* **18,** 160 (1963).

6.7. Bismuth

Although bismuth is situated in column V/1 of the periodic system, in solution it is stable only in the trivalent form. From an analytical aspect its complexes with S-containing ligands and with the halides (primarily I^-) are of importance. Its anionic complexes, e.g. BiI_4^-, can be extracted with amines or other organic cations. No selective reagent is known, but in most cases the effects of interfering ions can be eliminated by masking. Besides precipitation with NH_4OH, extraction of the halide complexes, and ion-exchange, procedures that are very suitable for its separations are the extractions with dithizone and dithiocarbamate.

Because of its low Clark value, bismuth cannot generally be determined photometrically in rocks. Photometry is of importance for sulphide ores and in metallurgy, and our examples are taken from these fields.

TABLE 6.4.
More important photometric reagents for bismuth

Reagent	Optimum pH	λ_{max} nm	$\varepsilon \times 10^{-3}$	Interfering ions	Refs
Dithizone (CCl_4)	2.8–3.2	490	8.6	Ag^+, Hg^{2+}, Au^{3+}, Cu^{2+} (not if extracted at pH 0); Tl^+, Sn^{2+} (not if oxidized)	[4]
Thiourea	0.5 M H^+	470	9	Sb^{3+} (can be masked with tartaric acid), Fe^{3+} (can be reduced)	[5]
KI	0.5 M H^+	460	10	Oxidants; much Sn^{4+}, Sb^{5+}, Pt^{4+}, Cu^{2+}	[2]
Dimercaptothiopyrone	2.5 M H_2SO_4	360	2.2	As^{3+}, Hg^{2+}, Cu^{2+}	[3]

The tendency of Bi^{3+} and the main constituents of such samples to undergo hydrolysis means that some organic oxyacid is generally required to keep them in solution.

The most suitable photometric reagents for bismuth are given in Table 6.4, together with the conditions of determination and the relative sensitivities.

Other procedures used to determine bismuth include ion-pair formation with basic dyes, extraction, and trihydroxyfluorone derivatives [1].

Surveys of the photometric determination of bismuth have been given by Akatsevich and Stolyarov [2] and Usatenko [3].

Determination of the bismuth content of copper ores with dithizone [4]

The interfering ions are partly masked and partly removed by separation with DDTC. Bi(III) dithizonate can be extracted selectively at pH 2.8–3.2 and determined.

Analytical procedure: 0.5–1.0 g sample is digested with $HNO_3 + HF + + H_2SO_4$ in a platinum vessel and, with the aim of separation from copper, the bismuth is precipitated, together with the metal hydroxides, with NH_4OH. The precipitate is dissolved in HCl and tartaric acid, and complexone III solution is added to mask the lead and other disturbing ions (Ni and Co traces, etc.). The pH is adjusted to 8–9 with NH_4OH. KCN solution is added to mask the Cu traces, then NaDDTC is added and the complex $Bi(DDTC)_3$ is extracted with $CHCl_3$. Since the resulting sensitivity of the determination is generally not satisfactory, the Bi^{3+} is re-extracted with 0.5 M HNO_3. The pH of the solution is adjusted to 2.8–3.2 in the presence of methyl orange indicator. Extraction is carried out with a 0.002% solution of dithizone in CCl_4 or $CHCl_3$, added in small portions. The absorbance of the solution is measured at 490 nm. The sensitivity of the method is 1 ppm Bi.

Determination of bismuth in galena and arsenopyrite with thiourea [5]

Galena is boiled with 1 : 1 HNO_3, and the dissolution is completed with tartaric acid. Arsenopyrite is boiled with HCl, in the presence of hydrazine to precipitate the arsenic. After decomposition of the N_2H_4 excess with

165

$HNO_3 + H_2SO_4$, tartaric acid is added, followed by sufficient thiourea to give a concentration of 2 g/100 cm³, and then HNO_3 to be in a final concentration of 0.5 M. The absorbance is measured at 450 nm, or visual comparison is made with a standard series. The sensitivity is 20 ppm Bi.

The bismuth content of lead and tin alloys can similarly be determined with thiourea [6]. In order to keep the lead and tin in solution, the alloys are dissolved in a mixture of dilute HNO_3, tartaric acid and citric acid. The background colour of the solution may disturb the determination, and therefore a portion of the digestion solution is used for reference purposes. Measurements are performed as above.

Determination of bismuth in concentrates and in non-ferrous alloys by the iodide method [7]

To eliminate interfering effects, the bismuth is extracted in the form of its diethyldithiocarbamate (DDTC) complex. After re-extraction with HCl, the bismuth is once more extracted, this time as the xanthate. The complex is decomposed by evaporation with acid, and evaluation is carried out after complex formation with iodide. The method is suitable for the determination of the bismuth content of alloys based on Pb, Sn and Cu, and concentrates of the sulphides of Cu, Mo, Pb, Zn and Ni.

Analytical procedure for concentrates: 0.2–0.5 g sample is fused with 3 g Na_2O_2 in a zirconium vessel. The melt is dissolved out with a mixture of 80 cm³ water+25 cm³ 1 : 1 H_2SO_4 in a teflon vessel. After the addition of HF, the solution is evaporated until the appearance of SO_3 fumes. 40 cm³ water, 5–10 g NaCl, 20 cm³ citric acid–tartaric acid (25% solutions) and 50 cm³ EDTA–NaOH (12% solutions) are added to the residue. The pH of the solution is adjusted to 11.5–12.0 with 50% NaOH. After the addition of 30 cm³ 20% KCN and 5 cm³ 1% Na-DDTC, the solution is extracted with 10 cm³ $CHCl_3$ for 1 min. The process is repeated with 10+5 cm³ $CHCl_3$. The bismuth is re-extracted from the combined $CHCl_3$ phase with 16 cm³ conc. HCl. 2 cm³ 20% NH_4Cl is added to the acidic solution, and the volume is made up to 75 cm³. 10 cm³ $CHCl_3$ and 1 cm³ 20% potassium ethylxanthate are added to the solution, which is then shaken for 1 min. The extraction is repeated with the addition of 10+5+5 cm³ $CHCl_3$ and 1+0.5+0.5 cm³ xanthate solution, respectively. The combined $CHCl_3$

phase is shaken with 10 cm³ washing liquid (215 cm³ conc. HCl+25 cm³ 20% tartaric acid+200 cm³ 25% NH_4Cl in 1 dm³). The $CHCl_3$ phase is evaporated with 10 cm³ 1 : 1 HNO_3 and 1 cm³ conc. $HClO_4$. 1 cm³ 1 : 1 HNO_3 and 3 cm³ 1 : 1 H_2SO_4 are added to the dry residue, and the mixture is evaporated until the appearance of dense fumes. 2 cm³ 15% NaH_2PO_2 and 5 cm³ 20% KI are added to the residue, and the volume is made up to 25 cm³. After 5 min, photometry is carried out in 4 cm cells at 337 or 460 nm. The calibration plot is produced in the concentration range 5–500 μg Bi/25 cm³. Alloys are dissolved in a mixture of $HNO_3 + H_2SO_4$.

Vokhrysheva and Gayun [8] recommend extraction separation with di--2-ethyl-hexylphosphoric acid, and evaluation with KI in the presence of thiourea, for the determination of the bismuth contents of lead smelting residues and copper-bismuth concentrates.

References

[1] Antonovich, V. P.: *Zh. Anal. Khim.* **30**, 1566 (1975).
[2] Akatsevich, I. N., Stolyarov, K. P.: *Instrumentalnye i khimicheskie metody analiza.* Leningrad University, 1973, p. 60.
[3] Usatenko, Yu. I.: *Zav. Lab.* **31**, 788 (1965).
[4] Filippova, N. A., Korosteleva, V. A., Yurovskaya, S. B.: *Spektralnye i khimicheskie metody analiza materialov.* Metallurgiya, Moscow 1964.
[5] Senderova, V. M.: *Khimicheskii analiz mineralov i ikh khimicheskikh sostav.* Nauka, Moscow 1964.
[6] Cheng, K. L., Bray, R. H., Melsted, S. W.: *Anal. Chem.* **27**, 24 (1955).
[7] Donaldson, E. M.: *Talanta* **25**, 131 (1978).
[8] Vokhrysheva, L. E., Gayun, M. G.: *Zav. Lab.* **49**, 8, 23 (1983).

6.8. Boron

Boron is the first element in column III/1 of the periodic system. Its behaviour results from this and from its small ionic radius. In solution it is trivalent and tetracoordinated. In aqueous solution, for example, besides the oxyanions of boric acid it tends to form complex ions of the type BF_4^-; it also forms iso- and heteropolyacids in which the boron is the central atom, e.g. borosalicylic acid. Photometry is often carried out in concentrated H_2SO_4 medium, where the boron is present in the forms B^{3+} and BO^+.

The most widespread procedure for the separation of boron is its selective dissolution with alcohol, distillation of the methyl ester (possibly with micro-diffusion), and extraction of BF_4^- as an ion-pair with the tetraphenylarsonium cation or with a basic dye [1–4].

The following are the main possibilities for the photometric determination of boron:

1. With α-hydroxyanthraquinone derivatives, or with other compounds containing adjacent OH and quinone groups, in concentrated H_2SO_4. In addition to the best-known, but very unsuitable quinalizarin, mention may also be made of carmine, 1.1-dianthrimide and chromotrope 2 B.

2. Another group of reagents permit determination in weakly acidic, possibly basic or aqueous alcoholic medium. The use of these is much more convenient. They include curcumin, various azomethine derivatives, H-resorcin and numerous other reagents.

TABLE 6.5.

Characteristic data on some reagents used for determination of boron

Reagent	Medium used in determination	λ_{max} of complex nm	$\varepsilon \times 10^{-3}$	Interferences	Refs
Carminic acid	conc. H_2SO_4	610	7.0	NO_3, NO_2^-, Zr^{4+}, Ti^{4+}, Ce^{4+}, Nb^{5+}, Sc^{3+}, change in H_2SO_4 concentration	[1, 4] [10]
Curcumin	Aqueous alcoholic, oxalic acid soln.	550 530	180	NO_3^-, NO_2^-, F^-, Ce^{4+}, oxidants	[1, 4]
Crystal violet	Extraction with benzene from acidic medium, with BF_4^- as ion-pair	~600	2.6	CSN^-, NO_3^- and other anions forming an ion-pair with the reagent	[1, 2]
H-resorcin	Acetic acid, pH 5.5	530	3.0	F^-, reductants, oxidants, Fe^{3+}, Cu^{2+}, Al^{3+} and other cations	[11]
DHNS	pH 10.2; extraction with 1,2-dichloro-ethane	341	24	Al, Cu, Fe, Ti, etc.	[8]

3. The third group of determinations is based on the extraction of BL_4^- with basic dyes. Mainly crystal violet, malachite green and methylene blue are used here.

The properties of some of the more important reagents are given in Table 6.5.

Detailed data on the photometric reagents for boron and on its determination are to be found in the publications by Nemodruk and Karalova [1], Eristavi and Brouchek [2], Boltz [3] and Babko and Pilipenko [4].

Two fundamental difficulties must be reckoned with in the determination of the element in small amounts, one of which is the volatility of boron compounds. As a result of this, it is permitted to work with acidic solutions only at lower temperature or for only a short time, e.g. warming-up. Solutions must be evaporated in basic medium. The other difficulty is the boron content of laboratory glassware. Platinum, quartz or plastic vessels must be used; even stirring with a glass rod must be avoided.

Determination of boron content of steels with H-resorcin and thionine [5]

The basis of the method is that boron forms a complex with H-resorcin. Development of the colour is accelerated by the addition of a basic dye, thionine (see Appendix).

Analytical procedure: 0.1–0.5 g steel is dissolved up with 10 cm³ conc. HCl in a quartz vessel fitted with a reflux condenser. The solution is transferred to a beaker, together with the aqueous washings from the digestion vessel and the condenser; its volume should be about 40 cm³. 18 M NaOH is added dropwise to precipitate the metal hydroxides, and then a 0.5 cm³ excess of the alkali is added. The filtrate from the precipitation, after washing with water, is transferred to a 50 cm³ volumetric flask, and the volume is made up to the mark with water. A 5–15 cm³ aliquot of this solution is taken in a 25 cm³ volumetric flask, and is neutralized with 6 M HCl in the presence of 1 drop 0.1% *p*-nitrophenol (until the yellow colour disappears). 4 cm³ 1×10^{-3} M H-resorcin, 1 cm³ 1×10^{-3} M thionine (in ethanol) and 0.5 cm³ 0.05 M EDTA are added, and the solution is made up to the mark with ammonium acetate buffer of pH 5. The absorbance is measured at 510 nm, in 3 cm cells, after 2 hours. A solution containing the reagents is used for comparison. The calibration plot is recorded for 1–10 μg B. The sensitivity of the method is 5 ppm B.

169

Determination of hot water-soluble boron content of soil, with quinalizarin [6]

Pretreatment of sample: 10 g soil sample is taken in a 250 cm^3 conical quartz flask, 50 cm^3 0.1% MgSO$_4$ is added, the vessel is covered with a small funnel, the mixture is boiled for 5 min, and it is then filtered on a fluted filter paper, the initial filtrate being discarded.

Analytical procedure: 10–20 cm^3 aqueous extract is taken in a polyethylene beaker, 2 cm^3 H$_2$O$_2$ is added to destroy organic matter, the solution is heated on a water-bath for 1 min, 2 drops of 2 M NaOH are added, and the mixture is evaporated to dryness. 1 cm^3 10% calcium hypophosphite and 9 cm^3 quinalizarin (15 mg dissolved in 1 dm^3 conc. H$_2$SO$_4$) are added to the residue. The solution is carefully mixed, then left in the dark for 2 hours, and photometry is performed at 620 nm, against a comparative solution of 1 cm^3 distilled water +9 cm^3 quinalizarin solution. The calibration plot is produced with 0.5–10 μg B, in accordance with the above description.

Determination of boron in plant matter with azomethine-H [7]

Analytical procedure: 100 mg plant matter is ashed at 550°C, and the residue is dissolved in 10 cm^3 0.1 M HCl. Portions of the solution are fed by means of a peristaltic pump into an automatic analyzer. 1 cm^3 solution is injected into a flow of 0.1 M HCl, in which a buffer–masking solution (132 g [NH$_4$]$_2$HPO$_4$+25 g EDTA/500 cm^3) and the reagent (0.5 g azomethine H+2 g ascorbic acid/100 cm^3) are mixed. The range of determination is 0.1–6.0 ppm B.

Determination of traces of boron in river water [8]

The boron traces to be found in water are extracted in the form of an ion-association with a solution of 1.8-dihydroxynaphthalene-4-sulphonic acid (DHNS) and tetradecyldimethylbenzylammonium chloride (zephyramine) in 1.2-dichloroethane (DCE).

Analytical procedure: After filtration through a membrane filter, an aliquot of at most 50 cm^3 of the water sample is taken. 5 cm^3 acetate buffer (0.25 M acetic acid–sodium acetate in 0.01 M EDTA, pH 3.8) and 5 cm^3 0.01 M DHNS (2.62 g DHNS Na-salt in 1 dm^3) are added to the polyethylene separating-funnel, followed after 30 min by 5 cm^3 carbonate buffer (0.8 M NaHCO$_3$–Na$_2$CO$_3$ in 4 M NaCl, pH 10.2) and 5 cm^3 1.6 M Na$_2$SO$_4$ (to accelerate separation of the phases). Extraction is performed for 30 min with 5 cm^3 2×10^{-3} M zephyramine solution in DCE. The organic phase is washed with 10 cm^3 washing liquid (1 M NaCl, 0.2 M NaHCO$_3$, 0.2 M Na$_2$CO$_3$; pH 10.2) to remove the reagent excess. After shaking for 10 min and a separation time of 30 min, the absorbance is measured in 1 cm quartz cells at 341 nm. The calibration plot is recorded in the range 0–400 μg B. The sensitivity of the determination is 1 μg B/dm^3 water sample.

Determination of boron in rocks with crystal violet or carmine [9]

Analytical procedure: 0.25 g sample is fused with 3 g KNaCO$_3$ in a platinum crucible. The crucible is placed in a porcelain beaker containing 50 cm^3 water and 20.0 cm^3 2 M H$_2$SO$_4$. After dissolution of the melt, the crucible is removed, 2.0 g NaF is added to the solution, the vessel is covered, and the solution is boiled for exactly 10 min, and then made up to 250 cm^3 in a plastic volumetric flask. On the following day, a 2–25 cm^3 aliquot is transferred to a plastic separating-funnel already containing 25 cm^3 benzene. Since the partition of BF$_4^-$ depends sensitively on temperature, the benzene must be thermostated at 20°C. After the addition of 0.5 cm^3 1% crystal violet, the mixture is shaken for 30 s, and the benzene solution is then run off into a dry plastic beaker. Its absorbance is measured at 584 nm after 40–60 min. A parallel blank analysis is recommended. The calibration plot is recorded with 0–120 μg B$_2$O$_3$. Care must be taken that the H$_2$SO$_4$ concentration in the extraction should be 0.025–0.035 M, since a deviation from this can cause a large error.

The determination can also be carried out with carmine. In this case the sample is fused with Na$_2$CO$_3$ or Eschka mixture, the melt is leached with water, and the solution is filtered, acidified and made up to the mark in a volumetric flask. An aliquot of not more than 25 cm^3, containing at most 300 μg B, is made alkaline with Na$_2$CO$_3$ and evaporated to dryness. 10–15 cm^3 conc. H$_2$SO$_4$ and 5 cm^3 0.025% carmine in conc. H$_2$SO$_4$ are added to the residue, and the solution is made up to the mark with conc. H$_2$SO$_4$ in a volumetric flask. After 4 hours or on the next day, the solution is evaluated by comparison with a series containing 0–2 mg B. (The complex does not obey Beer's law.)

171

References

[1] Nemodruk, A., Karalova, Z. K.: *Analiticheskaya khimiya bora*. Nauka, Moscow 1964.
[2] Eristavi, D. I., Brouchek, F. I.: *Analiticheskie metody opredelniya bora*. Metsniereba, Tbilisi 1965.
[3] Boltz, D. F.: *Volumetric Determination of Nonmetals*. Wiley, New York 1978.
[4] Babko, A. K., Pilipenko, A. T.: *Fotometricheskii analiz. Metody opredelenie nemetallov*. Khimiya, Moscow 1974.
[5] Nazarenko, V. A., Flyantikova, G. B., Chekidra, T. N.: *Zav. Lab.* **47**, No. 1, 19 (1981).
[6] Derzharin, L. M. (ed.): *Metodicheskie ukazaniya po kolorimetricheskomu opredelenie podvizhnym form mikroelementov v pochvakh*. TSINAO, Moscow 1977.
[7] Krug, F. J., Mortatti, J., Pressenda, L. C. R., Zagatto, E. A. G., Bergamin, F. H.: *Anal. Chim. Acta* **125**, 29 (1981).
[8] Korenaga, T., Motomizu, S., Toei, K.: *Analyst* **105**, 955 (1980).
[9] Popov, N. P., Stolyarov, I. A. (eds.): *Khimicheskii analiz gornykh porod i mineralov*. Nedra, Moscow 1974.
[10] Sámsoni, Z., Szeleczky, M. A.: *Microchim. Acta* **I**, 445 (1980).
[11] Grizo, V. A., Poluektova, E. N.: *Zh. Anal. Khim.* **13**, 435 (1958).

6.9. Cadmium

Cadmium is situated in column II/2 of the periodic system. In solution it is only divalent. In its analytical properties it displays similarities to Zn^{2+}, and Hg^{2+}, i.e. elements from the same column, and also to Cu^+, In^{3+}, Tl^{3+} and Ag^+, which have the same electronic configuration and a similar ionic radius.

Its sulphide is precipitated in mildly acidic medium, and its hydroxide at pH 7.0. A complex of composition $Cd(NH_3)_3^{2+}$ is formed with NH_4OH. Accordingly, it yields complexes with ligands containing S, O and N donor atoms. Its halide complexes of type CdX_4^{2-} are suitable for its separation, by extraction or by ion-exchange. Extraction of the halide complexes as ion-pairs with basic dyes provides a possibility for the separation and photometric determination of cadmium.

A survey of the analysis of cadmium is to be found in the book by Shcherbov and Matveets [1].

The characteristics of some photometric reagents are listed in Table 6.6.

TABLE 6.6.

Characteristics of some reagents used for photometric determination of cadmium

Reagent	Medium	Extractant	λ_{max} nm	$\varepsilon \times 10^{-3}$	Interferences	Refs
Dithizone	5% NaOH	CHCl$_3$	520	80	Fe, Mn, Ni, Zn, Cr, Co, Cu	[1]
Br-benzthiazo	3% NaOH	Xylene	580	52	Co, Cu, Ni, Zn	[4]
Cadion	pH 9.4, acetone, tartrate		480	119	Ag, Fe, Co, Hg, Cu, Zn, etc.	[3]
Chromepyra-zole II	2 M HCl+ +0.8 M HBr	Benzene	560	24	many ions	[5]
Sulphoarsazene	pH 9:5–10.0		510	51	Hg, Zn	[6]
Rhodamine B	1–2 M H$_2$SO$_4$	1 : 1 TBP– benzene	575	100	Cu, Hg, Ag, Si, Sb, Fe, Sn, Pb	[2]

Electroanalytical methods too are of great importance in the determination of cadmium.

Determination of cadmium content of zinc concentrates with rhodamine B [2]

Principle of method: The iodide complex of cadmium is extracted at pH 6–10 with TBP in benzene; this leads to its separation from interfering ions. KBr and an aqueous solution of rhodamine B are added to the extract, in the course of which an iodo–bromo anionic complex is produced; with the rhodamine B cation this yields an ion-association, which passes into the organic phase and can be measured by photometry.

Analytical procedure: 0.1–0.2 g zinc concentrate is dissolved in 10 cm^3 conc. HCl by boiling, 3 cm^3 conc. HNO$_3$ is added, and the volume is evaporated to 1–2 cm^3. After the addition of 10 cm^3 1 : 1 H$_2$SO$_4$, the contents are evaporated until SO$_3$ fumes are observed, and the solution is then transferred to a 100 cm^3 volumetric flask and made up to the mark with water. An aliquot containing 5–10 µg Cd is taken in a shaking-funnel, 0.5 cm^3 5% ascorbic acid and 0.5 cm^3 1 M KI are added, the pH is adjusted to 7 with 1 M NaOH, the solution is diluted to 10 cm^3 with water, and the Cd is extracted with a 1 : 4 mixture of TBP–benzene. The aqueous phase is poured off, 0.5 cm^3 4 M KBr, 0.5 cm^3 2×10^{-3} M rhodamine B and 4 cm^3

2.5 M H_2SO_4 are added to the organic phase, which is then shaken for 1 min and centrifuged. Its absorbance is measured in 0.3 cm cells at 575 nm. The calibration plot is produced with 1–30 μg Cd, in an analogous manner.

Determination of cadmium in sewage water with cadion [3]

This determination is more selective and more sensitive than the dithizone procedure. The reagent used, cadion (see Appendix), forms a complex which is kept in solution with Triton X-100. Disturbing effects are eliminated by the addition of a complex former, and by the demasking of Cd with formalin.

Analytical procedure: The water sample (<8 μg Cd) is measured into a 25 cm³ volumetric flask, and 50 mg ascorbic acid, 2 cm³ 20% $KNaCO_3$ solution, 3 drops saturated KOH solution, 1 cm³ 1 M KCN, 1 cm³ 1 M KF, 5 cm³ cadion reagent (1 cm³ 20% $KNaCO_3$, 12 cm³ 4 M KOH, 20 cm³ ethanol, 1 cm³ 10% Triton X-100 and 10 cm³ 0.02% cadion in 0.02 M KOH in alcohol, diluted to 50 cm³ with water) and 1 cm³ 1 : 1 formalin are added. The mixture is made up to the mark. The absorbance is measured in 2 cm cells at 480 nm. The calibration plot is recorded for 0–8 μg Cd/25 cm³.

References

[1] Shcherbov, D. P., Matveets, M. A.: *Analiticheskaya khimiya kadmiya*. Nauka, Moscow 1973.
[2] Kiss, P. P., Balog, T. S.: *Zh. Anal. Khim.* **34**, 326 (1979).
[3] Hsu Chung Gin, Hu Chao-sheng, Jing Ji-hong: *Talanta* **27**, 676 (1980).
[4] Shkrobot, E. P., Bakinovskaya, L. M.: *Zav. Lab.* **32**, 1452 (1966).
[5] Zhivopistsev, V. P., Chelnokova, M. N.: *Zh. Anal. Khim.* **18**, 717 (1963).
[6] Partoshnikova, M. Z., Shafran, I. G.: *Tr. IREA* **25**, 258 (1963).

6.10. Calcium

Calcium belongs in column II/1 of the periodic system, in the group of alkaline earth metals. With oxalate and carbonate it forms compounds that are only poorly soluble in water. It displays only a slight tendency to form complexes. Its complexes with complexones and with organic acids containing OH groups, e.g. tartaric acid, are of relatively low stability.

There is no selective photometric reagent for calcium. Coloured complexes can be formed with metal indicators in general, but the latter also give colour-reactions with other cations.

The properties of the reagents used most frequently are reported in Table 6.7.

TABLE 6.7.

Characteristic data on some reagents used for calcium determination

Reagent	Medium	Reagent λ_{max} nm	Complex λ_{max} nm	$\varepsilon \times 10^{-4}$	Interferences	Refs
Murexide	pH 12.5	552	514	1.4	Cations of groups I–III, Ba, Sr	[7]
Calcion	pH 12.5, 15% acetone	500	600	0.76	Hydrolyzable ions, inorganic salts (e.g. 0.1 g NaCl/5 cm^3)	[1, 2]
Chlorphosphonazo III	pH 2.5–3	540	665	~7	Many ions	[3]
Glyoxal-bis(2--hydroxyanil)	0.04 M NaOH	450	516	1.8	Cations of groups I–III, PO_4^{3-}, F^-, oxalate, tartrate, citrate, complexone III	[7]

In every case when samples of complicated composition are to be analyzed, the photometric determination of calcium must be preceded by separation. The most appropriate reagent is selected in accordance with the nature of the sample under examination.

The determination of calcium with calcion (calcichrome) [1, 2] is not disturbed by a 1000-fold amount of alkali metal, Al, Ge and Se, a 100-fold amount of Bi, Cd, Hg, As, Mo, Zn, V and Pb, a 50-fold amount of Ba, Mg, Ga, La, Sn, U and Pt, and a 25-fold amount of Sr. Slight quantities (25–50 µg) of hydrolyzable ions can be masked with triethanolamine. Inorganic salts inhibit development of the complex.

175

When chlorphosphonazo III (see Appendix) is used, the determination of calcium is disturbed by very many ions, but these can generally be separated by ion-exchange after masking with complexone III [3].

Azo-azoxy BN reacts with calcium to yield a cherry-red complex that can be extracted with TBP. It has the advantage that Ba^{2+} and Sr^{2+} are not extracted. The reaction is suitable for photometric measurement too, but it is being increasingly more widely used to separate calcium from interfering ions (mainly Ba and Sr). The calcium can be re-extracted from the organic phase with acid, and can be determined by another method.

The reagent most frequently recommended for the determination of calcium is glyoxal-bis(2-hydroxyanil). This ensures the highest sensitivity and selectivity. The determination can be performed in aqueous solution and after extraction too [4]. Small amounts of cations of groups I–III can be masked with cyanide (hydroxylamine too is added in the presence of iron) and with sodium sulphide. The disturbing effects of Ba and Sr can be eliminated by the addition of a little sulphate or carbonate. Magnesium and the alkali metals do not interfere in the determination.

Calcium separation:

1. Precipitation of cations of groups I–III as hydroxides at around pH 8.

2. Oxalate precipitation at pH 2–3 (separation from Fe^{3+}, Al^{3+}, Ti^{4+}) or at pH 4 (separation from PO_4^{3-}).

3. Extraction of calcium with azo-azoxy BN [4]. The solution to be examined (0.5–1 M in NaOH) is extracted for 1 min with a mixture of azo-azoxy BN + TBP (preparation: 0.01% azo-azoxy BN solution in CCl_4; before extraction, 1 cm³ TBP is added for each 10 cm³ of this solution). After clearing, the calcium can be re-extracted from the organic phase with 0.1 M HCl.

Determination of calcium in silicate minerals with GBHA [5]

The calcium to be found in silicate minerals in determined in alkaline medium with glyoxal-bis(2-hydroxyanil) (GBHA). The aluminium is then present in the form of aluminate and does not interfere. The calcium is kept in solution with mannitol.

Analytical procedure: 1 cm³ HF and 2 cm³ 9 M H_2SO_4 are added to 50 mg material, and the mixture is heated until dense SO_3 fumes appear. 100 cm^3 stock solution (0.6 M in H_2SO_4) is prepared from the residue. A 20 cm^3 aliquot is taken and 10 cm^3 10% mannitol is added, followed 30 min later by 5 cm^3 30% NaOH. 1 cm^3 0.5% GBHA in ethanol is added to the cooled solution, followed by 20 cm^3 1 : 1 $CHCl_3$–isopentanol, and the mixture is shaken for 30 s in a separating-funnel. After centrifugation to promote separation, the organic phase is subjected to photometry at 531 nm. The calibration plot is obtained for 0–80 μg Ca. The usefulness of the method has been demonstrated by the accurate analytical values (with low scatters) for standard rocks G-I, BCR-I and W-I.

Determination of calcium content of alloyed steels
with chlorphosphonazo III [6]

The alloying elements of steel can be separated from Ca^{2+} by cation-xchange after complex formation with EDTA; the residual amounts do not disturb the determination.

0.5 g steel sample is dissolved up with $HCl + HNO_3$ in a fluoroplast vessel, and the mixture is then evaporated with $HF + HClO_4$ to remove SiO_2. 70 cm^3 5% EDTA of pH 5.3 is added, the pH is adjusted to 2, and the solution is boiled. After cooling, the pH is adjusted to 3.5 with NH_4OH, and the solution is passed through a cation-exchange resin in NH_4^+ form. The column is washed with water, and the Ca^{2+} is then eluted with 120 cm^3 3 M HCl. The eluate is evaporated in a quartz dish until the NH_4^+ salts separate out; these are fumed off with aqua regia, the residue is dissolved up, and the volume is made up to 50 cm^3 in a volumetric flask.

For Ca^{2+} determination, a 25 cm^3 aliquot is taken in a 50 cm^3 volumetric flask, and 3.0 cm^3 0.1 M HCl, 2.5 cm^3 0.05 M EDTA and 10 cm^3 0.03% chlorphosphonazo III are added. The absorbance is measured at 665 nm. A blank test is carried throughout the analysis in parallel with the sample. Despite this, however, it is still necessary to use purified reagents, e.g. doubly-distilled water. This procedure allows the determination of 10 ppm calcium. The method is also suitable for the analysis of alloys of Cr, W, Co, V, Ni and Mo.

References

[1] Lukin, A. M, Smirnova, K. A., Zavarikhina, G. B.: *Zh. Anal. Khim.* **18**, 444 (1963).
[2] West, T. S.: *Analyst* **87**, 630 (1962).
[3] Yakovlev, P. Ya., Zhukova, M. P.: *Zav. Lab.* **36**, 1169 (1970).
[4] Gorbenko, F. P., Sachko, V. V.: *Zh. Anal. Khim.* **18**, 1198 (1963).
[5] King, H. G., Pruden, G : *Analyst* **94**, 39 (1969).
[6] Yakovlev, P. Ya., Zhukova, M. P.: *Zav. Lab.* **37**, 1292 (1971).
[7] Frumina, N. S., Kruchkova, E. S., Mushtakhova, S. P.: *Analiticheskaya khimiya kaltsiya*. Nauka, Moscow 1974.

6.11. Chloride

Similarly as for most anions, there is no direct method for the photometric determination of chloride. The indirect procedures are based on ligand-exchange, i.e. on the loss in colour of a coloured complex [1,2]. The determination of Cl_2 and ClO^- is likewise of importance in water analysis; the procedures here make use of coloured complexes developing as a result of oxidation or chlorination.

The indirect procedures applied for the determination of chloride have the common error that, on the action of other anions, decreases occur in the concentrations of the complexes, which are not too highly stable; interference is also caused by metal ions which form complexes with the ligand or with the chloride ion.

Determination of chloride in water with mercury diphenylcarbazone [3]

Analytical procedure: To decrease the disturbing effects of metal ions which react with diphenylcarbazone, their complexes are extracted with $CHCl_3$ before the determination. Next (or directly if the sample does not contain interfering ions), an aliquot containing 10–60 μg Cl^- is transferred to a 50 cm³ volumetric flask, and 5 cm³ 0.1 M sodium formate (adjusted to pH 3.4 with HNO_3), 10 cm³ 0.02% diphenylcarbazone in 10% alcohol solution, and 1 cm³ 2% gum arabic solution are added. The mixture is diluted with 45 cm³ water, 2.0 cm³ 1×10^{-3} M $Hg(NO_3)_2$ solution is added, and the volume is made up to the mark. The temperature of the solution is maintained at a constant value, e.g. 20°C, and the absorbance is measured at 520 nm. The calibration plot is produced with 10–60 µg Cl^-. The method is very sensitive to variations in pH and temperature.

Determination of chloride in plants [4]

The chloride content of plants is determined in an indirect spectrophotometric method. For this purpose, use is made of mercury(II) thiocyanate, from which the chloride in the sample liberates thiocyanate. This is extracted in the form tris(1.10-phenanthroline)-iron(II) thiocyanate with nitrobenzene, and photometry follows.

Analytical procedure: 0.1–1 g plant sample is ashed at 500°C. The residue is dissolved in 50 cm³ 5×10^{-3} M H_2SO_4. An aliquot is taken, and 5 cm³ 0.25 M KH_2PO_4 (adjusted to pH 2.5 with H_2SO_4), 5 cm³ 10^{-3} M mercury(II) thiocyanate, and 5 cm³ 6×10^{-4} M tris-(1,10-phenanthroline)-iron(II) solution [0.3922 g iron(II) ammonium sulphate hexahydrate+0.5947 g 1,10-phenanthroline monohydrate are dissolved in 100 cm³ 5×10^{-3} M H_2SO_4; this solution is diluted 100-fold before the analysis] are added, and the solution is diluted to 25 cm³ and extracted with 10 cm³ nitrobenzene for 1 min. The extract is subjected to photometry at 516 nm, in a 1 cm cell. The calibration plot is recorded in the interval 20–120 µg Cl^-/10 cm³. The accuracy is equal to that of the Volhard method, but in the present procedure the quantity of sample can be decreased to even 2–200 mg.

Determination of active chlorine in water with o-tolidine [5]

Free chlorine (hypochlorite, chlorite) dissolved in water oxidizes the colourless o-tolidine (see Appendix) to a yellow product. In the most common cases, the disturbing effect of Fe^{3+} is eliminated with H_3PO_4, and the interference occurring in the joint presence of Fe^{3+}, Mn^{3+} and NO_2^- is prevented by the addition of AsO_2^-.

Analytical procedure: 50 cm³ of the water to be examined is taken, and 2.5 cm³ conc. H_3PO_4 and 1 cm³ o-tolidine solution (1.35 g o-tolidine in 1 dm³ 3.6 M HCl) are added. Photometry is carried out on the solution at 440 nm. The calibration curve is obtained for 0.01–1.5 mg chlorine/dm³.

179

Evaluation can also be performed visually, by comparison with a standard simulating solution of $K_2Cr_2O_7 + CuSO_4$ [6].

Toyoaki Aoki et al. [7] absorb the free chlorine content of water in dilute alkali, after release of the chlorine in a microporous tube. The hypochlorite formed is measured photometrically at 290 nm in a continuously flowing medium.

References

[1] Babko, A. K., Pilipenko, A. T.: *Fotometricheskii analiz. Metody opredelenie nemetallov.* Khimiya, Moscow 1974.
[2] Boltz, D. F.: *Colorimetric Determination of Nonmetals.* Wiley, New York 1978.
[3] Kemula, W., Hulanicki, A., Janowski, A.: *Talanta* **7**, 65 (1960).
[4] Fessel Graner, C. A., Paulucci, J. B.: *Anal. Chim. Acta* **123**, 347 (1981).
[5] Chovanecz, T.: *Ipari vízvizsgálatok (Industrial water examinations).* Műszaki Könyvkiadó, Budapest 1977.
[6] Lure, Yu. Yu. (ed.): *Unifitsirovannie metody analiza vod.* Khimiya, Moscow 1973.
[7] Toyoaki Aoki, Makoto Munemori: *Anal. Chem.* **55**, 209 (1983).

6.12. Chromium

Chromium belongs in column VI/2 of the periodic system. In its compounds it occurs with valencies of between 2 and 6; in aqueous solution the di-, tri- and hexavalent forms are stable. Cr^{2+} is a strong reductant (E_0 $Cr^{2+}/Cr^{3+} = -0.41$ V), while $Cr_2O_7^{2-}$ is a strong oxidant (E_0 $Cr_2O_7^{2-}/Cr^{3+} = +1.36$ V).

A detailed review of the general analytical properties of chromium, and of the possibilities for its determination, is to be found in the book by Lavrukhina [1].

The behaviour of Cr^{3+} is similar to that of Al^{3+}, while $Cr_2O_7^{2-}$ behaves in some ways like the highest oxidation state ions of elements in the same or adjacent columns of the periodic table (MoO_4^{2-}, WO_4^{2-}, VO_3^-).

The hydroxide of the amphoteric Cr^{3+} precipitates out at pH 5, but dissolves up again at high basicity. Aqueous leaching following the oxidation of Cr^{3+} with Na_2O_2 is more suitable for quantitative separation. Further compounds providing a possibility for the separation of chromium by precipitation are $PbCrO_4$ and $BaCrO_4$. Cr^{3+} may be extracted with acetylacetone or oxine, but greater importance is attached to the extraction of $Cr_2O_7^{2-}$ or CrO_4^{2-} from HCl medium with extractants containing oxygen donor atoms (TBP, methyl isobutyl ketone, etc.), and with amines. The volatility of $CrOCl_2$ (b.p. 117°C) is primarily made use of to remove chromium interfering in the determination of other elements.

Methods suitable for the separation of small amounts of chromium are discussed by Beyermann [2], and the photometric reagents for Cr^{3+} are compared by Tataev and Abdullaev [3]. The role of the reagents in the analysis of chromium is generally not a great one. The simplest procedure is measurement of the absorbance of $Cr_2O_7^{2-}$ or CrO_4^{2-}, but this is only of limited applicability, because of the low sensitivity. The most widespread reagent for the determination of chromium is 1,5-diphenylcarbazide [4]. With CrO_4^{2-} this forms a purple compound, probably as a consequence of oxidation to diphenylcarbazone. Especially Mo and W disturb the determination of chromium with diphenylcarbazide. Fe, Cu and Hg also interfere in 10–100-fold amounts. The disturbances can be eliminated partly by extraction with acetylacetone, and partly by ion-exchange.

DETERMINATIONS WITH DIPHENYLCARBAZIDE

Determination of chromium in rocks [5]

The optimum acidity for the reaction is 0.005–0.1 M H_2SO_4. Only Mo, Hg, Fe, V and oxidants interfere in the determination. $\lambda_{max} = 546$ nm; $\varepsilon_{546} = 3.4 \times 10^4$. The complex can be extracted with alcohols immiscible with water.

Analytical procedure: 0.1 g rock sample is fused for 45 min at 900°C with 0.6 g $Na_2CO_3 + MgO$ (4 : 1) in a covered platinum crucible. After cooling, a little water and alcohol are added, and the mixture is boiled until the melt breaks down and is then kept hot for 1 hour. (The MgO ensures the precipitation of silica, while the alcohol reduces MnO_4^-.)

The precipitate is filtered off and washed with a little Na_2CO_3 solution (20 g/dm³). The volume of the filtrate is made up to 20 cm³, its pH is adjusted to 1.3–1.7 with H_2SO_4, and the CO_2 evolved is removed by shaking. Diphenylcarbazide solution is added (prepared by dissolving 0.25 g in 100 cm³ acetone), and the volume is made up to the mark in a 25 cm³ volumetric

181

flask. The absorbance of the solution is measured at 540 nm. The calibration plot is produced with 1–8 μg chromium.

Easton [6] reports that the treatment with alcohol is not completely successful in the presence of 0.1% manganese, and the results are then low. To avoid this, he recommends the reduction of MnO_4^- with EDTA.

Determination of chromium content of ilmenite [7]

A fundamental step in the method is elimination of the disturbing effect of iron. For this purpose, the Cr^{3+} is oxidized to $Cr_2O_7^{2-}$ (with MnO_4^-) after digestion of the sample with $KHSO_4$, and the $Cr_2O_7^{2-}$ is extracted from 2 M HCl with methyl isobutyl ketone. A little H_3PO_4 too is added to decrease the extraction of Fe^{3+}. After washing of the organic phase with 1 M HCl, the $Cr_2O_7^{2-}$ is re-extracted with water, and a little H_3PO_4 and diphenylcarbazide are added, together with sufficient H_2SO_4 for its final concentration to be 1 M. Photometry is carried out at 546 nm.

Determination of chromium content of saline waters [8]

Before the determination the Cr^{3+} is oxidized with NaOCl. If the iron content of the sample is higher than 1 mg/dm³, the iron must be preliminarily extracted at pH 3–4 with a 1 : 1 acetylacetone–chloroform solution.

Analytical procedure: 50 cm³ of the saline water to be analyzed is measured into a 250 cm³ conical flask and neutralized. 25 cm³ distilled water, 1 cm³ 20% NaOH and 0.5 cm³ NaOCl solution (75–100 g/dm³) are added, and the mixture is boiled for 15 min. After cooling, the pH of the solution is brought to 5–6 with 2.5 M H_2SO_4, an extra 1.5 cm³ H_2SO_4 is added, and the solution is boiled for 45 min to decompose the NaOCl and to expel the chlorine. The mixture is cooled and transferred to a separating funnel, 1 cm³ freshly prepared 1% diphenylcarbazide in acetone is added, and the complex is extracted for 2 min with 10 cm³ isoamyl alcohol. The absorbance of the organic phase is measured in 2 cm cells at 540 nm. The calibration plot is obtained with 1–5 μg Cr, carried through the process in an analogous way. The sensitivity of the method is 20 μg Cr/dm³.

182

Analysis of steels and alloys [9]

Analytical procedure: 0.1–1.0 g sample is evaporated almost to dryness with a little aqua regia, and the evaporation is then continued until the appearance of fumes of SO_3 after the addition of 1–2 cm^2 1 : 1 H_2SO_4. After cooling, the solution is made up to the mark with water in a 100 or 250 cm^3 volumetric flask. A 2–2.5 cm^3 aliquot is taken, 0.2–0.5 g KIO_4 is added and the solution is boiled for a few minutes to oxidize the chromium. After cooling, the solution is transferred to a separating-funnel, 10 cm^3 1 M H_2SO_4 and 3 cm^3 saturated NaCl solution (to mask Mo) are added, and the mixture is diluted to 50 cm^3 with water. The CrO_4^{2-} is extracted with 5.0 cm^3 5% TOMA (trioctylmethylamine) or TOA (trioctylamine) in $CHCl_3$, and the absorbance is measured at 450 nm. Tungsten would interfere in the determination, but is left undissolved in the acidic treatment.

SOME OTHER POSSIBILITIES FOR THE PHOTOMETRIC DETERMINATION OF CHROMIUM

It has been found [3] that the pH range 4.0–6.0 is the most suitable for the formation of Cr(III) complexes; it is necessary to boil the solution. However, the resulting complexes (with glycine thymol blue, glycine cresol red, or PAR) are also stable at higher acidity (confirming the inert nature of the complexes) and thus Fe^{3+}, Ni^{2+}, Cu^{2+} and V^{5+} no longer interfere. This provides the possibility of using otherwise non-selective reagents, without separation, and in this way 0.1% Cr in steels can be determined with PAR in 0.5 M H_2SO_4 [10]. The disturbing effect of Co^{2+} can be taken into account via a comparison solution prepared without boiling.

Tataev and Guseinov [11] report that arsenazo III is suitable for determination of the chromium content of steels, at chromium levels above 0.5%.

For purposes of metallurgical analysis where the sensitivity is lower, the absorbance of Cr(III) complexonates may be measured ($\varepsilon = 500$–1500), possibly after extraction with TOMA [12].

183

References

[1] Lavrukhina, A. K.: *Analiticheskaya khimiya khroma*. Nauka, Moscow 1979.
[2] Beyermann, K.: *Z. Anal. Chem.* **190**, 4 (1962).
[3] Tataev, O. A., Abdullaev, R. R.: *Zh. Anal. Khim.* **25**, 930 (1970).
[4] Van der Walt, C. F. J., van der Merwe, A. J.: *Analyst* **63**, 809 (1938).
[5] Fruge, R.: *Chem. Geol.* **2**, 289 (1967).
[6] Easton, A. J.: *Anal. Chim. Acta* **31**, 189 (1965).
[7] Pilkington, E. S., Smith, P. R.: *Anal. Chim. Acta* **39**, 321 (1967).
[8] Lozovik, A. S.: *Zav. Lab.* **45**, 305 (1979).
[9] Adam, J.: *Talanta* **18**, 91 (1971).
[10] Akhmedov, S. A., Tataev, O. A., Abdullaev, R. R.: *Zav. Lab.* **37**, 756 (1971).
[11] Tataev, O. A., Guseinov, V. K.: *Zh. Anal. Khim.* **30**, 935 (1975).
[12] Kinhikar, G. M., Dara, S. S.: *Talanta* **21**, 1208 (1974).

6.13. Cobalt

Cobalt is a characteristic transition metal, from column VIII/2 of the periodic system. It exists in solution in di- and trivalent forms. Co^{3+} is a very strong oxidant (E_0 $Co^{3+}/Co^{2+} = 1.8$ V), and cobalt is generally found in the Co^{2+} form in aqueous solution. The tendency to complex formation is very strong in both oxidation states. Particularly the Co(III) complexes are of high stability; the Co(III) state is stabilized by ligands containing N and O donor atoms. Thus, the oxidation state $+2$ is favoured in the ionic form, and the oxidation state $+3$ in the complexes.

The analytical value of these complexes is increased by the fact that, similarly to Cr(III) complexes, they are (kinetically) inert. Accordingly, if the acidity is increased to 2–5 M after cobalt complex formation at lower acidity, the other metal complexes decompose, and consideration need be given to the removal or masking of interfering ions (Fe^{3+}, etc.) only if they are present in high concentration.

Co^{2+} forms complexes suitable for separation or determination with NH_4OH, halides (e.g. $CoCl_4^{2-}$), thiocyanate and other S-containing ligands.

Co^{3+} primarily forms complexes with N and O-containing ligands. The extraction of $Co(NH_3)_6^{3+}$ and of chelate complexes of Co(III) provides a possibility for its separation.

The photometric reagents for cobalt can be classified into three main groups:

1. nitrosophenols (nitroso-R salt, nitrosonaphthols, etc.);
2. reagents containing N donor atoms (terpyridyl, pyridyl-azo derivatives, etc.);
3. reagents containing S donor atoms (dithizone, dithiocarbamates).

Properties of the reagents and complexes are reported in Table 6.8.
A detailed review of the analysis of cobalt is presented in the monograph by Pyatnitskii [1].

TABLE 6.8.

Characteristics of some photometric reagents for cobalt

Reagent	Medium	Extractant	Complex λ_{max} nm	$\varepsilon \times 10^{-3}$	Refs
SCN$^-$ (Co^{2+})	0.1–1.0 M HCl, aqueous + organic	Alcohols, acetate esters, MIBK, etc.	620	1.7	[8]
Nitroso-R salt	Weakly acidic; subsequently 1–2 M H$^+$		500	15	[1]
1-Nitroso-2- -naphthol	pH 4–5	Toluene, CHCl$_3$, etc.	425	30	[1]
PAR	Weakly acidic; subsequently 2–5 M H$^+$		510	56	[9]
PAN	Weakly acidic; subsequently 2–5 M H$^+$		625	21	[10]
3,5-Br-PADAB	Weakly acidic; subsequently 2–5 M H$^+$		590	120	[6]

DETERMINATION WITH NITROSO-R SALT

Determination of cobalt in rocks [2]

Co^{2+}, Cu^{2+}, Ni^{2+} and Zn^{2+} are all determined in the stock solution obtained via precipitation with NH$_4$OH after digestion by the procedure described in Section 6.43 for zinc. The ammine complex of Co(II) is of low

185

stability, but immediately after its formation it is transformed to the Co(III) complex as a result of the strong ligand field. This permits separation from the hydroxide precipitate in ≥ 2 M NH_4OH $+2$ M NH_4^+ solution. If the complexes formed with nitroso-R salt at pH 5.5–6.0 are subsequently boiled in HNO_3, all the other metal complexes decompose, leaving the Co(III) complex, and thus the determination is selective.

Analytical procedure: Depending on the cobalt content expected, 5–20 cm^3 of the stock solution obtained after digestion of the sample and precipitation with NH_4OH is measured into a beaker, and the bulk of the NH_4OH is removed by boiling. The solution is acidified with 1 M HCl, 1.0 cm^3 33% sodium acetate is added to adjust the pH to 5.5–6.0, the solution is boiled, and 0.5 cm^3 0.1% nitroso-R salt is added to form the Co(II) complex. A few minutes later, after the addition of 0.5 cm^3 1 : 1 HNO_3, the solution is again boiled. The volume is made up to 25 cm^3 and the absorbance is measured at 520 nm. The reagent solution is used for comparison purposes. The calibration plot is recorded for 1–10 µg cobalt. The sensitivity of the method is a few ppm.

Determination of cobalt in steels and iron ores [3]

A condition for the method to be used without separation is that the Co content should be more than 0.2%, and the Cr content less than 2%. If this condition does not hold, separation by hydrolysis is applied. The minor Co loss occurring is established by the addition method and is taken into account.

Analytical procedure: A steel sample is dissolved in HCl+HNO_3. An iron ore is decomposed with a mixture of HF+H_2SO_4 until the appearance of fumes of SO_3; after cooling, HNO_3 is added to the residue, and the mixture is heated to boiling. 300 or 400 µg Co is added to the parallel samples. The strong acidity of the solution is decreased with NaOH, the mixture is heated to boiling, and ZnO is added in slight excess. The precipitation-containing solution is boiled, cooled, made up to 100 cm^3, and filtered.

An aliquot corresponding to 40–400 µg Co is taken from the stock solution. 5 cm^3 citrate solution (40 g sodium citrate/100 cm^3, neutralized to pH 7) and 0.5 cm^3 indicator (0.1% bromothymol blue in 50% alcohol) are added, and the solution is neutralized until the indicator changes colour.

186

10 cm³ 0.5% nitroso-R salt is then added, and the solution is boiled. The boiling is continued for 1 min after the addition of 5 cm³ 1 : 1 HNO_3. After cooling, the volume is made up to 100 cm³. The absorbance is measured at 500 nm, in 1 cm cells. The calibration plot is obtained for the range 60–600 μg Co. If a 5 cm cell is used, a concentration of even 5 μg Co/100 cm³ can be determined.

The procedure is also suitable for the analysis of Ni and Cu alloys, and silicates (glass, enamel, cement).

Determination of cobalt content of alkali metal hydroxides with diphenylguanidine extraction [4]

The sensitivity of the determination can be increased by extraction of the cobalt complex into $CHCl_3$ in the presence of diphenylguanidine. 0.1 ppm cobalt can then be determined in a 10 g sample.

Determination of cobalt in high-purity lanthanide metals and their compounds, after preliminary concentration [5]

Cobalt can be extracted from a 7–8 M HCl medium with a 1 : 3 : 5 mixture of trioctylamine+dioxan+CCl_4. With an acetate solution it can be re-extracted quantitatively and subsequently determined with nitroso-R salt. If the medium contains nitrate (in the case of the dissolution of CeO_2), the cobalt must be extracted from a thiocyanate medium.

Analytical procedure: 1–5 g of the sample to be analyzed (with the exception of CeO_2) is dissolved in 30 cm³ 8 M HCl, and the cobalt is then extracted for 2 min with 2×3 cm³ of the trioctylamine+dioxan+CCl_4 mixture. The organic phases are combined, and the cobalt is re-extracted by intensive shaking for 1 min with 10 cm³ 10% sodium acetate. The extract is transferred to a 50 cm³ beaker, 3 cm³ 1 M HCl (pH ca. 5.5) and 2 cm³ 0.2% nitroso-R salt are added, and the mixture is boiled for 1 min. After the addition of 3 cm³ H_2O_2 and 2 cm³ 1 : 1 HCl, the volume is evaporated down to 5 cm³. The colour of the resulting solution is compared with the colours of a standard series containing 0.05, 0.10, 0.25 and 0.50 μg Co.

In the analysis of CeO_2, the material is dissolved in a mixture of HNO_3+ +H_2O_2. The solution is evaporated, 30 cm³ water and 0.5 g NH_4SCN are added, the cobalt is extracted at pH 1–1.5 with the above extractant mixture, and the organic phase is washed with water containing ascorbic acid. The re-extraction and the subsequent steps are the same as described above.

The analysis is not disturbed by $10^4–10^5$-fold amounts of Al, Cr, Cu, Fe, Hg, Mn, Mo, Bi, Ni, Pb, Sn, Sb, Ti, V, W, Th and U. The sensitivity of the method is 1×10^{-5} % Co.

DETERMINATION WITH 3,5-Br-PADAB

Determination of cobalt in rocks [6]

3,5-Br-PADAB (see Appendix) is a very sensitive and selective cobalt reagent. Similarly as with the other pyridylazo reagents, the high selectivity is ensured by the strongly acidic medium after the complex formation; the slight interference caused by Fe^{3+} is eliminated with F^-.

Analytical procedure: 0.5–1.0 g sample is evaporated with 5 cm³ H_2O + 5 cm³ 1 : 1 H_2SO_4 + 15 cm³ HF in a platinum dish on a water-bath. This operation is repeated after the addition of 5 cm³ HF + 3 cm³ HNO_3, and the vessel is then heated on a hot-plate until fumes of SO_3 appear. The residue is washed with water into a 250 cm³ volumetric flask, and the volume is made up to the mark.

An aliquot of at most 25 cm³, containing 0–25 μg cobalt, is transferred to a 50 cm³ volumetric flask, and 5 cm³ 20% NaF, 2 cm³ 2×10^{-3} M Br-PADAB (in alcohol), and then 5 cm³ 4 M ammonium acetate are added. The mixture is heated for 10 min on a water-bath, then cooled, and 10 cm³ conc. HCl is added. The solution is made up to the mark, and the absorbance is measured by photometry at 590 nm. The comparison is made with distilled water containing the reagents. The calibration plot is produced in a similar way with 1–5 μg cobalt. With a layer thickness of 5 cm, 1 ppm cobalt can be determined.

The selectivity of the method can be increased by ion-pair extraction with anthraquinone-2-sulphonic acid (AQSA) into $CHCl_3$, and subsequent re-extraction of the cobalt complex with acid [7]. 2 cm³ 0.01 M AQSA (0.62 g/100 cm³) is added to the coloured solution formed with Br-PADAB in the presence of the complex former. The aqueous solution is extracted for 3 min with 10 cm³ $CHCl_3$, and the cobalt complex is then re-extracted by shaking with 10 cm³ 2.4 M HCl. The absorbance of the aqueous solution is measured at 573 nm. The calibration plot is recorded in the interval 0.6–25 μg Co/50 cm^3.

References

[1] Pyatnitskii, O. V.: *Analiticheskaya khimiya kobalta.* Nauka, Moscow 1965.
[2] Görbicz, M., Upor, E.: *Acta Chim. Acad. Sci. Hung.* **66,** 373 (1970).
[3] Wünsch, G.: *Talanta* **26,** 177 (1979).
[4] Shustova, M. V., Nazarenko, V. A.: *Zav. Lab.* **39,** 18 (1973).
[5] Glinskaya, I. V., Merisov, Yu. B.: *Zav. Lab.* **45,** 987 (1979).
[6] Kiss, E.: *Anal. Chim. Acta* **77,** 320 (1975).
[7] Baudino, O., Marone, C. B.: *Anal. Chim. Acta* **119,** 393 (1980).
[8] Jackwerth, E., Schneider, E. L.: *Z. Anal. Chem.* **207,** 188 (1965).
[9] Radovskaya, T. L.: *Zav. Lab.* **42,** 398 (1976).
[10] Ersova, N. S., Orlov, V. V., Ivanov, V. M., Busev, A. I.: *Zav. Lab.* **41,** 913 (1975).

6.14. Copper

Copper belongs in group I/2 of the periodic system. It occurs in mono- and divalent forms in its complexes; otherwise, in solution Cu^+ has a tendency to undergo disproportionation. In its analytical properties, Cu^+ is primarily similar to Ag^+, Au^+ and Tl^+, while Cu^{2+} rather resembles Zn^{2+} and Co^{2+}. The complexes formed with N and S-containing ligands are characteristic. Similarly to Fe^{2+}, Cu^+ gives stable complexes with reagents containing several N atoms.

In the separation of Cu^{2+}, besides the soluble ammine complexes, primarily the extractions involving dithizone and dithiocarbamate are of importance.

The most frequently used reagents are listed in Table 6.9.

(A) DETERMINATION WITH SODIUM DIETHYLDITHIOCARBAMATE

This reagent forms complexes with numerous other elements besides Cu^{2+}, and the selectivity must therefore be ensured with separations, pH adjustment and masking. The compound formed in aqueous solution can

189

TABLE 6.9.

More important photometric reagents for copper

Reagent	λ_{max} nm	$\varepsilon \times 10^{-3}$	pH of medium	Interferences	Note	Refs
For Cu^{2+} ions						
Sodium diethyldi-thiocarbamate	436	12.8	9	Many elements	Separa-tion	[1, 7, 8]
NH$_4$OH	578	0.055	10	Ni^{2+}, Co^{2+}, CN$^-$	Separa-tion	[6, 8, 9]
5-(2-Hydroxy-3,5-dinitrophenylazo)-N-oxide-8-hydroxy-quinoline	570	16	5.5–6	Al^{3+}, Fe^{3+}, Cr^{3+}		[11]
For Cu$^+$ ions						
Cuproin	546	6.3	4–7.5	Ag, Cd, Co, Hg, Sb, Sn, much PO$_4^{3-}$, oxalate		[8, 9]
Neocuproin	460	7.9	3–10	Ag, Cd, Co, Hg, Sb, Sn, much Fe, oxalate		[8]
Bathocuproin	480	13	4–10	Ag, Cd, Co, Hg, Sb, Sn, much Fe, oxalate		[14]
8,8'-Biquinolyl-di-sulphide	432	9.5	2–13	Au		[12, 13]

be kept in solution only with protective colloids, and therefore extraction is performed (with CHCl$_3$, CCl$_4$, amyl alcohol, etc.). The resulting compound is photosensitive, and direct sunlight should be avoided. Of the metals, Ag, Co, Cr, Fe, Hg, Mn, Ni and V cause interference, whereas the anions Cl$^-$, SO$_4^{2-}$, NO$_3^-$, PO$_4^{3-}$, citrate, tartrate and oxalate have no perceptible influence.

Determination of copper in rocks after separation with ammonia and extraction with dithizone [1]

The steps involved in the digestion of the rock and the preparation of the stock solution are identical with those described for zinc in Section 6.43. A 5–20 cm^3 aliquot is taken from the NH$_4$OH+NH$_4$Cl-containing stock

190

solution, and is diluted to 50 cm³. The Ni is extracted with dimethylglyoxime solution, the organic phase is washed with 5 cm³ NH_4OH solution (dilution 1 : 49), and the washing liquid is combined with the base solution. The aqueous phase is extracted with 0.01% dithizone. The zinc is removed from the dithizone phase with 0.1 M HNO_3 and is determined. The copper is next transferred to the aqueous phase by shaking with 2×5 cm³ 6 M HCl. The solution is neutralized with conc. NH_4OH and the pH is adjusted to 8.5. The cobalt also passing into the aqueous phase is bound with 3 cm³ 0.1% nitroso-R salt. 5 cm³ 0.5% DDTC solution is added to the cooled solution, extraction is carried out with 2×4 cm³ CCl_4, and the volume is made up to 10 cm³. The absorbance of the solution is measured at 436 nm, in 1 cm or 5 cm cells. The calibration plot is obtained with 1–10 μg Cu^{2+}/10 cm³, measurements being made in 5 cm cells.

If only copper is determined, the dithizone step following the NH_4OH separation is omitted. The course of the analysis is then as follows:

The sample is evaporated with 30 cm³ aqua regia and then with 10 cm³ 10% HCl. The residue is taken up in a little distilled water, the solution is filtered and the solid is washed. The volume is adjusted to 100 cm³, this medium containing 2 M NH_4OH+2 M NH_4Cl. 10 cm³ 10% $Na_4P_2O_7$, 3 cm³ 0.1% nitroso-R salt and 2 cm³ 1% dimethylglyoxime (in alcohol) are added to an aliquot of the stock solution. The solution volume should then be 50–60 cm³. The solution is extracted with 2×5 cm³ $CHCl_3$, whereby the nickel is removed. 5 cm³ 0.5% DDTC is next added to the solution, and extraction is performed with CCl_4 as described above.

Determination of copper traces in plant material [2]

The plant material is decomposed with a mixture of $HNO_3+H_2SO_4+$ $+HClO_4$, and the copper and zinc are extracted together with dithizone at pH 9 in the presence of buffer. The zinc is transferred from the dithizone phase to the aqueous phase with 0.02 M HCl, and is determined.

The dithizone phase containing the copper is evaporated to dryness, and the residue is heated with 2 drops H_2SO_4 and 5 cm³ HNO_3 for 10 min after the first fumes of SO_3 have appeared. An EDTA–citrate mixture is prepared by dissolving 17.28 g citric acid in water, neutralizing with NH_4OH, adding 5 g EDTA, and making the volume up to 1 dm³. 10 cm³ of this EDTA–

191

citrate mixture is added to the cooled residue, together with thymol blue as indicator, and 3.5 M NH_4OH is added until a green colour develops. 1 cm^3 1% DDTC and 10 cm^3 CCl_4 are added, the solution is extracted for 2 min, and the absorbance is measured at 436 nm.

Determination of copper traces in natural waters and industrial waste waters [3, 4]

The copper in natural waters is to be found in the form Cu^{2+}, while in industrial waste waters it may also occur in a form bound to cyanide. The cyanide complex is then decomposed by evaporation with $H_2SO_4 + H_2O_2$.

Analytical procedure: 1 cm^3 20% citric acid, 2 cm^3 10% NH_4OH and 0.5 cm^3 20% NH_4Cl are added to 100 cm^3 of the water to be examined. If the water is initially yellow in colour, several extractions with 10 cm^3 CCl_4 are performed to remove this colour. 1 cm^3 1% DDTC is added, and the $Cu(DDTC)_2$ is extracted with 30 cm^3 CCl_4. The absorbance is measured at 435 nm. In the presence of nickel, 1 μg Cu is deducted per 10 μg nickel.

(B) DETERMINATION OF COPPER BY MEASUREMENT OF ABSORBANCE OF $[Cu(NH_3)_4]^{2+}$

When NH_4OH is added to Cu^{2+}, a precipitate is first obtained containing the *trans*-diamminecopper(II) complex; this dissolves in excess NH_4OH to give an intensely blue solution of $[Cu(NH_3)_4]^{2+}$ [5]. The determination is disturbed in high concentration by Ni and Co, which similarly form coloured complexes with NH_3, and also by anions such as CN^-, citrate and tartrate, which react with copper to give complexes that are more stable than the ammine complex.

Determination of copper content (>0.2%) of rocks [6]

The preliminary steps here too are the same as those described for zinc in Section 6.43, up to the preparation of the 2 M $NH_4OH + 2$ M NH_4Cl medium. In this medium the separation of copper from metal hydroxides is loss-free. The blue colour is subjected to photometry at 620 nm. The calibration plot is recorded up to a concentration of 20 mg $Cu^{2+}/100$ cm^3.

As a consequence of its simplicity and speed, the method based on measurement of the tetramminecopper(II) complex is used in metal-analysis practice too, e.g. for pig iron and cast iron samples [7, 8].

(C) DETERMINATION OF COPPER WITH 2,2'-BIQUINOLYL (CUPROIN)

In the pH range 4–7.5, Cu^+ reacts with 2,2'-biquinolyl to give a compound that is insoluble in water, but that can be extracted into organic solvents, e.g. CHCl$_3$, CCl$_4$, amyl alcohol, etc. The complex is stable and is not sensitive to atmospheric oxygen. The reagent is selective; only certain complex-forming anions interfere (e.g. CN$^-$, oxalate, EDTA).

Determination of copper content of rocks [9]

Digestion with $K_2S_2O_7$ in a glass tube heated by an alcohol flame is a suitable procedure for the determination in the field of the copper content of pyroxene-andesite rock samples. This procedure permits the on-site detection of a few μg Cu. Parallel digestions under laboratory conditions with HF+HClO$_4$, with Na$_2$CO$_3$ and with NaOH provide practically the same result as the $K_2S_2O_7$ technique.

Analytical procedure: 0.5 g rock sample is fused for 20 min with 2.5 g $K_2S_2O_7$. The melt is dissolved in 20 cm^3 water. 1 cm^3 15% NH$_2$OH.HCl and 2 cm^3 20% sodium acetate are added to a 1 cm^3 aliquot of this solution in a separating-funnel. After the addition of 3 cm^3 buffer (8.2 g sodium acetate+5.8 cm^3 acetic acid in 200 cm^3), when the pH should be 4–5, the solution is extracted with 2×1 cm^3 0.02% 2,2'-biquinolyl in iso-amyl alcohol, for 1 min on each occasion. The volume is made up to 3 cm^3 with iso-amyl alcohol. Evaluation is performed either by means of a comparative series or spectrophotometrically at 546 nm. The individual solutions in the comparative series should contain 0–5 μg Cu.

(D) DETERMINATION OF COPPER WITH 8,8-BIQUINOLYL-DISULPHIDE

8,8'-Biquinolyl-disulphide is a specific reagent; masking agents, e.g. tartrate, are needed only if the solution also contains ions that are readily hydrolyzed, e.g. Sb^{3+}, Bi^{3+}.

Determination of copper traces in soils [10]

The pretreatment and digestion of the soil sample (ignition at 500°C, digestion with $HF + H_2SO_4$) are the same as described in connection with zinc in Section 6.43. Aliquots containing 5–60 µg Cu^{2+} are pipetted from the 50 cm³ stock solution. The HCl solution is brought to pH 4 with 50% sodium acetate. 5 cm³ 20% ascorbic acid (at pH 4) is added to reduce Cu^{2+}, followed by 10 cm³ 0.2% 8,8'-biquinolyl-disulphide in $CHCl_3$ for extraction. The absorbance of the Cu^+ complex is measured in 1 cm cells at 453 nm. The calibration plot covers the concentration range 5–60 µg Cu^{2+}/10 cm³.

(E) DETERMINATION OF COPPER WITH 5-(2-HYDROXY-3,5-DINITROPHENYLAZO)-N-OXIDE-8-HYDROXYQUINOLINE

This compound (reagent I) is a sensitive reagent for Cu(II). The resulting complex is poorly soluble in water, but dissolves well in oxygen-containing organic solvents, and can be extracted with CCl_4. The determination is not disturbed by a 200-fold amount of Zn, a 100-fold amount of Mn, Ag and Pb, a 25-fold amount of Ni and Co, a 1000-fold amount of Cl^- and $NO_3^-/$, an 800-fold amount of SO_4^{2-}, and a 125-fold amount of F^-. Al, Fe(III) and Cr(III) do interfere; the disturbing effects of Al and Fe can be eliminated by the addition of NaF.

Determination of copper content of zinc, zinc alloys and steel [11]

Analytical procedure for zinc and zinc alloys: 0.1–0.5 g zinc or zinc alloy (Cu content 0.5–1.6%) not containing Fe or Al is dissolved in 8–10 cm³ 1 : 1 HNO_3, and the solution is evaporated until salts crystallize out. The residue is dissolved in water and the volume is made up to 250 cm^3. An aliquot corresponding to 4–12 μg Cu is transferred to a separating-funnel, 2 cm^3 acetate buffer of pH 5.5 (28.7 cm^3 1 M CH$_3$COOH+25 cm^3 1 M NaOH, made up to 250 cm^3) is added, and the volume is made up to 5 cm^3. 2 cm^3 2.5×10^{-4} M reagent I in acetone and 3 cm^3 acetone are added to the solution, which is extracted for 1 min with 4 cm^3 CCl$_4$. Anhydrous Na$_2$SO$_4$ is added to the organic phase, and the absorbance of the solution is measured in 1 cm cells at 530 nm. An extract prepared in the same way from the reagents is used as comparative solution.

Analytical procedure for steel: 0.5 g sample is dissolved in 15 cm^3 1 : 1 HNO$_3$, the oxides of nitrogen are removed by boiling, and the solution is evaporated to low volume. The residue is diluted to 100 cm^3. An aliquot (4–12 μg Cu) is taken, 0.6 cm^3 saturated NaF is added, the pH is adjusted to ~ 3 with 1 M NaOH, 2 cm^3 acetate buffer of pH 5.5 is added, and the volume is made up to 5 cm^3. The subsequent steps are analogous to those for zinc and zinc alloy samples. The calibration plot is produced for 2–14 μg Cu(II).

References

[1] Görbicz, M., Upor, E.: *Acta Chim. Acad. Sci. Hung.* **66**, 373 (1970).
[2] Page, E. R.: *Analyst* **90**, 435 (1965).
[3] *The Testing of Water.* 5th Ed. Merck, Darmstadt 1964.
[4] Lure, Yu. Yu., Rybnikova, A. I.: *Khimicheskii analiz proizvodstvennykh stochnykh vod.* 4th Ed. Khimiya, Moscow 1974.
[5] Grant, D. M., Kollrack, R.: J. *Inorg. Nucl. Chem.* **23**, 35 (1961).
[6] Upor, E.: Unpublished work.
[7] Spauszus, S.: *Methoden der Chemischen Stahl- und Eisenanalyse.* VEB Deutscher Verlag für Grundstoffindustrie, Leipzig 1968.
[8] Dragomirecky, A., Mayer, V., Michal, J., Rericha, K.: *Photometrische Analyse anorganischer Roh- und Werkstoffe.* VEB Deutscher Verlag für Grundstoffindustrie, Leipzig 1968.
[9] Simó, B.: *Magy. Kém. Lapja* **26**, 408 (1971).
[10] Vazhenina, I. G. (ed.): *Metody opredeleniya mikroelementov v pochvakh, rasteniyakh i vodakh.* Kolos, Moscow 1974.
[11] Klemoneva, O. K., Nemodruk, A. A., Gibalo, I. M.: *Zh. Anal. Khim.* **34**, 1485 (1979).
[12] Fotiev, A. A. (ed.): *Analiticheskaya khimiya neorganicheskii soedinenii.* Akad. Nauk., Sverdlovsk Vol. 27, 1974.
[13] Bankovskii, Yu. A., Evinsh. A. F., Luksha, E. A., Bochkhaus P. Ya.: *Zh. Anal. Khim.* **16**, 150 (1961).
[14] Smith, G. F., Wilkins, D. H.: *Anal. Chem.* **25**, 510 (1953).

6.15. Fluorine

Fluorine belongs in column VII/1 of the periodic system. It forms the lightest and smallest halide ion. In aqueous solution it occurs partly as F^- and partly as HF_2^-, because of its strong tendency to take part in hydrogen-bonding. It differs somewhat from the other halogens as regards its chemical properties. This is due in part to the fact that the fluorine atom has no d orbital. Fluoride forms precipitates with Ca^{2+} and other multivalent metal ions. As a ligand of moderate donor strength, it yields complexes of use in analysis with tri- and tetravalent metal ions.

The relatively high stability of BF_3 (BF_4^-) and SiF_4 (SiF_6^{2-}) is of great importance in separations involving extraction or distillation. A systematic listing of the inorganic and organic complexes of fluorine, and also the fields of application of fluorine, are to be found in the book by Ishikawa and Kobayashi [23].

At present two groups of methods are known for the photometric determination of fluoride (as a colourless complex former). The more widespread ones are indirect methods based on the complex-forming ability of fluoride. In these procedures the concentration (and hence the absorbance) of some coloured complex is decreased as a consequence of ligand-exchange with fluoride.

The procedure elaborated by Belcher and West [1] for the direct determination of fluoride is based on measurement of the absorbance of an alizarin-complexone-fluoride mixed ligand complex of a lanthanide. The critical article by Deane et al. [2] provides a valuable comparison of the properties of the La complexes of alizarin fluorine blue (AFB) and of sulphonated alizarin fluorine blue (AFBS). Supplementary examinations were carried out with a F^- ion-selective electrode. It was found that the AFBS reagent is undoubtedly advantageous in comparison with the AFB reagent, while the F^--selective electrode can be recommended as an alternative possibility. The method may be used for the analysis of water samples.

The photometric determination of fluorine is treated in a number of monographs and reviews [3–8].

Multivalent anions and cations that form complexes with fluoride interfere in the determination. Separation from such disturbances can primarily be achieved by means of distillation, including pyrohydrolysis (HF) [9], steam-distillation [10] and microdiffusion [11] (H_2SiF_6, SiF_4). In some cases, ion-exchange, extraction, and precipitation of the interfering metal ions may be considered.

(A) INDIRECT METHODS FOR DETERMINATION OF FLUORIDE

Fluorides can be determined indirectly with coloured complexes (sometimes lakes) the metal ion of which forms a more stable complex with fluoride than with the original ligand. A solution of pH 2–3 is generally favourable for the determination; the disadvantage of a medium more alkaline than this is that the metal ion may undergo hydrolysis (competition by OH^-), while in more acidic medium there is a decrease in the dissociation of HF. The main features of some indirect analytical procedures are given in Table 6.10.

TABLE 6.10.

Main features of some indirect procedures for fluorine determination

Sample	Means of separation	Reagent and conditions of determination	Refs
Fluorinated water ($\cong 1$ mg F/dm^3)		Zr^{4+} eriochrome cyanine R	[8]
Rocks and minerals	Digestion with $ZnO + Na_2CO_3$; distillation from $HClO_4$ or H_2SO_4	Zr^{4+}-eriochrome cyanine R or Zr^{4+}-alizarin S	[3,19,22]
Phosphate-containing solutions	Precipitation of $Ba_3(PO_4)_2$ at pH 10	Zr^{4+}-pyrocatechol violet	[13]
Air		Fe(III)-thiocyanate, extracted into amyl alcohol	[21]

Determination of fluoride in rocks (based on [3] and the investigations of the authors)

Fluoride-containing silicate rocks are fused with a mixture of $KNaCO_3 + + ZnO$, and the melt is leached with water. The solution is filtered, and the

197

filtrate is distilled. The fluoride in the distillate is determined with the Zr-alizarin S reagent. The calibration plot is negative in slope.

Analytical procedure: 4 g $KNaCO_3$ and 0.6 g Zn are mixed with 0.1–0.3 g sample, and the mixture is fused. The melt is leached with 30–40 cm^3 hot water, the mixture is filtered, and the residue is washed with 1% Na_2CO_3 solution in water. The alkaline filtrate is neutralized with 1 : 1 H_2SO_4 and transferred to a steam-distillation apparatus (the 0.5 dm^3 boiling flask is connected to a 2 dm^3 steam-producing flask, a condenser, a collector and a feeding head). 0.1 g powdered quartz, 30 cm^3 2 : 1 H_2SO_4 and 2 cm^3 conc. H_3PO_4 are added to the solution in the flask. The mixture is boiled over a gas burner. The temperature of distillation is $140\pm5°C$. 2×250 cm^3 distillates are collected (the main and the secondary distillate). 10–25 cm^3 aliquots are taken from the distillates. 2.5 cm^3 zirconium solution (0.354 g $ZrOCl_2.8H_2O+33.3$ cm^3 conc. H_2SO_4+100 cm^3 conc. HCl, in 1 dm^3) and 2.5 cm^3 alizarin S (0.75 g in 1 dm^3) are added, and the volume is made up to 50 cm^3. Photometry is performed in 4 cm cells at 520 nm, after a waiting period of 1 hour, against a solution containing only alizarin S. The calibration plot is recorded in the range 5–50 µg $F^-/50$ cm^3.

Notes:
(a) The alizarin S solution must be freshly prepared every 1–2 days.
(b) A calibration plot must be recorded in every determination.
(c) The colour of the solution is practically constant in the period 1–1.5 hours; the calibration plot is linear.
(d) Evaluation via addition to the distillate yields results agreeing with those read off the calibration plot.
(e) Determinations were also made with an F^--selective electrode on the leached solution after $NaKCO_3+Zn$ fusion (neutralization with HCl; 1 : 1 dilution with 1 M Na-citrate; measurement at pH 5.3). The photometric and ion-selective electrode measurements yielded practically identical results.
(f) In the F concentration range 0.08–0.12%, the scatter in the determinations was within 0.01%.

Determination of fluoride content of air samples [16]

The air sample to be examined (100–200 dm^3) is bubbled at a rate of 1 dm^3/min through 15 cm^3 2×10^{-3} M NaOH solution. In the presence of

198

disturbing ions (PO_4^{3-}, chlorine $\cong 0.2$ mg/m^3, SO_2 $\cong 0.6$ mg/m^3), the sample is distilled in a microdistillation apparatus [5]. The distillate solution is made up to 15 cm^3.

Analytical procedure: 0.5 cm^3 eriochrome cyanine R solution (3.6 g/dm^3) and 1 cm^3 zirconyl chloride solution (0.25 g $ZrOCl_2.8H_2O + 300$ cm^3 conc. HCl in 1 dm^3) are added to 9.5 cm^3 sample solution. The absorbance is measured at 550 nm. The calibration plot, which has a negative slope, is produced for the range 1–9 μg F.

(B) DIRECT DETERMINATION OF FLUORIDE

As mentioned in the introduction, there is a possibility for the direct determination of fluoride via the photometry of the alizarin-complexone-fluoride mixed ligand complex of some metal ions. Fluoride will enter complexes of lanthanides with alizarin-complexone (also known as alizarin fluorine blue). This is indicated by an increase in the molar absorbance (6.0×10^3) and a shift in λ_{max} (from 570 nm to 610 nm) [1]. The determination is disturbed by Cr^{3+}, Cu^{2+}, Fe^{3+}, Fe^{2+}, Ni^{2+}, Pb^{2+}, PO_4^{3-}, Ca^{2+}, etc. The disturbing effects can be eliminated by masking or separation; they otherwise strongly limit the sensitivity of the determination (to $\cong 0.1\%$ in rock analysis, for instance [12]). Detailed data on the properties of the reagent are reported by Minin and Filippova [13]. It was demonstrated more recently by Dedkova *et al.* [14] that such triple complexes are also formed with other systems, e.g. with Zr^{4+}-arsenazo III, when the determination may be more sensitive and more selective.

Fast method for determination of fluorine content of silicate rocks and minerals [15]

Analytical procedure: 1 g rock sample is fused with Na_2CO_3 in a platinum crucible, and the melt is leached out with hot water and filtered. 2 g $(NH_4)_2CO_3$ is added to the filtrate, the mixture is left on a water-bath for 30 min, a further 1 g is added, and the mixture is allowed to stand until the next day. The resulting precipitate, which contains the bulk of the interfering metal ions, is filtered off, the filtrate is just acidified with 1 : 1 HCl, and the volume is made up to 500 cm^3 in a polyethylene flask. A 1.0

199

cm³ aliquot is pipetted into a 50 cm³ volumetric flask, 20.0 cm³ reagent solution (25 g lanthanum–alizarin complexone dissolved in 150 cm³ isopropanol + 350 cm³ distilled water; filtered before use) is added, and the solution is made up to the mark. Photometry is performed at 630 nm after 1 hour. A comparison and standard fluoride solutions are measured simultaneously. The method is suitable for the analysis of samples containing $> 0.3\%$ F.

Determination of fluoride in environmental water, after distillation enrichment with hexamethyldisiloxane (HMDS) [17]

The method is suitable for water samples with low silica contents. The spectrophotometric evaluation of the fluoride is based on the loss in colour of the La–alizarin complexone (La–AC, Dotite–Alfusone).

Analytical procedure: A 40 cm³ aliquot is pipetted into a 200 cm³ distillation vessel. The two receivers (connected in series) each contain 30 cm³ 0.1 M NaOH. About 0.05 cm³ HMDS solution, 2 cm³ conc. H_3PO_4 and 25 cm³ conc. H_2SO_4 are added to the aliquot. The solution is heated up to about 90°C, and N_2 is bubbled through it for 30 min at a rate of 50–90 cm³/min. The contents of the two receivers are combined and made up to 100 cm³. A 5–10 cm³ aliquot of this solution is neutralized with 1 M HCl, and after the addition of 2 cm³ La–AC reagent (4.75 g of a 1 : 1 : 1 mixture of La–AC chelate + potassium hydrogen phthalate + hexamethylenetetramine, dissolved in 50 cm³ water) and 10 cm³ acetone the volume is made up to 25 cm³. The absorbance is measured at 620 nm. It is very significant that the calibration curve is linear for low F levels (2–5 µg).

References

[1] Belcher, R., West, T. S.: *Talanta* **8**, 853 (1961).
[2] Deane, S. F., Leonard, M. A., McKee, V., Svehla, G.: *Analyst* **103**, 1134 (1978).
[3] Nikolaev, N. S.: *Analiticheskaya khimiya ftora*. Nauka, Moscow 1970.
[4] Babko, A. K., Pilipenko, A. T.: *Fotometricheskii analiz. Metody opredeleniya nemetallov*. Khimiya, Moscow 1974.
[5] Boltz, D. F.: *Colorimetric Determination of Nonmetals*. Wiley, New York 1978.
[6] Moraru, L., Suten, A.: *St. cerc. Chim.* **19**, 757 (1971).
[7] Tusl, J.: *Chem. listy* **61**, 1302 (1967).
[8] Crosby, N. T., Dennis, A. L., Steves, J. G.: *Analyst* **93**, 643 (1968).

[9] Lorec, S.: *Chim. Anal.* **49**, 557 (1967).

[10] Richter, F.: *Z. Anal. Chem.* **124**, 161 (1942).

[11] Tusl, J.: *Chem. listy* **63**, 777 (1969).

[12] Prilutskaya, T. A., Arapova, G. A.: *Khimicheskie i fizikokhimicheskie metody analiza rud, porod i mineralov.* Nauka, Moscow 1974.

[13] Minin, A. A., Filippova, R. P.: *Alizarinovii komplekson. Assortiment reaktivov na ftor.* NIITEKHIM, Moscow 1970.

[14] Dedkova, V. P., Ozhashchi, D. O., Savvin, S. B.: Organicheskie reagenty v analiticheskoi khimii. Naukova Dumka, Kiev, **1**, 129, 1976. (Lectures at the 4th All-Union Conference.)

[15] Hall, A., Walsh, J. N.: *Anal. Chim. Acta* **45**, 341 (1969).

[16] Várkonyi, T., Cziczó, T.: *A levegőminőség vizsgálata (Air-quality investigations).* Műszaki Könyvkiadó, Budapest 1980.

[17] Yoshida, M., Kitami, M., Murakami, N., Katsura, T.: *Anal. Chim. Acta* **106**, 95 (1979).

[18] Cooke, J. R.: *Proc. Soc. Water Treat. Exam.* **14**, 145 (1965).

[19] Huang, P. M., Jackson, M. L.: *Am. Miner.* **52**, 1053 (1967).

[20] Frambe, A.: *Vom Wasser* **42**, 161 (1974).

[21] Moucmanova, A., Malinovsky, M.: *Chem. Zvesti* **19**, 287 (1965).

[22] Huang, W. H., Johns, W. D.: *Anal. Chim. Acta* **37**, 508 (1967).

[23] Ishikawa, N., Kobayashi, Y.: *Fluorine Compounds—Chemistry and Application.* Kodansha Scientific, 1979.

6.16. Gallium

Gallium is situated in group III/1 of the periodic system. It is present in solution in the trivalent form. In alkaline medium it can be dissolved in the anionic form GaO_2^- (gallate). Chemically, it resembles aluminium. In nature it is always to be found in small quantities as an accompanying element of aluminium.

The analysis of gallium is dealt with in the publications by Dimov and Savostin [1] and Savostin [2].

Numerous reagents are recommended for the photometric determination of gallium; though none of these are selective, after appropriate separation the most suitable reagent can be chosen in accordance with the type of sample to be analyzed. The organic reagents most frequently suggested for gallium determination are listed in Table 6.11.

Of the tabulated reagents, malachite green [3–5] and rhodamine B [6–8] are most suitable for the analysis of rocks, bauxites, and alloys containing Al, Zn, Cu and In.

TABLE 6.11.

Some photometric reagents for gallium

Reagent	Medium	λ_{max} nm	$\varepsilon \times 10^{-3}$	Extractant	Interfering elements
Quinalizarin	pH 5	500	11		Many elements
Aluminon	pH 3.5–4	490	18		Many elements; Zn and In do not interfere
Brilliant green	6 M HCl	635	15	CHCl$_3$, benzene	W, Ag, Hg(I), Ir, Sn(II), Fe(III), Au(III), Tl(III), Sn(IV), Mo(VI)
Malachite green	6 M HCl+ TiCl$_3$	635	~76*	benzene	Sn, oxidants, Au, Pd, Se, Te
Methyl violet	6 M HCl	580	27	CHCl$_3$	Many elements
Rhodamine B	6 M HCl+ TiCl$_3$	565	110*	benzene+ acetone (9:1)	Sb(V), Au(III), Tl(III), Fe(III), oxidants
Gallion	pH 3.6	600	25		Many elements

*Results of present authors

Malachite green forms a complex with gallium in 6–6.5 M HCl medium; this complex can be extracted with benzene. The reagent itself is not extracted, but its oxidation product is, and the extraction must therefore be carried out in the presence of TiCl$_3$.

Interference is caused by elements which are reduced to the metal by TiCl$_3$ (Au, Pd, Se, Te), and by ions which are prone to hydrolysis in acidic medium and which hamper the separation of the phases. The determination is not disturbed by Cu, Mo, Sb and Tl, even when present in 100–1000-fold excess.

Similarly to malachite green, rhodamine B forms a complex with gallium, which can be extracted with benzene. It has the advantage that Al, Zn and Cu do not give coloured compounds that are extracted with the reagent.

Possibility of separation of gallium from interfering elements by extraction

The extraction of GaCl$_4^-$ is convenient for the separation of gallium from aluminium and many other elements. Ether saturated with HCl is applied most often. It can be extracted from 6 M HCl medium: $D_{Ga} > 100$ [9, 10]. The

202

separation factor of Ga and Al is $>10^5$ (according to the authors' own examinations). Fe^{3+}, Pd^{2+} and Au^{3+} pass into the extract, while As, Sb, Mo, Te, V, In, Cu, Ni, Hg, Se, Ge and Sn do so partially. On the addition of $TiCl_3$, Fe^{3+} is reduced to Fe^{2+}, which is practically not extracted. Gallium can be re-extracted quantitatively from the organic phase with water.

Isopropyl ether, methyl isobutyl ketone and butyl acetate can also be used as extractants.

Gallium can be extracted from 6 M HCl medium with $CHCl_3$ in the presence of diantipyrylmethane. In this way, Al, Zn, Pb, Cd, Mg, Bi, As, In, Sn and Ge are not extracted; nor is Te if it is first reduced with $TiCl_3$ [11].

Determination of gallium content of aluminate liquor with malachite green ([1, 4, 12] and present authors)

Since the gallium content of aluminate liquor is relatively high (100–200 mg/dm³), there is no need for the preliminary separation of the gallium from the disturbing elements.

Analytical procedure: A 50–100-fold dilution of the solution to be examined is prepared (the acidity of the solution should be 6 M HCl). A 2 cm³ aliquot of the stock solution is transferred to a separating-funnel, 2 cm³ 15% $TiCl_3$ in 6 M HCl and 2 cm³ 2% malachite green in 6 M HCl are added, and extraction is performed with 20 cm³ benzene. The volume of the organic phase is made up to 25 cm³ with benzene (the mildly opalescent solution then becomes clear). The absorbance is measured in 1–5 cm cells at 635 nm, against pure benzene. The interval covered by the calibration plot is 1–10 μg Ga.

Rapid determination with rhodamine B of gallium content of rocks, bauxites and solutions obtained by acidic digestion of bauxites [13]

Principle of method: Our experiments have demonstrated that digestion with $HF + H_2SO_4$, followed by dissolution in HCl, is sufficient for the digestion of rocks and bauxites of average composition. Separation from interferences can be achieved even without preliminary ether separation, if the gallium-rhodamine B extract is washed with a washing liquid of appropriate composition.

Analytical procedure: A mixture of 1 g sample and 10–15 cm^3 HF+
+0.5 cm^3 1 : 1 H$_2$SO$_4$ is evaporated to dryness in a platinum dish on a
sand-bath. The residue is dissolved up in 1 : 1 HCl by heating on a water-
bath, and the volume is then made up to 25–50 cm^3 with 1 : 1 HCl. An
aliquot of at most 4 cm^3 is transferred to a separating-funnel from the sedi-
mented-out solution (if the volume is less than 4 cm^3, it is made up to this
value with 1 : 1 HCl), 1 cm^3 rhodamine B in 1 : 1 HCl is added, and the
solution is extracted with 7 cm^3 of a 9 : 1 mixture of benzene+acetone. If
the iron content of the solution to be analyzed is very high (Fe : Ga$>$10^4),
the organic phase is washed for 20 s with a washing liquid (4 cm^3 1 : 1
HCl+2 cm^3 TiCl$_3$+1 cm^3 rhodamine B). Depending on the concentration,
0.5–5 cm cells are used to measure the absorbance of the organic phase at
565 nm, against the 9 : 1 benzene+acetone mixture. The calibration plot is
obtained with 0.1–2 μg Ga, the operation being started from the extraction
in all cases. The sensitivity of the method is 1 ppm Ga.

In the analysis of solutions originating from the acidic digestion of
bauxites, a 2 cm^3 aliquot is taken and its volume is adjusted to 4 cm^3 so
that the acidity should be 6 M. The subsequent procedure is the same as
for rocks and bauxites. In this way, 50 μg Ga/dm^3 can be determined well.

Notes:

(a) The absorbance of the solution depends on the concentration and
quality of the reagent.

(b) Under the described conditions, the partition coefficient of the GaCl$_4^-$-
rhodamine B complex is 10.8. Washing of the organic phase therefore leads
to a Ga loss. This error can be eliminated by recording the calibration
curve in the same way. At least one point on the calibration plot should be
recorded simultaneously with analysis of the sample.

Determination of gallium content of copper alloys with chrome azurol S [14]

With this method, a few per cent of Ga can be determined in copper
alloys without preliminary separation.

Analytical procedure: 0.1 g alloy is dissolved in 5 cm^3 HF+a few drops
of conc. HNO$_3$ in a platinum dish. 12 cm^3 1 : 1 H$_2$SO$_4$ is added, and the
mixture is evaporated until fumes of SO$_3$ appear. The residue is transferred

to a 100 cm³ volumetric flask and the solution is made up to the mark with water. A 0.1–0.2 cm³ aliquot is taken in a 50 cm³ volumetric flask, and diluted to about 10 cm³ with water. 1 cm³ 0.5% gelatin, 1 cm³ 0.1 M $Na_2S_2O_3$, 1 cm³ chrome azurol S (10 g is dissolved in 500 cm³ ethanol, and after the addition of 6 cm³ 1 : 1 HNO_3 and 0.75 g urea the volume is made up to 1000 cm³) and 5 cm³ acetate buffer of pH 4.6 are added. The solution is made up to the mark with water, and 15 min later the absorbance is measured in 2 cm cells at 575 nm. A gallium-free copper alloy treated in the same way is used for comparison. The same procedure as above is followed to prepare the calibration plot for 0.4–10 μg Ga.

References

[1] Dimov, A. M., Savostin, A. P.: *Analiticheskaya khimiya galliya*. Nauka, Moscow 1968.
[2] Savostin, A. P.: *Zav. Lab.* **9**, 1045 (1964).
[3] Jankovsky, J.: *Talanta* **2**, 29 (1959).
[4] Klein, P., Skrivanek, V.: *Chem. Prumysl.* **13**, 250 (1963).
[5] Urbanek, E.: *Paliva* **41**, 88 (1961).
[6] Onishi, H.: *Anal. Chem.* **27**, 832 (1955).
[7] Culkin, F., Riley, J. P.: *Anal. Chim. Acta* **24**, 413 (1961).
[8] Parissakis, G., Issopoulos, P. B.: *Microchim. Acta* **28** (1965).
[9] Grahame, D. C., Seaborg, T. J.: *J. Amer. Chem. Soc.* **60**, 2524 (1938).
[10] Irving, H. M., Rossotti, F. J. C.: *Analyst* **77**, 801 (1952).
[11] Busev, A. I., Skrebitskova, L. M., Talipova, L.: *Zh. Anal. Khim.* **17**, 801 (1962).
[12] Thakur, R. S., Sant, B. R.: *J. Sci. Ind. Res.* **31**, 384 (1972).
[13] Mohai, M.: Unpublished work.
[14] Pyatiletova, N. M., Korneva, V. I., Ershova, N. S., Orlov, V. V., Cherkasov, A. A., Ivanov, V. M., Bushev, A. I.: *Zav. Lab.* **45**, 797 (1979).

6.17. Germanium

Germanium belongs in column IV/1 of the periodic system; it is a metalloid. In solution it is stable only in the tetravalent state. Similarly to other tetravalent ions, it is strongly hydrolyzed in solution; Ge^{4+} is virtually non-existent in aqueous solution, but the element can be present in the dissolved state in the form of the germanate anion (GeO_3^{2-} or $Ge_5O_{11}^{2-}$ depending on the circumstances). Among the important analytical properties are the volatility of its halides, the high stability of the anionic halide complexes, and the tendency to form complexes with organic acids and O-containing compounds.

The germanium-molybdenum heteropolyacid is suitable for the photometric determination of germanium [1]; other good reagents include various diphenols, e.g. dithiol, certain triphenylmethane derivatives and trihydroxyfluorones [2, 3]. Most important of all is phenylfluorone (see Appendix) [4].

Germanium(IV) reacts with phenylfluorone to give a $1:2$ polymeric complex as a colloid poorly soluble in water. The reagent itself is also poorly soluble at lower acidities (< 0.3 M); the solubility can be improved by the addition of alcohol. Even so, a protective colloid, such as poly(vinyl alcohol), gelatin or gum arabic is needed to keep the complex in solution. Complex formation occurs in a very wide range of acidity (from pH 3–3.5 M H_2SO_4). However, the reaction is slow at pH < 1. It is therefore advisable to add the reagent first at pH 4–5 (acetate medium), and only subsequently transform the medium to one more favourable from the aspect of the stability of the colour (0.5–1 M HCl) [5].

The absorbance maximum for the reagent is at 462 nm, while that for the germanium complex is at 510 nm. The very low absorbance of the reagent means that measurements can be made at the latter wavelength. Depending on the medium and the reagent concentration, ε_{510} varies in the range 4–8×10^4; this value can be increased to 1.7×10^5 by forming the mixed ligand complex also containing the cetyltrimethylammonium ion [6].

The elements primarily causing interference in the determination of germanium are the multivalent metal ions (Ga^{3+}, Fe^{3+}, Sn^{4+}, Sb^{5+}, Nb^{5+}, Ti^{4+}, W^{6+}, etc.) and higher concentrations of ions which are themselves coloured (Cu^{2+}, Cr^{3+}, Ni^{2+}). The disturbing effects can sometimes be eliminated by masking, but at times only by separation.

The most suitable possibilities for separation are distillation of germanium halides (in certain cases the evolution of GeH_4, which is gaseous even at room temperature) and extraction of anionic halide complexes.

Primarily $GeCl_4$ is separated by distillation (b.p. 86°C), usually from 6 M HCl. In this way it is accompanied by As^{3+}, Sb^{3+} and Sn^{4+}; As can be retained after oxidation with $KMnO_4$.

$GeCl_4$ can be well extracted from 9 M HCl with CCl_4 or other apolar solvents [2, 7]. From 9 M HCl, $D_{GeCl_4} \sim 500$, whereas from 8 M HCl it is only 50. Although many chloro complexes are otherwise extracted from such a solution, if CCl_4 is used then the co-extraction of As^{3+} is appreciable [5].

Determination of germanium in rocks and in coal ash with phenylfluorone
(based on [2, 8] and the authors' modifications)

Analytical procedure: 1 g of the sample to be analyzed is heated for 90 min on a water-bath with a mixture of 4 cm³ 1 : 1 H_2SO_4+2 cm³ conc. HNO_3+ +10 cm³ HF in a platinum dish, followed by careful evaporation almost to dryness on a sand-bath. The residue is dissolved in 25 cm³ hot 2% boric acid with gentle boiling, 5 cm³ conc. HCl is added to promote complete dissolution, and the mixture is immediately cooled. The solution is transferred to a 200 cm³ separating-funnel, 70 cm³ conc. HCl is added, and the germanium is extracted with 20 cm³ CCl_4. The organic phase is washed with 20 cm³ 9 M HCl, and the germanium is re-extracted with 2×9 cm³ distilled water (1 min). The re-extract, in a 25 cm³ volumetric flask, is mixed with 2 cm³ conc. HCl, a little solid $Na_2S_2O_3$, 1 cm³ 1% gelatin and 3 cm³ 0.05% phenylfluorone (0.25 g phenylfluorone is dissolved in a little hot water, the volume is made up to 500 cm³ with 96% ethanol containing 3 cm³ conc. HCl, and the solution is filtered on the next day), and the solution is made up to the mark with water. After 60 min, the absorbance is measured in 1-5 cm cells at 510 nm, against a blank prepared from the reagents. The calibration plot is recorded for 0.5-10 μg Ge, taken through the procedure from the extraction. The sensitivity of the method is 0.5 ppm Ge. If the sample contains Ge in excess of 10 ppm, a stock solution must be prepared from the re-extract and an aliquot taken for the determination.

Pretreatment of coal and coal-processing products for analysis

The germanium content of mineral coals is higher than that of rocks, and thus such coals may be examined as potential industrial raw materials.

The organic matter is removable by dry-ashing at 600°C, but to ensure the avoidance of germanium loss this step is best carried out after mixing with Na_2CO_3 or some other alkaline additive ($CaCO_3$, MgO+$MgCO_3$, etc.) [2]. This is followed by separation by distillation of $GeCl_4$. The ashing can also be carried out at 160° C in the wet, in mixtures with $K_2Cr_2O_7$, H_2SO_4 and H_3PO_4, for instance [9].

Coal tar is first digested with HNO_3 in order to avoid sputtering during ashing. The ashing is performed at 450°C in the presence of CaO + Ca(NO$_3$)$_2$ [10]. After ignition of the residue at 750°C, and then dissolution, the GeCl$_4$ is extracted and determined as described in connection with germanium analysis in rocks.

Determination of germanium in silver alloys with lumogallion [11]

For the analysis of silver alloys used in electromechanics, which contain a small quantity (0.1–1%) of germanium, lumogallion is recommended [4]. The determination can be carried out in 3–3.5 M H$_3$PO$_4$. Silver is removed by the precipitation of AgCl.

Analytical procedure: 0.1–0.2 g sample is dissolved in HNO$_3$, and the silver is precipitated as AgCl. From the filtrate, an aliquot corresponding to 10–50 µg Ge is pipetted into a 25 cm^3 volumetric flask, 10 cm^3 water, 2 cm^3 1 : 1 HCl, 5 cm^3 H$_3$PO$_4$ and 2.5 cm^3 0.02% lumogallion are added, and the volume is made up to the mark. After 30 min, the absorbance is measured at 560 nm.

References

[1] Mirzoyan, F. V., Tarayan, V. M., Hairyan, E. K. H.: *Anal. Chim. Acta* **124**, 185 (1981).
[2] Nazarenko, V. A.: *Analiticheskaya khimiya germaniya.* Nauka, Moscow 1973.
[3] Marczenko, Z.: *Spectrophotometric Determination of Elements.* Ellis–Horwood, London 1976.
[4] Lukin, A. M., Efremenko, O. A., Podolskaya, B. L.: *Zh. Anal. Khim.* **21**, 970 (1966).
[5] Luke, C. J., Campbell, M. E.: *Anal. Chem.* **28**, 1273 (1956).
[6] Shijo, Y., Takeuchi, J.: *Japan Analyst* **16**, 51 (1967).
[7] Schneider, W. A., Sandell, E. B.: *Microchim. Acta* 263 (1954).
[8] Popov, N. P., Stolyarova, I. A.: *Khimicheskii analiz gornykh porod i mineralov.* Nedra, Moscow 1974.
[9] Ura, M.: *J. Chem. Soc. Jap., Pure Chem. Sect.* **78**, 316 (1957).
[10] Nazarenko, V. A., Lebedeva, N. V., Ravitskaya, R. V.: *Zav. Lab.* **24**, 9 (1958).
[11] Budyak, N. F., Ekimenkova, T. A.: *Analiz kontaktnykh i provodnikovykh splavov.* Metallurgiya, Moscow 1975.

6.18. Indium

Indium belongs in group III/1 of the periodic system. In its analytical properties, however, it displays similarities not only to the ions Ga^{3+} and Tl^{3+} from the same column, but also to the adjacent ions Cd^{2+} and Sn^{2+}. In solution it is exclusively trivalent. Its hydroxide is amphoteric; it dissolves in concentrated alkali. It forms anionic complexes with halides, these providing the main basis for separation. For instance, it can be extracted from 5 M HBr in this way, and it can be separated from Ga^{3+} by re-extraction with 6 M HCl [1]. The photometric reagents include ligands containing O and S donor atoms; there is no selective reagent. Of its reagents, mention may be made of oxine and its derivatives, hydroxyanthraquinones (e.g. quinalizarin), dithizone, various triphenylmethane derivatives (pyrocatechol violet, chrome azurol S, xylenol orange, rhodamines, methylthymol blue), PAN, PAR, fluorone derivatives and hydroxy-azo compounds (gallein, stilbazo) [2–4].

Certain reagents, e.g. rhodamines, are also used for its determination via luminescence.

The choice of the optimum reagent depends on the composition of the sample under examination. It is reported by Röbisch [4] that, of the 7 reagents tested, rhodamine 6G proved best for the determination of indium in $ZnSO_4$ formed in the course of lithopone production.

A survey of the photometric reagents for indium is provided in the book by Busev [2].

Use of rhodamine 6G to determine the indium content of rocks, sulphidic ores and the zinc sulphate solution formed for lithopone production [4, 4a]

In H_2SO_4 and HBr medium, indium forms a complex with rhodamine 6 G; this complex can be extracted with benzene. Separation from disturbing elements is achieved by extraction with di-2-ethylhexylphosphoric acid, followed by antagonistic re-extraction in the presence of TBP. In this way, the determination is not disturbed by 10^5-fold quantities of Pb^{2+}, Cu^{2+}, Cd^{2+}, Mn^{2+}, Fe^{2+}, Co^{2+} and Zn^{2+}.

Analytical procedure:

(a) Digestion for sulphidic ores: 1–2 g sample is boiled for 40–50 minutes with 10 cm³ conc. HCl and 15 cm³ conc. HNO_3, and the mixture is then evaporated almost to dryness. 10 cm³ conc. HCl and ca. 50 cm³ water are added, and the mixture is boiled and filtered. 5 cm³ 5 M H_2SO_4 is added to the filtrate, which is then evaporated until the appearance of SO_3 fumes. The residue is washed with water into a 50 cm³ volumetric flask, and the volume is made up to the mark. The acidity of the solution should be 0.5 M H_2SO_4.

(b) Digestion for silicates: 1–2 g sample in a platinum dish is evaporated with 15 cm³ HF + 3 cm³ conc. HNO_3. The residue is washed into a beaker with 10 cm³ conc. HCl, 15 cm³ conc. HNO_3 is added, and the mixture is boiled. The subsequent procedure is as described in point (a).

10 cm³ stock solution is pipetted into a separating-flask and extracted for 1 minute with 10 cm³ 0.2 M di-2-ethylhexylphosphoric acid in petroleum ether. The organic phase is washed with 0.5 M H_2SO_4. The aqueous phase is poured off, 10 cm³ TBP is added to the organic phase, and the indium is re-extracted for 2×1 minute with 2×5 cm³ 5 M H_2SO_4. The re-extract is transferred to a 25 cm³ volumetric flask, and the volume is made up to the mark with water (the acidity of the solution is 2 M H_2SO_4).

At most 5 cm³ of this solution is taken in a separating-funnel, 4 cm³ conc. HBr, 1 cm³ freshly prepared ca. 10% $Ti_2(SO_4)_3$ (metallic Ti is dissolved in 5 M H_2SO_4) and 0.5 cm³ 0.30% rhodamine 6G are added, and the mixture is extracted with 10 cm³ benzene–acetone (9.5 : 0.5). The extract is subjected to photometry at 630 nm in 1–5 cm cells, against an extract of the reagents taken throughout the analysis processes. A calibration plot is prepared with 1–50 μg In.

The indium content of $ZnSO_4$ prepared for lithopone production can be determined in the same way, the operations being begun from the extraction separation.

The sensitivity of the method is 20 ppm In.

Note:

If the indium content of the solution to be measured by photometry is less than 1 μg, evaluation may be performed by fluorimetry. An increase in sensitivity of an order of magnitude can be attained in this way.

Determination of indium in rocks and ores with disulphophenylfluorone
(see Appendix) [5]

Analytical procedure: The sample is digested with $HF + H_2SO_4$, the residue is filtered off, and the indium in solution is precipitated at $pH \sim 5$ with $Fe(OH)_3$ carrier. After dissolution of the precipitate, this procedure is repeated with a view to further separation from interfering components. The precipitate is dissolved in 50 cm^3 1 M H_2SO_4, 8 g KI is added to the solution (~ 1 M I^-), and sufficient $Na_2S_2O_3$ solution is added to attain decolorization. The indium is extracted with 2×30 cm^3 ethyl ether, and the combined organic phase is washed with 4×10 cm^3 washing liquid (8 g KI + a little $Na_2S_2O_3$, in 50 cm^3 0.5 M H_2SO_4). The re-extraction is performed with 3×15 cm^3 water, and the solution is made up to the mark in a 50 cm^3 volumetric flask.

The analysis can also be carried out [1] by evaporating the solution to dryness after the digestion with $HF + HNO_3$. 3–5 cm^3 HBr and 1–2 drops of H_2O_2 are added to the residue, and the mixture is heated on a water-bath and again evaporated to dryness. This is repeated several times, at the end without H_2O_2. The residue is dissolved in 10–15 cm^3 5 M HBr, and 0.1 g KI and a little solid $Na_2S_2O_3$ (for complete reduction of Fe^{3+}) are added.

The indium is extracted with the same volume of butyl acetate, and the organic phase is washed with 2×5 cm^3 5 M HBr. The re-extraction is performed with 6 M HCl + a little H_2O_2.

For the determination, an aliquot containing at most 40 µg indium is taken in a 50 cm^3 volumetric flask, 0.5 cm^3 0.1% NH_4F solution, 0.5 cm^3 2% ascorbic acid, 1 cm^3 0.25% o-phenanthroline (to mask Fe^{2+}), 2 cm^3 1% gelatin, 2 cm^3 0.05% disulphophenylfluorone, and 20 cm^3 acetate buffer of pH 4.6 are added, and the solution is made up to the mark. After 45 min, the absorbance of the solution is measured at 530 nm, against a comparison solution not containing the reagent.

Determination of indium content of silver wire with PAR [7]

0.1–0.2 g sample is dissolved in HNO_3, the silver is precipitated out electrolytically, and the remaining solution is diluted to 100 cm^3. Depending on the expected indium content (0.01–1.0%), two 1.0–5.0 cm^3 aliquots are taken from this stock solution. 1 cm³ 5% complexone III solution is added to one of these aliquots, which will serve as the comparison in the determination. To both of the aliquots, 10 cm³ buffer solution of pH 3.3 (90 cm³ glacial acetic acid + 10 cm³ conc. NH₄OH, diluted to 1000 cm³) and 2 cm³ 0.002 M PAR are added, and the volume is made up to 25 cm³ with buffer solution. The absorbance is measured at 510 nm.

Determination of indium content of lead- and zinc-containing products and alloys, with PAN [8]

Principle of method: In can be extracted selectively from strongly acidic medium in the presence of KI and dimethylformamide with a solution of PAN in benzene. The indium-PAN complex in the organic phase is colourless. However, if the extract is shaken with a weakly acidic (pH 3–4) solution containing KI, it is transformed into a red complex suitable for photometry. Fe(III), Ni and Co interfere in the determination. The absorbance maximum for the complex is at 570 nm, where $\varepsilon = 1.9 \times 10^4$.

Analytical procedure: 1 g of the sample to be analyzed is heated with 20 cm³ conc. HCl for 10 min on a sand-bath, 10 cm³ conc. HNO_3 is added, and the mixture is evaporated nearly to dryness. After the addition of 20 cm³ 0.5 M H_2SO_4, the mixture is filtered, and the filtrate is made up to 100 cm³ with 0.5 M H_2SO_4. A 2–3 cm³ aliquot (2–400 μg In) is pipetted into a 25 cm³ glass-stoppered test-tube, 9 cm³ of a mixture (0.5 M H_2SO_4, 0.5 M KI and 0.15% ascorbic acid) and 1 cm³ dimethylformamide are added [together with 100 mg NaF, thiourea and tartaric acid in the presence of Sn(IV), Cu and W(VI)], and the mixture is shaken for 1–1.5 min with 10 cm³ 0.03% PAN solution in benzene. The whole is transferred to a separating-funnel and the aqueous phase is run off; the funnel is carefully washed (without shaking) with 4–5 cm³ 0.05 M H_2SO_4 (also 2.5 M in KI). The aqueous phase is carefully run off, and the organic phase is gently shaken for 5–10 s with 2–3 drops of 0.5 M NH₄OH (2.5 M in KI). 2 cm³ ammonium acetate buffer (2.5 M in KI) is added to the mixture, which is gently shaken for 10–15 s. The aqueous phase is poured off, and the organic phase is filtered. Its absorbance is measured in 2, 1 or 0.3 cm cells at 570 nm, against an extract of the reagents. The calibration plot is obtained in an analogous way; for the ranges 2–10 µg, 10–100 µg and 100–400 µg In, the absorbance is measured in 2, 1 and 0.3 cm cells, respectively.

References

[1] Popov, N. V., Stolyarova, I. A. (eds.): *Khimicheskii analiz gornykh porod i mineralov*. Nedra, Moscow 1974.
[2] Busev, A. J.: *Analiticheskaya khimiya indiya*. AN SSSR, Moscow 1958.
[3] Marczenko, Z.: *Spectrophotometric Determination of Elements*. Ellis–Horwood, London 1976.
[4] Röbisch, G.: *Z. Chem.* **13**, 64 (1973).
[4a] Mohai, M.: Fotometriás és fluorimetriás módszerek összehasonlítása az In és Ga meghatározására (Comparison of photometric and fluorimetric methods for the determination of In and Ga). Paper presented at the 8th Rare-Metal Conference. Vol. III, p. 713, Budapest, 1982.
[5] Nazarenko, V. A., Ravitskaya, R. V.: *Zav. Lab.* **31**, 1301 (1965).
[6] Levin, I. S., Nazarenko, T. G.: *Zav. Lab.* **28**, 1313 (1962).
[7] Budyak, N. F., Ekimenkova, T. A.: *Analiz kontaktnykh i provodnikovykh splavov*. Metallurgiya, Moscow 1975.
[8] Rakhmatullayev, K., Zakhirov, B. G.: *Zh. Anal. Khim.* **34**, 469 (1979).

6.19. Iron

Iron belongs in column VIII/2 of the periodic system. In aqueous solution it is to be found in the forms Fe^{2+} and Fe^{3+}. The transformation from one of these oxidation states to the other proceeds readily, and this is utilized in various ways analytically. Fe^{2+} forms stable complexes primarily with ligands containing N donor atoms. Some characteristic atomic groups in the chelate-forming ligands are selective as regards colour formation, e.g. compounds with the ferroin structure in the determination of Fe^{2+}. The phenolic hydroxy group is selective in the determination of Fe^{3+}, and a second hydroxy group, e.g. in the *ortho* position, increases the sensitivity.

The properties of some reagents for iron are to be seen in Table 6.12.

In the following we shall deal in more detail with the practical application of iron(III) thiocyanate, and the complexes of "ferroin".

TABLE 6.12.

More important photometric reagents for iron

Reagent	λ_{max} nm	$\varepsilon \times 10^{-3}$	Medium	Interferences	Notes	Refs
For Fe^{3+}						
SCN$^-$	475	7.4	0.5–1 M HNO$_3$	Many elements	Separation	[1]
SCN$^-$-extracted	500	20	0.2 M HNO$_3$	Many elements	Separation	[2, 4]
Sulphosalicylic acid	420	5.6	pH 8.5–11.5	Cu, Ni, Co, Mn	ZnO precipitation	[11]
For Fe^{2+}						
2,2'-Bipyridyl	522	8.6	pH 3–9	Many elements		[12]
1,10-Phenanthroline	512	11	pH 2–9	Many elements		[13]
Bathophenanthroline	535	20	pH 4–7	Cu	Masking with thiourea	[7, 8, 14, 15]
PPDT-DAS	570	29.3	pH 3–8	F$^-$, EDTA, CN$^-$, Co^{2+}, Cu^{2+}		[6]
BAPH	520	44	pH 6–7	Cu, Zn, Co		[5]
Pyrogallol red	560	75	pH 8.5–10	Al, Cu, Ge	Masking with NH$_2$OH, KF	[10]

(A) APPLICATION OF IRON(III) THIOCYANATE

Iron(III) forms thiocyanate complexes in a stepwise manner. Of the anions, fluoride and oxalate, for example, disturb the development of the colour; sulphate and phosphate decrease the intensity of the colour only when present in higher concentrations. As concerns the cations, high concentrations of Ag$^+$, Hg$^+$ and Cu$^+$ interfere by forming precipitates with thiocyanate, while Cu^{2+}, Bi^{3+}, Ti^{4+}, U^{6+} and Mo^{6+} give a yellowish-orange coloration with the reagent. The iron(III)-thiocyanate complex can be extracted into the organic phase, e.g. with methyl isobutyl ketone, whereby the sensitivity of the method can be increased [1–4].

Determination of a small quantity of iron in yttrium oxide via thiocyanate extraction [2]

Analytical procedure: 0.5 g of the examined yttrium oxide is dissolved in sufficient 1 M HNO_3 to give an acidity of 0.2 M after dilution to 25 cm^3. 5 cm^3 0.5 M NH_4CNS is added to the digestion solution, and the volume is made up to 25 cm^3. The resulting red complex is extracted with 10 cm^3 methyl isobutyl ketone, and the absorbance of the solution is measured in 5 cm cells at 500 nm. The calibration plot is produced under similar conditions, with 0.5–5 μg Fe^{3+}. The sensitivity of the method is 1×10^{-4} %.

(B) APPLICATION OF FERROIN-TYPE COLOUR REACTIONS

In connection with the colour-producing reagents of ferroin type, mention will be made of the disturbing effects observed when 1,10-phenanthroline is used. Three ligands each coordinate via two N atoms to one Fe^{2+}, and $[Fe(phen)_3]^{2+}$ (ferroin) is formed. Interfering effects are caused by Bi, Cd, Zn, Ag and Hg, which precipitate the reagent. The presence of oxalate, tartrate and fluoride decreases the intensity of the colour. In the presence of strong oxidants, e.g. Ce^{4+}, Mn^{3+}, MnO_4^-, $Cr_2O_7^{2-}$, the ferroin is transformed to the weakly blue $[Fe(phen)_3]^{3+}$ (ferriin), and the original colour intensity weakens. The oxidation of Fe^{2+} is promoted by higher pH, and it is therefore advisable to maintain a weakly acidic medium.

Determination of iron content of manganese ores with BAPH [5]

BAPH (see Appendix) is a suitable reagent for the extraction and photometric determination of iron-containing substances. $\varepsilon_{520} = 4.4 \times 10^4$.

Analytical procedure: 0.02 g manganese ore is dissolved in 10 cm^3 conc. HCl. The solution is evaporated to about 5 cm^3, 5 cm^3 1 : 1 H_2SO_4 is added, and the mixture is evaporated until the removal of H_2SO_4. The residue is taken up in 10 cm^3 water, filtered, washed with water, and the filtrate is made up to 100 cm^3 with water. A 1 cm^3 aliquot is taken in an extraction vessel, 1 cm^3 1% NH_2OH.HCl and 5 cm^3 acetate buffer of pH 6.2 are added, and the solution is diluted with water to 10 cm^3. It is then

215

mixed intensively for 20 min with BAPH reagent dissolved in 10 cm³ iso-amyl alcohol. The extract is centrifuged, and its absorbance is measured at 520 nm. The method is suitable for the determination of an iron content of $\sim 10^{-1}\%$. The iron content of aluminium alloys and various pure salts, such as $CaCO_3$, $(NH_4)_2C_2O_4$, etc. can be determined in an analogous way.

Determination of Fe^{2+} and total iron contents in rocks with PPDT–DAS reagent [6]

The outstanding sensitivity, stability and acid-resistance of PPDT–DAS (see Appendix) make it a suitable reagent for the photometry of Fe^{2+}. Of the cations, Cu^+ and Co^{2+} form coloured complexes with the reagent, but the intensities of these are much lower. The development of the colour is delayed by much fluoride and oxalate, and inhibited by EDTA and cyanide.

Preconditions of a good result are that the oxidation of the Fe^{2+} should be avoided in the powdering and digestion, and that the dissolution should be quantitative.

Analytical procedure: A small amount (1–5 mg) of sample is weighed into a polyethylene thimble. A few drops of 20% acetone and 1 cm³ reagent mixture (10% HF + 5% H_2SO_4 + 2% PPDT–DAS) are added, and the contents are homogenized. The thimble is covered and left to stand in the dark for 24 hours. 10 cm³ saturated boric acid is taken in a 100 cm³ volumetric flask, and the digestion solution is added. More intensive colour development is promoted by elevation of the solution pH with 10 cm³ 2 M sodium acetate buffer, and the volume is made up to the mark. The absorbance is measured in 1 cm cells at 570 nm.

Silicate rocks that are more difficult to digest are dissolved in a N_2 atmosphere in a microbomb, under pressure at 100–200° C.

For determination of the total iron, 10–30 mg sample is moistened with a few drops of water in a platinum vessel, and 1 cm³ 1 : 1 H_2SO_4 and 4 cm³ HF are added. The mixture is evaporated almost to dryness on a water-bath, and is then heated on a sand-bath up to the appearance of fumes of SO_3. The cooled residue is dissolved in 5 cm³ water, with warming, and a 100 cm³ stock solution is prepared. A 5–10 cm³ aliquot is pipetted from this into a 100 cm³ volumetric flask, and 10 cm³ 10% NH_2OH.HCl is added. After

15 min, 10 cm³ 0.005 M PPDT–DAS reagent and 10 cm³ 2 M sodium acetate buffer are added for colour development, and the volume is made up to the mark. Photometry is as above.

Determination of iron in blood serum with bathophenanthroline-disulphonic acid, without removal of protein [7, 8]

In the determination, the iron bound to the transferrin is decomplexed with Teepol reagent and reduced at the same time (Teepol reagent: 5 g sodium dithionite is dissolved in 100 cm³ 1% $MgSO_4.7 H_2O$, 100 cm³ Teepol 710 (Serva) detergent is added, and the solution is made alkaline by the addition of 50 cm³ 3% NaOH; the pH of the iron-freed reagent is adjusted to 5.4–6.2 with glacial acetic acid; the reagent is stable for 2 months at 4°C). The resulting Fe^{2+} forms a coloured complex with the iron reagent, and photometry follows (iron reagent: 1 g bathophenanthroline-disulphonic acid disodium salt is made up to 100 cm³ with distilled water; the reagent can be kept for a prolonged period).

0.5 cm³ Teepol reagent is added to 0.2 cm³ serum, the mixture is shaken for 20 s, reagent is added, and the absorbance is measured at 541 nm. In parallel with the sample, absorbance measurements are also made on the Teepol reagent (0.2 cm³ distilled water+0.5 cm³ Teepol reagent)+iron reagent, and on standard samples (0.2 cm³ standard sample+0.5 cm³ Teepol reagent)+iron reagent.

(C) APPLICATION OF OTHER REAGENTS

Determination of iron content of brass with benzimidazole-oxime derivative [9]

Principle of method: With Fe^{2+} at pH 9.3, N-methyl-2-(α-isonitroso-α-cyano)-methylbenzimidazole forms a stable complex which can be extracted with a 3 : 1 mixture of benzene+iso-amyl alcohol. Photometry is performed at 560 nm, where $\varepsilon = 1.4 \times 10^3$.

Analytical procedure: 0.5 g brass is dissolved in 5 cm³ 1 : 1 HNO_3, the solution is evaporated to remove oxides of nitrogen, and the volume is then made up to 250 cm³ in a volumetric flask. A 3–5 cm³ aliquot is taken, 2 cm³

2% ascorbic acid, 2 cm^3 5% thiourea and 2 cm^3 5×10^{-3} M reagent in acetone are added, and the pH is adjusted to 9.3 with NH$_4$OH. The coloured complex formed is extracted for 30 s with a 3 : 1 mixture of benzene+isoamyl alcohol. The extract is dried by the addition of Na$_2$SO$_4$ (anhydr.), and the absorbance is measured at 560 nm. The method is suitable for determination of $\sim 10^{-2}\%$ Fe.

Determination of iron content of natural waters with pyrogallol red [10]

Pyrogallol red forms a coloured complex with iron traces in water. In the solution containing excess reagent, the Fe(II)-pyrogallol red complex is reacted with zephyramine (tetradecyldimethylbenzylammonium chloride), and the resulting ternary complex is extracted with CHCl$_3$.

Analytical procedure: 5–20 cm^3 sample is acidified to pH 2 with H$_2$SO$_4$ and filtered, and 0.5 cm^3 0.5 M hydroxylammonium sulphate, 1 cm^3 masking solution (5.81 g KF+4.41 g sodium citrate+1.65 g potassium pyrophosphate/100 cm^3), 1 cm^3 5×10^{-4} M purified pyrogallol red, 0.5 cm^3 5×10^{-3} M zephyramine solution, 1 cm^3 NH$_4$OH+(NH$_4$)$_2$SO$_4$ buffer (pH 9.6), and 1 cm^3 5.75 M NaCl are added. The volume of the solution is made up to 25 cm^3, and the mixture is shaken for 30 min with 5 cm^3 CHCl$_3$. Following this, 0.5 cm^3 0.02 M EDTA solution is added, and shaking is continued for a further 30 min. After standing for 30 min, the absorbance of the organic phase is measured in 1 cm cells at 560 nm, against a blank solution prepared with the reagents. The calibration plot covers the range 0.2–4 µg Fe. The sensitivity of the method is 20 µg Fe/dm^3.

References

[1] Babko, A. K., Pilipenko, A. T.: *Photometric Analysis.* Mir, Moscow 1971.
[2] Szalai, J., Upor, E.: *Pécsi Műszaki Szemle* **16**, 1 (1971).
[3] Upor, E.: *Acta Chim. Acad. Sci. Hung.* **73**, 133 (1972).
[4] Luke, C. L.: *Anal. Chim. Acta* **36**, 122 (1966).
[5] Pustovar, P. Ya.: *Zh. Anal. Khim.* **34**, 511 (1979).
[6] Kiss, E.: *Anal. Chim. Acta* **72**, 127 (1974).
[7] Richterich, R.: *Klinische Chemie.* 3rd Ed. Karger, Basel 1971.
[8] Jobst, K.: Unpublished work.

[9] Bagdasarov, K. N., Chernovyants, M. C., Chernoivanova, T. M., Tsupakh, E. B., Seifulina, A. M.: *Zav. Lab.* **44**, 387 (1978).
[10] Korenaga, T., Motomizu, S., Toei, K.: *Anal. Chim. Acta* **104**, 369 (1979).
[11] Moizhes, I. B. (ed.): *Rukovodstvo po analizu v proizvodstva fosfora, fosfornoi kisloty i udobrenii.* Khimiya, Leningrad 1973.
[12] French, W. J., Adamo, S. J.: *Analyst* **97**, 828 (1972).
[13] Fadrus, H., Maly, J.: *Anal. Chim. Acta* **77**, 315 (1975).
[14] Gahler, A. R., Hamner, R. M., Shubert, R. C.: *Anal. Chem.* **33**, 1937 (1961).
[15] Clark, L. J.: *Anal. Chem.* **34**, 349 (1962).

6.20. Lanthanides and yttrium

Lanthanum belongs in group III/2 of the periodic system. It is observed that the properties of lanthanum and the following elements, with atomic numbers 58–71 (lanthanides or rare earth metals), vary only slightly with change in the atomic number. The reason for this lies in the electronic configurations of their atoms: the outermost electronic orbitals of lanthanum and the lanthanides are uniformly filled, with one electron in the $5d$ orbital and two in the $6s$ orbital, the only differences being in the degree of occupation of the $4f$ orbital. The great similarity in their physical and chemical natures stems from this. All of the lanthanides form trivalent ions; europium, samarium and ytterbium can also be divalent, while cerium, praseodymium and terbium can also be tetravalent. Yttrium, which likewise belongs in group III/2 of the periodic system, displays analytical properties similar to those of the lanthanides. A condition for the combined determination of the lanthanides and yttrium is the use of a reagent which forms complexes of similar molar absorbance with all these elements.

With the exception of arsenazo III, the reagents recommended in the literature [1] (alizarin S, pyrocatechol violet, bromopyrogallol red, arsenazo I, arsenazo III, etc.) form complexes with the lanthanides and yttrium only in nearly neutral medium. This restricts their use merely to the analysis of pure solutions, or very complicated separation procedures must be introduced. The complexes of arsenazo III have much higher molar absorbances than those of the other reagents; below, therefore, the discussion of the determination of the sum of the lanthanides and yttrium is confined to the application of arsenazo III.

(A) ANALYTICAL PROPERTIES OF ARSENAZO III COMPLEXES OF LANTHANIDES AND YTTRIUM

The arsenazo III prepared by Savvin [2] forms stable green complexes with the lanthanides at pH 2–3. The absorbance maxima lie in the range 650–655 nm.

$$\Delta\lambda = 120 \text{ nm}; \quad M:L = 1:1.$$

It may be seen from Table 6.13 that the molar absorbance values vary between 62.5×10^3 and 74×10^3, whereas the absorbances for identical quantities of the individual lanthanides are almost completely the same.

TABLE 6.13.

Molar absorbances (ε) of arsenazo III complexes of lanthanides and yttrium, and absorbances for equal weights of lanthanides and yttrium (0.003 M HCl medium, 2 cm^3 0.1% arsenazo III, 50 cm^3 volume, 5 cm layer thickness, $\lambda = 650$ nm)

Element	La	Ce	Pr	Nd	Sm	Eu	Gd	Tb
$\varepsilon \times 10^{-3}$	62.5	64.5	65.0	65.0	70.5	71.5	72.0	72.5
Absorbance of 10 μg	0.450	0.460	0.460	0.450	0.470	0.470	0.460	0.455

Element	Dy	Ho	Er	Tm	Yb	Lu	Y	
$\varepsilon \times 10^{-3}$	73.0	73.5	73.5	72.0	72.5	70.0	74.0	
Absorbance of 10 μg	0.450	0.455	0.440	0.430	0.420	0.400	0.830	

Yttrium is an exception: it has a similar molar absorbance to those of the lanthanides, but its lower atomic weight means that it has a higher absorbance than the lanthanides for an identical amount. The yttrium : lanthanide ratio is known from the Clark values for samples of average composition, and this can therefore be taken into account in the recording of the calibration plot. For informatory purposes, the average occurrences of the individual lanthanides in the Earth's crust are listed in Table 6.14 [1]. In samples with different compositions, the yttrium must be determined with another method (possibly emission spectrum analysis), and the average absorbance must be given on this basis. The molar absorbance value depends appreciably on the purity of the arsenazo III reagent. The tabulated data relate to a reagent prepared in our laboratory. The authenticity of the values must be checked for every preparation, and it is advisable to record the calibration plot in the knowledge of the concentrations of the lanthanides and the yttrium.

TABLE 6.14.

Average occurrence of lanthanides in the Earth's crust

Element	Atomic no.	ppm in the Earth's crust	Relative occurrence within the group, %
La	57	18	11.4
Ce	58	45	28.6
Pr	59	7	4.5
Nd	60	25	15.0
Pm	61	—	—
Sm	62	7	4.4
Eu	63	1.2	0.8
Gd	64	10	6.4
Tb	65	1.5	1.0
Dy	66	4.5	2.9
Ho	67	1.3	0.8
Er	68	4.0	2.5
Tm	69	0.8	0.5
Yb	70	3.0	1.9
Lu	71	1.0	0.6
Y	39	28.0	17.8

Interfering elements. Although the determination may be performed in 0.005–0.001 M acidic medium, and thus the selectivity is higher than for the other reagents, it is nevertheless necessary to reckon with the disturbing effects of many ions.

The determination is strongly disturbed by Th^{4+}, U^{6+}, Fe^{3+}, Ti^{4+}, Nb^{5+} and Sc^{3+}. Large quantities of Ca^{2+}, Al^{3+}, Fe^{2+}, ZrO^{2+}, Cu^{2+}, etc. also cause errors. Although the effects of certain interfering ions can be eliminated, e.g. Al^{3+} with sulphosalicylic acid, Cu^{2+} with thiourea, ZrO^{2+} with phenyl-arsonic acid, Fe^{3+} by reduction, etc. [3], the determination cannot be carried out without appropriate separation.

Determination of total lanthanides and yttrium in rocks, various ores (monazite, apatite, bauxite, molybdenite, tungsten ore, etc.), cast iron and misch metal [4]

Course of separation:

(a) Alkaline separation (pH>13). This dissolves up Al^{3+}, W^{6+}, Mo^{6+}, V^{5+}, CrO_4^{2+}, and the bulk of the UO_2^{2+} and silica.

(b) Oxalate precipitation in the presence of a calcium carrier. Nb^{5+}, Mg^{2+}, Fe^{3+}, TiO^{2+} and ZrO^{2+} predominantly remain in solution (pH~2).

(c) Extraction with 0.2 M di-*n*-butylphosphoric acid (HDBP) dissolved in petroleum ether, from 0.1 M HCl. The bulk of the Ca^{2+} and the Fe^{2+} remain in the aqueous phase ($D_{Ca}<0.1$; $D_{Ln+Y}>100$).

(d) Washing of the organic phase with 0.1 M HCl (separation from Ca).

(e) 8 M HCl antagonistic re-extraction with addition of TBP. Th^{4+} and Sc^{3+} remain in the organic phase. The slight amounts of TiO^{2+} and Ca^{2+} still present are re-extracted quantitatively together with the lanthanides; Their disturbing effects can be eliminated completely if photometry is performed twice, the lanthanides being masked with hexametaphosphate before the second measurement, when the absorbance due to the Ca^{2+} and TiO^{2+} is obtained.

In the event of necessity, the separations may be repeated, e.g. the oxalate precipitation in the case of a high titanium content, and the alkaline separation in the case of a high titanium content, and the alkaline separation in the case of much tungsten or molybdenum.

Analytical procedure: Depending on the expected lanthanide content, 0.1–1 g rock sample (powdered to a grain size of 100 μm) is weighed into a silver, nickel or iron dish, and fused with ca. 10 g NaOH over a gas flame. After cooling, the melt is washed into a beaker with hot water (if a small hydroxide precipitate is obtained, 50–100 mg Fe^{3+} is added as carrier), the mixture is boiled, allowed to cool, and filtered on a Whatman No. 30 filter paper, and the precipitate is washed well with hot water.

In the case of the analysis of metals, an acid mixture is used for digestion; after total dissolution the metal hydroxides are precipitated with 5 M NaOH (pH>13), the mixture is boiled and filtered, and the precipitate is washed with hot water.

The precipitate is washed back into the beaker, and sufficient conc. HCl is added for the precipitate to dissolve. 4–5 g solid oxalic acid and a few drops of H_2O_2 are added, the solution is boiled, and the pH is adjusted to 2 with 5 M NaOH (universal indicator paper). If there is not much precipitate, 4 cm^3 5% $CaCl_2$ solution can be added as a carrier. After cooling, the precipitate is filtered off, washed with 1% oxalic acid, and ignited at 700°C in a platinum or porcelain crucible. The oxides are dissolved in 1 : 1 HCl in the presence of a little H_2O_2, the solution is evaporated to 1–2 cm^3 on a sand-bath, the beaker is washed round with a little water, the solution is neutralized with 1 M NaOH in the presence of methyl orange, 2 cm^3 1 M HCl is added, and the volume is made up to 20 cm^3 with distilled water. The solution is transferred to a separating-funnel and extracted for 30 s with 20 cm^3 0.2 M HDBP in petroleum ether (preparation: see [5]). The aqueous phase is poured off, and the organic phase is washed with 3×10 cm^3 0.1 M HCl for total removal of calcium.

To return the lanthanides to the aqueous phase, 5 cm^3 TBP is added and re-extraction is performed for 1 min with 2×20 cm^3 8 M HCl. The combined HCl solutions are evaporated to 0.5–1 cm^3, this is washed with water into a 50 cm^3 volumetric flask, a little ascorbic acid is added, the solution is neutralized in the presence of phenolphthalein, 5.0 cm^3 0.1 M HCl is added, and the volume is made up to the mark with water. Depending on the expected lanthanide content, an aliquot of at most 10 cm^3 of this stock solution is pipetted into a 50 cm^3 volumetric flask, sufficient 0.1 M HCl is added to give a final acid concentration of 0.003 M, 2 cm^3 0.1% arsenazo III is added, and the volume is made up to the mark with water. The absorbance of the solution is measured in 2–5 cm cells at 650 nm, against distilled water. 2–3 drops 1% sodium hexametaphosphate solution is added to the cell, and the photometry is repeated after mixing. The difference between the two measurements is the desired absorbance value.

The calibration plot is recorded directly, with photometry on 3×10^{-3} M HCl solutions of 5–50 µg lanthanide or 2–20 µg yttrium. (If possible, the absorbance of all the lanthanide-arsenazo III complexes is measured. It is sufficient to perform this with 10 µg lanthanide. The average absorbance is calculated in accordance with the occurrence ratios of the individual lanthanides and yttrium. Alternatively, the calibration plot is produced with solutions of an artificial mixture prepared in the proportions of the Clark values.) The sensitivity of the method is 10 ppm Ln+Y. The error is ±10%.

Determination of total lanthanides+yttrium in thorium compounds and in samples containing much thorium [6]

In the presence of macro amounts of thorium, the dibutylphosphoric acid extraction and antagonistic re-extraction described in connection with rock analysis cannot be used for separation of the lanthanides from thorium. The thorium–dibutylphosphate complex is only slightly soluble in petroleum ether, and hence the organic phase contains a precipitate. The copolymer also containing the lanthanides does not dissolve on addition of TBP either, and thus the lanthanides are only partially re-extracted.

The bulk of the thorium can conveniently be removed by precipitation of Th^{4+} as iodate.

Analytical procedure: Depending on the expected lanthanide content, at most 0.1 g sample is taken and dissolved appropriately, e.g. $Th(NO_3)_4$ is soluble in dilute acid; fusion with alkali metal hydroxide or pyrosulphate is necessary with certain products. Sufficient HNO_3 and H_2SO_4 are added for their solution concentrations to be 6 M and 1 M, respectively. The thorium is precipitated in the hot with the same volume of 15% KIO_3 and, after cooling, the precipitate is filtered off. Gaseous H_2S is introduced to reduce the IO_3^-. The iodine partly vaporizes, and partly separates out in solid form; the latter is filtered off, and the solution is neutralized with NaOH, and then adjusted to be 0.1 M in HCl. The dibutylphosphoric acid extraction and the antagonistic re-extraction, with the subsequent photometric determination of the lanthanides can next be carried out as described for rocks.

Determination of total lanthanides and yttrium in solutions with high uranium and nitrate contents [7]

Principle of method: Uranium present in an excess of about 100–500-fold compared to the lanthanides is separated by extraction or ion-exchange, and the lanthanides remaining in the aqueous phase are determined with arsenazo III.

Analytical procedure for extraction separation: An aliquot of the HNO_3 solution to be examined is taken (U content 1–3 mg, lanthanide content 10–100 μg) and the volume is made up to 10 cm³. Sufficient HNO_3 is added for the acidity of the solution to be 3 M, and the uranium is extracted with

2×10 cm^3 0.2 M di-2-ethylhexylphosphoric acid in petroleum ether. 0.5 cm^3 of the aqueous phase is pipetted into a 50 cm^3 volumetric flask, the solution is neutralized with 1 M NaOH in the presence of 1 drop methyl orange, it is just acidified with 0.1 M HNO$_3$, and 2 cm^3 0.1 M HNO$_3$ is added in excess. After the addition of 1 cm^3 0.1% arsenazo III, the volume is made up to the mark with water, and the absorbance is measured at 650 nm. The same solution is used for comparison, after masking of the lanthanides with 2–3 drops 1% hexametaphosphate. The calibration plot is produced with 0.5–20 µg lanthanide mixture. Care must be taken in the recording of the calibration plot that the salt (NaNO$_3$) concentration should be the same as that resulting from neutralization of the solution under examination.

Analytical procedure for ion-exchange separation: The solution to be analyzed is diluted with 0.5 M H$_2$SO$_4$+0.5 M Na$_2$SO$_4$ solution so that the NO$_3^-$ concentration of the solution is ≤ 0.03 M, the lanthanide content is 10–100 µg/cm^3, and the U content is 1–3 mg/cm^3. During this operation the uranium is transformed into the anionic complex [UO$_2$(SO$_4$)$_3$]$^{4-}$. 10 cm^3 of the solution is passed at a rate of 30–40 cm^3/hour through a strongly basic anion-exchange resin in NO$_3^-$ form (10 cm^3 resin, column diameter 50 mm, Varion AP, Wofatit SBU, etc.). After the total quantity of solution has dripped through, the solution is made up to 100 cm^3 with water in a volumetric flask. This solution contains the lanthanides without loss, whereas the uranium content is negligible. 0.5 cm^3 of the solution is pipetted into a 50 cm^3 volumetric flask, 2 cm^3 0.1 M HNO$_3$ and 1 cm^3 0.1% arsenazo III are added, the volume is made up to the mark, and the absorbance is measured at 650 nm, against the same solution, but in which the lanthanides have been masked with hexametaphosphate. The calibration plot is recorded with 0.5–20 µg lanthanide mixture, care being taken that the sulphate content of the solutions should be the same as that of the solution analyzed.

The method is also suitable for the determination of the lanthanide content of burned-out [235]U-based heating elements [8].

Determination of cerium and yttrium subgroup in silicate rocks [9]

The principle of the method is that the total lanthanides+yttrium are separated from interfering ions by the application of extractant mixture and by variation of the acidity of the solution; the residual aqueous phase is subjected to repeated extraction to separate the cerium and yttrium subgroup elements from one another, and these are determined individually.

Analytical procedure: A mixture of 1 g sample and 10 cm³ HF+20 cm³ conc. HCl+3 cm³ 1 : 1 H_2SO_4 is evaporated to dryness in a platinum dish, the residue is dissolved in 10 cm³ conc. HCl, and the volume is made up to 25 cm³. 5 cm³ is pipetted into a separating-funnel, the iron(III) is reduced with 50 mg ascorbic acid, and extraction is performed for 3 min with 2 cm³ 0.1 M di-2-ethylhexylphosphoric acid+0.04 M mono-2-ethylhexylphosphoric acid+0.1 M tributylphosphate in petroleum ether (extractant I) (Th, Zr, U, Ti and Sc pass into the organic phase).

4 cm³ 10% NH_4Cl is added to the aqueous phase, the pH is adjusted to 1, and the yttrium subgroup (Y, Gd, Tb, Dy, Ho, Er, Th, Yb, Lu) is extracted with 2 cm³ extractant I. (The organic phase is washed three times with 0.1 M HCl.) The organic phase is transferred to a 25 cm³ volumetric flask, 4 cm³ 0.25% arsenazo III dissolved in a 0.1 M trioctylamine solution in benzene and 10 cm³ ethanol are added to the organic phase, which is then well mixed. The absorbance is measured at 650 nm. From the aqueous phase remaining after the extraction, the elements of the cerium subgroup are extracted with 2 cm³ 0.1 M di-2-EHPA+0.1 M TBP in petroleum ether (extractant II), and the lanthanides are determined with arsenazo III in TOA as above. The results are calculated from the calibration plot recorded for a lanthanide mixture as discussed in the introduction.

(B) DETERMINATION OF INDIVIDUAL LANTHANIDES IN MONAZITE, IN LANTHANIDE CONCENTRATES PRODUCED FROM MONAZITE AND APATITE, AND IN MIXED LANTHANIDE OXIDES, ON THE BASIS OF THE ABSORBANCE OF THE AQUO IONS [10]

Most of the lanthanides have coloured ions, their absorption spectra being composed of narrow bands. The sequence of the colours is characteristic. The sequence of colours for the first seven ions if repeated in the reverse order for the second seven ions: La^{3+} and Lu^{3+}, Ce^{3+} and Yb^{3+}, and Gd^{3+} are colourless; Pr^{3+} and Tm^{3+} are greenish; Nd^{3+} and Er^{3+} are red; Ho^{3+} is brownish-yellow; Sm^{3+} and Dy^{3+} are pale-yellow; and Eu^{3+} and Tb^{3+} are pale-pink. Thus, on proceeding from either end of the lanthanide series towards gadolinium, the absorption bands shift towards the ultraviolet. Since the spectra of the lanthanide ions are very characteristic,

226

there is a possibility for their individual determination on the basis of the absorbances of the aquo ions [1, 11].

The values of ε for the individual lanthanides lie in the range 0.1–10. These not too large values determine the field of application. The most favourable sensitivity attainable is of the order of 10^{-2} %.

The wavelength interval suitable for measurement of the individual lanthanides is 200–1400 nm. La^{3+}, Lu^{3+} and Y^{3+} do not absorb in this interval, while the spectra of Ce^{3+}, Gd^{3+} and Tb^{3+} lie in the far UV and their sensitivities are very low; the determination of these ions by this means is therefore of little practical importance.

Pr^{3+} and Nd^{3+} display the most sensitive lines ($\varepsilon_{max} = 7$–10); the other ions absorb much more weakly. For certain samples, however, this method is very useful. It has the advantages that it is fast, and that it is nearly always possible to find undisturbed absorption lines for the individual lanthanide ions.

The positions of the maxima are changed somewhat by the anions. Because of the high absorbance of NO_3^- in the UV range in HNO_3 medium, the method is not suitable for measurements at low wavelengths. It is best to use 1 M HCl or $HClO_4$ medium for analysis. The free acid contents of the sample solution to be analyzed and the standard lanthanide stock solutions must be the same. Exact maintenance of the measurement wavelengths is very important, because of the steepness of the absorption spectrum.

The solution subjected to photometry cannot contain any coloured ions which absorb in the appropriate parts of the spectrum (Co^{2+}, Cr^{3+}, Ni^{2+}, Cu^{2+}, Fe^{3+}, etc.).

Analytical procedure: For the examination of lanthanide oxides and lanthanide concentrates obtained from the processing of monazite or apatite, a solution with a concentration of at most 1 M can be prepared by boiling with HCl or $HClO_4$. If necessary, H_2O_2 too is added to promote dissolution. In this case the excess H_2O_2 must be removed by boiling. This is often not completely successful, and that UV portion of the spectrum where H_2O_2 absorbs can then not be evaluated. The acidity of the prepared solution should be 1 M.

A sample of at most 2 g can be taken for the analysis of monazite. This is digested with NaOH in a nickel or iron crucible. (In this case the cerium is not digested completely, but the determination of cerium is not the aim here. If the total digestion of the lanthanides is to be ensured, it is necessary to

227

apply the customary boiling with $H_2SO_4 + H_2O_2$, with evaporation, and subsequent precipitation of the lanthanides with alkali.) The melt is leached out with hot water, the solution is boiled, cooled and filtered, and the precipitate is washed back (80–100 cm³) into the beaker. A little conc. HCl is added and the precipitate is dissolved by boiling, a few cm³ H_2O_2 is added to keep titanium in solution, and the lanthanides are precipitated with 4–5 g solid oxalic acid. The acidity of the solution may be 0.1–0.2 M. After cooling, the precipitate is filtered off. It is advisable to repeat the oxalate precipitation twice more. The precipitate is ignited at 700°C, and the residue is dissolved by boiling with HCl (or $HClO_4$) + H_2O_2. The H_2O_2 excess is removed by boiling. If the solution is yellow from Fe^{3+}, a little ascorbic acid is added and the volume is made up to 15 cm³ (the acidity should be 1 M). The solution is filtered to make it optically pure, and the absorbance is measured in 1–5 cm cells at the appropriate wavelengths. Primarily the wavelengths listed in Table 6.15 are recommended for analytical use.

Evaluation: Spectrally-pure lanthanide oxides are used to prepare solutions with concentrations of 1–10 mg/cm³ in 1 M HCl or $HClO_4$. The concentration of the stock solution must be checked by titration or gravimetrically. The spectra of the solutions are obtained in 1–5 cm cells with a recording spectrophotometer.

The concentrations of unknown lanthanide solutions are established with the aid of calibration plots. The absorbance of the unknown solution is measured at the tabulated wavelength, and compared with the values for standard solutions. If the values obtained at the different wavelengths do not agree, an interfering effect is present. In such cases the available spectra of the pure lanthanide solutions are compared with the spectrum of the unknown sample, and it is established which ion is causing the disturbance. The result may be corrected on this basis.

The accuracy of the determination depends strongly on the nature of the sample being examined. The determination is most inaccurate if the sample contains all the coloured lanthanides in identical concentrations (the error can attain ±15–20%), or if the concentration of the desired lanthanide is low compared to those of other lanthanides in the sample that possess disturbing peaks. The error in the determination may then be very high, and it may even not be possible to carry out the analysis.

If the composition is relatively simple and the concentration ratios are

TABLE 6.15.

Absorbance data for solutions of lanthanides

Ion	λ, nm	ε_λ	Absorbance (5 mg Ln^{3+}/cm^3, $l=5$ cm)	% sensitivity of determination for 2 g sample/15 cm^3
Pr^{3+}	444	10.5	1.9	0.1
	467	4.64	0.830	0.2
	482	3.99	0.700	0.3
Nd^{3+}	510	1.9	0.300	0.6
	521	4.51	0.780	0.2
	575	7.11	1.250	0.1
	734	6.76	1.180	0.2
Sm^{3+}	402	3.49	0.575	0.3
	1085	1.8	0.296	0.5
Eu^{3+}	393	2.6	0.435	0.4
Dy^{3+}	364	2.3	0.350	0.5
	905	2.3	0.360	0.5
Ho^{3+}	412	2.4	0.350	0.5
	450	3.8	0.570	0.3
	453	2.0	0.300	0.6
	537	4.3	0.650	0.3
Er^{3+}	487	1.9	0.300	0.6
	523	3.0	0.500	0.4
	653	1.7	0.250	0.7
Tm^{3+}	683	2.5	0.375	0.5
	694	1.0	0.250	1.2
	774	1.1	0.265	1.1
Yb^{3+}	942	0.76	0.110	1.6
	973	1.8	0.260	0.7

favourable, the accuracy of the method may attain $\pm 2\%$. The accuracy of analysis is increased if the solution of an artificial mixture with a composition similar to that of the sample is used as the basis of the evaluation. The sensitivity of the method may be established from Table 6.15.

The determination may also be performed in the organic phase after extraction.

Extraction can be carried out from 0.5 M HNO_3 medium with tributyl phosphate in the presence of the salting-out agent $Al(NO_3)_3$ [12], or from an acetate-buffered medium of pH 6.5 with thenoyl-trifluoroacetone in benzene [13]. In these cases a slight shift in the wavelengths of the maxima may be observed. The sensitivity of certain lanthanide determinations can be increased several-fold by means of extraction.

The sensitivity of the method based on the absorbance of the lanthanide aquo ions can be improved by a factor of about 20 if the transmission of the solution is measured instead of the absorbance, with an extended scale at $T(80-100\%)$ [14]. A Specord UV–VIS recording spectrophotometer, or one of similar type, is excellently applicable for recording spectra. On this principle, a method has been elaborated for determination of the Pr, Nd and Sm contents of bauxites [15]. The digestion and separation are carried out as described in connection with the determination of the total lanthanides [14]. Since the concentrations of Pr, Nd and Sm in bauxite are only 50–150 ppm, a larger sample weight (4–5 g) must be taken. The re-extract obtained after the separations is evaporated to 10 cm^3, and the transmission of the solution is measured at a setting of $T(80-100\%)$. In this way, 20 ppm Pr, 20 ppm Nd and 40 ppm Sm may be determined. The method is also excellently suitable for the determination of other lanthanides in high-purity La_2O_3, Y_2O_3 and Gd_2O_3.

(C) POSSIBILITIES FOR DETERMINATION OF CERIUM

The analysis of cerium has been reviewed by Ryabchikov and Ryabukhin [1], Eremin et al. [16] and Kharlamova et al. [17].

The use of the following reagents is based on the oxidizing effect of Ce^{4+} and measurement of the absorbance of the dye formed: ferroin, methyl orange, o-tolidine, methylene blue. However, these determinations are disturbed by virtually all ions of variable valency, and they are sensitive to traces of oxidants and reductants.

Reagents forming complexes with Ce^{3+} are not selective either. These include bromopyrogallol, tiron, thenoyl-trifluoroacetone, arsenazo III and other organic reagents recommended for the determination of the lanthanides. The separation of cerium from the other lanthanides is not easy. If no other lanthanide is present in the sample (e.g. cerium-iron alloy), arsenazo III is convenient for cerium determination.

However, there is a possibility for the determination of cerium in the presence of the other lanthanides via measurement of the colour of Ce^{4+} in alkaline medium. The solution is yellow: $\lambda_{max} = 300-320$ nm. Ce^{3+} can be oxidized with H_2O_2 in alkali metal carbonate medium. Nevertheless, this procedure is a suitable one only at relatively low concentrations of the lanthanides, for their peroxocarbonates are poorly soluble and this may give rise to turbidity. Optical interference may similarly be caused by O_2 liberated from the H_2O_2 (bubble formation).

Ce³⁺ may also be oxidized by oxygen in the atmosphere, so that these disturbing effects can be avoided. It may be observed in Fig. 6.1 that the absorption spectra of the products formed in the two types of oxidation are not the same; it is obvious that in the former case H_2O_2 also participates in complex formation.

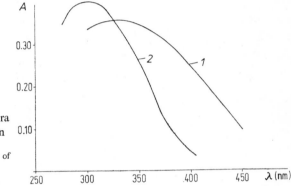

Fig. 6.1. Absorption spectra of Ce⁴⁺ in K_2CO_3 solution

1—oxidation in the presence of H_2O_2; 2—oxidation by atmospheric oxygen

Photometric determination in a carbonate medium of cerium in rocks and various ores (uranium ore, apatite) [18]

In the direct photometric determination of cerium, use is made of the process of aerial oxidation in alkaline medium: Ce³⁺ → Ce⁴⁺.

Analytical procedure: The course of the analysis is the same as that in the determination of the total lanthanides with arsenazo III, right up to the extraction with HDBP.

The cerium is re-extracted, together with the other lanthanides, with 2×10 cm³ 0.5 M K_2CO_3. The re-extract is filtered optically pure through a G4 glass filter, and the volume is made up to the mark with 3.0 M K_2CO_3 solution in a 25 cm³ volumetric flask. After 40 min (for oxidation of Ce³⁺), the absorbance of the pale-yellow solution is measured in 1–5 cm quartz

231

cells at 305 nm. If it is not wished to exchange the light source for the measurement, a bulb with a tungsten filament can be used for measurement at 320 nm; however, the sensitivity is then about 20% lower.

The calibration plot is obtained with 5–250 μg Ce, the operation being started at the extraction (1 and 5 cm cells). In a volume of 25 cm³, the absorbance of 100 μg Ce in a 1 cm cell is 0.22.

For a 1 g sample, the sensitivity of the method is 10 ppm; above 20 ppm, the error in the method is not more than ±7%.

The procedure in the determination of the total lanthanides can also be as described above. In the course of the analysis, therefore, 8 M HCl is used for re-extraction after the addition of TBP. This is of importance if both the total lanthanides and cerium are to be determined in the same analytical sample. In this case, the re-extract is evaporated to 0.5–1.0 cm³ on a sand-bath, and the residue is made up to 25 cm³ in a volumetric flask. An appropriate aliquot is taken, the K_2CO_3 concentration is adjusted to 1.0 M, and the absorbance is measured.

Determination of cerium content of high-purity lanthanide oxide and yttrium oxide (according to [19, 20] and the authors' investigations)

The principle of the method is that the lanthanides are kept in solution with a complexone (EDTA, nitriloacetic acid) in mildly alkaline medium (pH 9–10), and the absorbance of the complex formed in the presence of H_2O_2 is measured. Since H_2O_2 also absorbs at the absorption maximum of the complex, photometry is performed at 360 nm.

Analytical procedure: 2 g of the sample to be analyzed is dissolved in 1:1 HCl, and the solution is evaporated on a sand-bath until salt begins to separate out, and subsequently to dryness on a water-bath. The beaker is washed round with a little water, 70 cm³ 10% EDTA solution and 24 cm³ 1:1 NH_4OH are added, and the volume is made up to 100 cm³ in a volumetric flask. Three 25 cm³ aliquots are taken from this stock solution and placed in small beakers. 0.2 cm³ 10% H_2O_2 is added to two of these aliquots, followed by mixing. After 40 min, the absorbance is measured in 5 cm cells at 360 nm, against the solution not containing H_2O_2. The absorbance of 100 μg Ce is 0.23.

The sensitivity of the method is 40 ppm, and the determination is therefore suitable for the examination of lanthanide compounds with purities of 99.99%.

D) POSSIBILITIES FOR DETERMINATION OF EUROPIUM

As with the other lanthanides, there is no selective photometric reagent for europium. If it is separated from the other lanthanides, it can be determined with any dye suitable for the photometry of the lanthanides. The well-tried arsenazo III is the obvious choice for this purpose.

Eu^{3+} can be reduced to Eu^{2+} with sodium amalgam in citrate-containing alkaline solution, and passes into the amalgam by Na–Eu exchange [21]. For the determination of $1 \times 10^{-3}\%$ europium in various lanthanide compounds, the amalgam extraction of europium ensures a separation factor of the order of 10^8 from all the other lanthanides, including samarium, which is partially extracted with it under other conditions [22]. The europium can be dissolved up by decomposition of the amalgam with HCl; the slight amount of mercury also entering the aqueous phase does not disturb the photometry of europium with arsenazo III. If the separations described for the determination of the total lanthanide and yttrium content of rocks are applied, the solution obtained after extraction with HDBP and re-extraction with HCl contains practically only lanthanides and yttrium, and is suitable for the selective, amalgam extraction of europium.

Determination of europium in monazite, in apatite and in various lanthanide-containing industrial products [22]

Analytical procedure: At most 1 g of the sample to be analyzed is weighed out, and the procedure described for the determination of the lanthanide content of rocks is followed up to the re-extraction with HCl. The resulting solution is evaporated down to 0.5–1 cm^3. Depending on the expected europium content, either the total amount is taken, or a stock solution is prepared and an aliquot of this is examined.

At most 5 cm^3 solution is transferred to a separating-funnel, 5 cm^3 1 M sodium citrate and 1 drop of phenolphthalein are added, the solution is neutralized with 10% KOH, and if necessary the volume is made up to 10 cm³. The europium is extracted for 1 min with 2×1.5 cm³ 0.25% sodium amalgam, and the amalgam phase is transferred to another separating-funnel. The water drops on the combined amalgam phase are soaked into filter paper, the amalgam is washed with 2×10 cm³ 0.5 M citrate solution, poured into a small beaker, and again dried with filter paper. The europium is liberated from the amalgam by heating for a short time with 2.5 cm³ 2 M HCl. The aqueous phase containing the europium is washed into a 50 cm³ volumetric flask, the solution is neutralized with 1 M NaOH (phenol-phthalein indicator) and just re-acidified with 0.1 M HCl. 1.5 cm³ 0.1 M HCl is added in excess and, after the addition of 2.0 cm³ 0.1% arsenazo III, the volume is made up to the mark. The absorbance of the solution is measured in 2 or 5 cm cells at 650 nm, against distilled water. The photometry is repeated after the addition of 2 drops 1% hexametaphosphate. The difference between the two measurements gives the absorbance of the europium.

Evaluation of the analytical result: since the "active value" of the sodium amalgam can deteriorate during standing, a loss of even 5–10% of the europium can occur if the preparation used for the extraction is not fresh. The error is eliminated by means of an external standard method. Simultaneously with the sample, 10 µg europium too is extracted with the amalgam under the described conditions in every case. The absorbance of this standard is determined, and is used as the basis of the evaluation. The sensitivity of the method is 1 ppm, and its accuracy above 10 ppm is ±10%.

References

[1] Ryabchikov, D. J., Ryabukhin, V. A.: *Analiticheskaya khimiya elementov. Red-kozemelnye elementy i ittriya.* Nauka, Moscow 1966.
[2] Savvin, S. B.: *Arsenazo III.* Atomizdat, Moscow, 1966.
[3] Budanova, L. M., Pineeva, S. M.: *Zh. Anal. Khim.* **20,** 320 (1965).
[4] Mohai, M., Upor, E.: *Magy. Kém. Folyóirat* **74,** 270 (1968).
[5] Upor, E.: *Magy. Kém. Folyóirat* **73,** 479 (1967).
[6] Mohai, M., Klesch, K., Upor, E.: *Magy. Kém. Folyóirat* **79,** 79 (1973).
[7] Mohai, M., Csővári, M.: Paper presented at the 16th Transdanubian Analytical Conference, Győr, Hungary, 1981.
[8] Mohai, M., Csővári, M., Akopov, G. A., Malyshev, N. A.: „Kiégett fűtőelemek fel-dolgozása, valamint radioaktív anyagokat tartalmazó hulladékok megsemmisítése témakörében végzett kutatások". (Research into the processing of burned-out

heating elements, and the destruction of radioactive wastes.) Comecon Symposium, Marianske Lazne, Czechoslovakia, 5–7 April, 1981.

[9] Bikhovtsova, T. T., Cherkovnitskaya, I. A.: *Zh. Anal. Khim.* **35**, 1925 (1980).
[10] Mohai, M.: Unpublished work.
[11] Poluektov, N. S., Kononenko, L. I.: *Spektrofotometricheskie metody opredeleniya individualnykh elementov.* Naukova Dumka, Kiev 1968.
[12] Mishchenko, V. T., Lauer, R. S., Efryushina, N. P., Poluektov, N. S.: *Ukr. Khim, Zh.* **11**, 1188 (1965).
[13] Mishchenko, V. T., Lauer, R. S., Efryushina, N. P., Poluektov, N. S.: *Zh. Anal. Khim.* **20**, 1073 (1965).
[14] Poluektov, N. S., Mishchenko, V. T., Kovtun, V. D.: *Zav. Lab.* **42**, 1299 (1976).
[15] Mohai, M.: Paper presented at the 14th Transdanubian Analytical Conference, Pécs, Hungary, 1978.
[16] Eremin, Yu. G., Lavrova, L. A., Raevskaya, V. V., Romanov, P. N.: *Zav. Lab.* **30**, 1427 (1964).
[17] Kharlamova, L. I., Borcheva, T. A., Solomatin, V. T.: *Zav. Lab.* **40**, 1169 (1974).
[18] Mohai, M., Upor, E.: *Magy. Kém. Folyóirat* **73**, 489 (1967).
[19] Ganopolskii, V. I., Krivonozhnikova, L. G.: *Izvestiya vysshikh uchebnykh Zav. Khim. i Khim. Techn.* **6**, 913 (1963).
[20] Ganopolskii, V. I., Krivonozhnikova, L. G.: *Izv. AN BSSR. Ser. Khim. Nauki* **4**, 134 (1965).
[21] Poluektov, N. S., Nikonova, M. P.: *Redkozemelnye elementy.* AN USSR, Kiev 1959.
[22] Mohai, M., Upor, E., Nagy, Gy.: *Acta Chim. Acad. Sci. Hung.* **68**, 1 (1971).

6.21. Lead

Lead belongs in column IV/1 of the periodic system. It may be divalent or tetravalent in its compounds, but in aqueous solution it exists only as Pb^{2+}. Its hydroxide is amphoteric; at high alkalinity it dissolves as $Pb(OH)_4^{2-}$.

Its halides and sulphate are precipitated, but at high acidity and at high anion concentration they similarly form soluble complexes. The low solubilities of its sulphide, carbonate and oxalate may be utilized for its separation. It forms complexes of high stability with ligands containing O and S donor atoms; for example, acids (tartaric acid) are suitable for keeping lead in solution, while dithizone and dithiocarbamates are of use for extraction separation.

The most important photometric reagents for lead are given in Table 6.16.

TABLE 6.16.

More important photometric reagents for lead

Reagent	λ_{max} nm	$\varepsilon \times 10^{-3}$	pH of medium	Interferences	Notes	Refs
Dithizone	520	69	6.5–10.5	Many elements		[2, 4]
Sodium diethyl-dithiocarba-mate	440	12	10.5–12.5	Bi, Cd, Tl	Colour developed with Cu^{2+}	[5, 6]
Sulphoarsazene	500	45	8–10	Zn, Cu, Cd, Ni, Co, Mn, Hg, Ag	Extraction separation	[3, 7]

The importance of the determination of small quantities of lead is particularly emphasized by its poisonous nature and by the environmental pollution caused by lead from engine fuels.

(A) DETERMINATIONS WITH DITHIZONE

Since dithizone is not a selective reagent for lead, preliminary separation and masking are necessary to eliminate disturbances. The means of separation and masking are selected in accordance with the composition of the sample.

Determination of lead traces in silicate rocks (procedure of the authors)

A mixture of $HCl + HNO_3$ is suitable for the dissolution of sulphide minerals, but in rock analysis digestion with HF must be performed to ensure destruction of the silicate lattice; without this, an appreciable negative error may arise.

Analytical procedure: For sulphide ores, 1 g sample is boiled with 5 cm³ conc. $HCl + 15$ cm³ conc. HNO_3, and the solution is evaporated to dryness on a sand-bath. The residue is dissolved in 1:1 HCl, and the solution is filtered and made up to 50 cm³ with 1:1 HCl.

In the analysis of silicate rocks, 1 g sample is dissolved in 10 cm³ $HF + 1$ cm³ conc. HNO_3, and the solution is evaporated to dryness. Evaporation to dryness is repeated with 1 cm³ conc. HCl, and then with 1 cm³ conc. HCl+1 cm³ saturated boric acid. The residue is dissolved in 1:1 HCl, and the solution is filtered and made up to 50 cm³ with 1:1 HCl.

An appropriate aliquot, corresponding to at most 30 μg Pb, is pipetted from the stock solution into a separating-funnel, and if necessary the volume is made up to 10 cm³ with 1:1 HCl. Fe(III) is extracted with 2×10 cm³ methyl isobutyl ketone. After extraction, the residual extractant traces are removed from the aqueous phase by boiling. 2–3 drops 1% thymol blue indicator (in alcohol) is added to the cooled solution, and the mixture is neutralized with NH_4OH. 10 cm³ masking mixture is added to the solution to reduce Fe and Mn, and to eliminate the disturbing effects of traces of Cu, Ni, Zn, etc. (Masking mixture: 60 g sodium citrate and 8 g $NH_2OH.HCl$ are dissolved in 500 cm³ water, the pH is adjusted to 9.5 with NH_4OH, the solution is purified with dithizone, and 1 g KCN/100 cm³ is added before use.) The lead is extracted with 0.01% and then 0.001% dithizone solution portions in CCl_4 or $CHCl_3$. The volume of the organic phase is made up to 10 cm³, and its absorbance is measured at 520 nm. The calibration plot covers the range 1–30 μg Pb^{2+}/10 cm³, with measurements in 5–1 cm cells. The sensitivity of the method is 5 ppm Pb.

Determination of lead traces in the air [1]

An automatic air-sampler is used for sampling. The lead content of the air is absorbed on a membrane filter or in HNO_3 solution, the dithizone complex of lead is then formed, and this is extracted with $CHCl_3$. The air is passed through the filter for 1–8 hours at a rate of 1–15 dm³/min. In the other technique, the air is bubbled through two vessels each containing 10 cm³ 1:10 HNO_3, for 1–8 hours at a flow rate of 1–3 dm³/min.

Analytical procedure: In the case of the filtered sample, the filter paper is heated for 20 min with 10 cm³ 1:10 HNO_3 (if the paper does not dissolve, the solution is separated off by decantation). The solution is evaporated to dryness. The HNO_3 absorption solution is similarly evaporated to dryness. The evaporation residue is dissolved in 20 cm³ 1:100 HNO_3, and transferred to a separating-funnel with 1 cm³ 20% $NH_2OH.HCl$, 20 cm³ ammonium citrate (50 g citric acid is neutralized with conc. NH_4OH and made up to 1 dm³; it is purified with dithizone solution) and 5 cm³ 5% KCN

237

(purified with dithizone). The aqueous solution is shaken, for 1 min on each occasion, with 5×5 cm^3 0.01% purified dithizone in CHCl$_3$. The combined CHCl$_3$ phase is shaken intensively for 2 min with 30 cm^3 1:100 HNO$_3$, and after the separation the CHCl$_3$ phase is rejected. The HNO$_3$ solution is shaken with 5 cm^3 CHCl$_3$, 1 cm^3 conc. NH$_4$OH is added to the aqueous HNO$_3$ solution, and the lead is extracted with 5×5 cm^3 dithizone solution. The absorbance of the combined CHCl$_3$ phase is measured at 520 nm. The calibration plot is produced via the described procedure, with 0–10 µg Pb.

Determination of macro amounts of lead in plant materials [2]

The lead traces entering the atmosphere in the form of industrial fumes and in the exhaust gases from motor vehicles can be bound by plants. Lead serves as an indicator element to establish the degree of pollution.

Analytical procedure: For dissolution of the lead, 5 cm^3 water and 10–15 drops 0.005 M H$_2$SO$_4$ are added to 1–2 g dried sample, and it is again dried at 80°C. The material is next evaporated several times with 5 cm^3 portions of conc. HNO$_3$, until the residue is colourless. The residue is then dissolved in 2–3 cm^3 water and 2–5 cm^3 20% NH$_2$OH.HCl solution. The solution is heated until salts begin to separate out and, after the addition of 5 cm^3 citrate solution. it is filtered and the residue washed. (Citrate solution: 200 g citric acid is dissolved in 200 cm^3 water, the pH is adjusted to 9.5, with 5 M Na OH, dithizone purification is performed, and the volume is made up to 1 dm^3.) The resulting solution is neutralized to pH 9.5 with NH$_4$OH in the presence of 3 drops thymol blue indicator in a separating-funnel. After the addition of 20 cm^3 KCN-citrate–NH$_3$ solution (100 cm^3 citrate solution+200 cm^3 10% KCN, pH is adjusted to 9.5 with NH$_4$OH) the lead content is extracted with small portions of 0.02% dithizone solution in CCl$_4$. The combined organic phase is washed with 5 cm^3 KCN–NH$_4$Cl solution (1% KCN and 0.1% NH$_4$Cl) to remove the dithizone excess. The absorbance is measured at 520 mn.

(B) DETERMINATION WITH SULPHOARSAZENE

Determination of lead in fresh water and in mineral water [3]

Sulphoarsazene (see Appendix) can be used to advantage both as a photometric reagent and as a complexometric indicator. The metal ion: : reagent ratio in the complex is 1:1; Beer's law is obeyed up to the limit 90 μg $Pb^{2+}/100$ cm^3. Under the same conditions, coloured compounds are similarly formed with Zn, Cu, Cd, Ni, Co, Mn, Hg and Ag.

The procedure with this reagent can be employed under field conditions too. The first step in the analysis is extraction with dithizone; this is followed by re-extraction with HCl, development of the colour, and colorimetry with a comparative series.

Analytical procedure: 100 cm^3 of the water to be examined is taken in a separating-funnel. 1 cm^3 20% $NH_2OH.HCl$ and 1 cm^3 33% sodium potassium tartrate (both purified with dithizone) are added, the solution is neutralized with NH_4OH in the presence of 2–3 drops 0.1% phenol red indicator, and 2 drops NH_4OH are added in excess. Extraction is performed with 1–2 cm^3 portions of 0.01% dithizone in CCl_4 until the final portion remains green. In addition to the lead, the Zn, Cu, Mn, Ni and Co traces also enter the organic phase. The lead is re-extracted from the dithizone phase by intensive shaking with 3×2 cm^3 0.05 M HCl, for 2 min on each occasion. The aqueous phase is left to stand for 5–10 min, and the CCl_4 traces too are removed. 0.2 cm^3 1% freshly-prepared potassium hexacyanoferrate(II), 4.5 cm^3 0.05 M sodium tetraborate and 0.5 cm^3 0.05% sulphoarsazene are added, the mixture is left to stand for 30 min, and the colour is compared with those of a series containing 0, 0.5, 1.0, 2.0 and 3.0 μg Pb^{2+}, prepared under similar conditions. The sensitivity of the determination is 1–1.5 μg Pb^{2+}/dm^3.

References

[1] Várkonyi, T., Cziczó, T.: *A levegőminőség vizsgálata (Air-quality investigations).* Műszaki Könyvkiadó, Budapest 1980.
[2] Kerin, Z.: *Mikrochim. Acta* 927 (1968).
[3] Markova, A. J.: *Zh. Anal. Khim.* **117,** 952 (1962).

[4] Iwantscheff, G.: *Das Dithizon und seine Anwendung in der Mikro- und Spurenanalyse.* 2nd Ed. Chemie, Weinheim 1972.

[5] Keil, R.: *Z. Anal. Chem.* **229**, 117 (1967).

[6] Fries, J., Getrost, H.: *Organische Reagenzien für die Spurenanalyse.* Merck, Darmstadt 1975.

[7] Petrova, G. S., Yagodnitsin, M. A., Lukin, A. M.: *Zav. Lab.* **36**, 776 (1970).

6.22. Magnesium

Magnesium belongs in column II/1 of the periodic system, together with the alkaline earth metals. It occurs exclusively in the divalent form. Its hydroxide separates out at $pH > 9$. Its complex-forming tendency is slight, but it does form complexes of low stability with oxygen-containing ligands, tartrate, citrate and complexone III.

The photometric reagents for magnesium can be divided into two main groups: (1) those reacting with magnesium to give coloured complexes that are well soluble in water (chromotrope 2R, eriochrome black T, chlorophosphonazo III, magneson IREA); and (2) those which form coloured adsorption compounds with $Mg(OH)_2$ (titan yellow, phenazo).

The reagents in the first group are very sensitive and the results are well reproducible, but the disadvantage is the low selectivity. A certain degree of help here is offered by extraction of the complexes. The reagents forming adsorption compounds are more selective, but the reproducibility depends strongly on the quality of the reagent, the type and concentration of the protective colloid, the temperature, etc. Accordingly, the literature data in this respect are fairly contradictory.

The analysis of magnesium is dealt with in the comprehensive work by Tikhonov [1] and in an article on the photometric reagents [2].

The more important photometric reagents and their properties are listed in Table 6.17. The molar absorbance values vary between wide limits in the different publications, depending on the medium applied.

As the Table reveals, the photometric determination of magnesium is disutrbed by many elements; an appropriate separation or masking must be included, depending on the nature of the sample to be analyzed.

Precipitation with NH_4OH ($pH < 8$) is suitable for separation of interfering substances, as is precipitation with alkali metal hydroxide, when the $Mg(OH)_2$ is present in the precipitate (possibly with a carrier). Other

TABLE 6.17.
More important photometric reagents for magnesium

Reagent	Reagent λ_{max} nm	Complex λ_{max} nm	$\varepsilon \times 10^{-3}$	Medium	Interferences	Notes	Refs
Chromotrope 2R	507	570	37	pH 10.5-11	Many elements	Many elements can be masked with triethanolamine and CN^-	[1]
Eriochrome black T	610-620	550-560	22	pH 10-10.2	Many elements	Heavy metal traces can be masked with KCN	[1]
Chlorophosphonazo III	570	669	48	pH 7	Many elements		[1]
Magneson IREA	573	520	15.7	pH 9.8-11.2	Many elements	Al, Fe, Zn masked with triethanol-amine; Mn, Cu with CN^-; Ca effect avoided by known addition of Ca to sample and standard	[1, 2]
Titan yellow	405	540-550	1.8	0.6-0.8 M NaOH	Many elements		[1]
Phenazo	490	560	35.4	1-2 M NaOH	Many elements	No Ca disturbance up to Ca : Mg=100; Fe, Al masked with triethanolamine; Ti with H_2O_2	[1, 2]
Oxine/n-BuNH$_2$ in CHCl$_3$	—	380	5.2	pH 10.5-11.5	Many elements		[7]

possibilities used for separation are the precipitation or extraction of the diethyldithiocarbamates of certain metals, e.g. Mn. Otherwise, in extraction or ion-exchange separations the disturbing ions are generally removed from the Mg^{2+}, because of the low stability of the complexes of Mg^{2+}.

Since the complex-forming tendency of magnesium is slight, the importance of its photometric determination has been declining in recent years, whereas its atomic-absorption determination has become more important.

DETERMINATION OF MAGNESIUM WITH TITAN YELLOW [3–5]

Titan yellow forms a coloured adsorption product with $Mg(OH)_2$ above pH 12. Gelatin, gum arabic, starch and poly(vinyl alcohol) can be used as protective colloids. The best of these is poly(vinyl alcohol), for the others are usually contaminated with magnesium.

The reproducibility of the results is improved if the alkaline solution is added dropwise with stirring to the solution to be analyzed.

Between 10 and 30 min after development of the colour, there is hardly any variation in the absorbance of the solution.

It is preferable to keep the solution at a definite temperature, so that the dispersity of the colloid, and hence the absorbance, will be reproducible.

Ca^{2+} does not form a coloured compound with titan yellow, but in its presence the colour is intensified. Above a certain calcium concentration, this intensification is constant, and this can be utilized to compensate the disturbing effect.

Analytical procedure for solutions not containing interfering ions: An aliquot of the solution to be examined, containing at most 300 μg magnesium, is pipetted into a 50 cm^3 volumetric flask, and 2 cm^3 $CaCl_2$ solution, 5 cm^3 0.05% titan yellow, and 5 cm^3 1% poly(vinyl alcohol) are added. ($CaCl_2$ solution: 2% $CaCl_2.2H_2O$ is dissolved in 0.1 M NaOH, the insoluble residue is filtered off after 1 hour, and the solution is weakly acidified.) The resulting mixture is diluted to ca. 35 cm^3 with water, it is treated with 1 M NaOH until a precipitate separates out, and an excess of 5 cm^3 1 M NaOH is added. The volume is made up to the mark, and the absorbance is measured at 545 nm 15 min later. The reference solution is prepared from the mixed reagents in distilled water.

After appropriate dissolution and separation, the method is suitable for the analysis of plant materials [3], biological materials [4], silicate minerals [5], etc.

DETERMINATION OF SMALL AMOUNTS OF MAGNESIUM (0.01–0.06%) IN IRON ALLOYS WITH MAGNESON IREA [6]

Analytical procedure: 0.1 g sample is dissolved in 10 cm^3 1 : 1 HCl, a little HNO$_3$ is added to oxidize carbides and Fe^{2+}, and the mixture is evaporated to dryness several times with HCl to dehydrate the silica. The silica is filtered off, the aqueous solution is evaporated to 10 cm^3, 10 cm^3 HCl is added, and the mixture is shaken twice with 5 cm^3 TBP+5 cm^3 butyl acetate to remove the bulk of the Ti^{4+} and the Fe^{3+}.

The pH of the aqueous phase is adjusted to 5–6 with NaOH, and to remove the interfering ions (Mn^{2+}, Ni^{2+}, Cu^{2+}, Cr^{3+}, Mo^{5+}) 5 cm^3 3% NaDDTC solution is added, and the complexes are extracted with 10 cm^3 butyl acetate. If the residual aqueous phase is turbid because of the precipitation of sulphur, it must be filtered. The solution is evaporated to 20–25 cm^3, and transferred to a 50 cm^3 volumetric flask. 0.25 cm^3 10% NH$_4$Cl solution, 15 cm^3 acetone, 6 cm^3 0.02% Magneson IREA reagent and 0.25 cm^3 20% NH$_4$OH are added, and the volume is made up to the mark. The absorbance of the complex is measured at 490 nm, or with an appropriate colour filter. The calibration plot is recorded for 10–60 μg Mg^{2+}, up to which limit Beer's law holds. The developed colour is stable for 20 min.

References

[1] Tikhonov, V. N.: *Analiticheskaya khimiya magniya.* Nauka, Moscow 1973.
[2] Tikhonov, V. N.: *Zh. Anal. Khim.* **26**, 1616 (1971).
[3] Bradfield, E. G.: *Analyst* **85**, 667 (1960); **86**, 269 (1961).
[4] Butler, E. J., Forbes, D. H. S., Munro, C. S., Russel, J. C.: *Anal. Chim. Acta* **30**, 524 (1964).
[5] King, H. G., Pruden, G.: *Analyst* **92**, 83 (1967).
[6] Kulikovskaya, Zh. B., Gregorenskii, A. S.: *Zav. Lab.* **39**, 949 (1973).
[7] Fries, J., Getrost, H.: *Organische Reagenzien für die Spurenanalyse.* Merck, Darmstadt 1975.

6.23. Manganese

Manganese is situated in column VII/2 of the periodic system. In its aqueous solutions it is to be found in the divalent, trivalent and heptavalent forms. Mn^{2+} and Mn^{3+} form coloured compounds with organic reagents, while the absorbance of MnO_4^- can be measured directly. Mn^{2+} is stable in acidic solution, whereas Mn^{3+} is labile, readily being reduced to Mn^{2+}. In alkaline medium, on the other hand, $Mn(OH)_2$ is easily oxidized to $Mn(OH)_3$. Mn(III) is stabilized in certain complexes, e.g. those with formaldoxime and DDTC.

Table 6.18 lists the main possibilities for the photometric determination of manganese.

TABLE 6.18.

More important possibilities for photometric determination of manganese

Reagent	Medium	λ_{max}, nm	$\varepsilon \times 10^{-3}$	Mn : L	Extractant
MnO_4^-	0.1 M H_3PO_4	525	2.2	—	—
Formaldoxime	pH 10–13	455	11.2	1 : 6	—
PAN	pH 9.5	568	47	1 : 2	CCl_4, $CHCl_3$
DDTC	pH 6.5	500	4	1 : 3	$CHCl_3$

(A) DETERMINATION OF MANGANESE BY PHOTOMETRY OF MnO_4^-

In solutions for analysis, manganese is generally present in the form Mn^{2+}. Boiling of a H_3PO_4–HNO_3 medium of Mn^{2+} with a strong oxidant such as $S_2O_8^{2-}$ in the presence of Ag^+ as catalyst leads to its conversion to the purple-coloured MnO_4^-. Permanganate has two absorbance maxima: at 525 nm ($\varepsilon=2.2\times10^3$) and at 545 nm ($\varepsilon=2.4\times10^3$). The oxidation is a several-step complex process, in which a key role is played by the labile Ag^{3+} complex formed on the action of the persulphate. Of the various possibilities recommended for colour formation, one of the optimum media is: 0.75 M HNO_3+0.25 M H_2SO_4+0.001 M $AgNO_3$+0.01 M $(NH_4)_2S_2O_8$+0.3 M H_3PO_4 [1]. The phosphoric acid plays an important role in the avoidance of the formation of manganese dioxide hydrate.

244

After colour development by boiling, the solution is cooled rapidly, and photometry is carried out in the concentration range 25–500 μg Mn^{2+}/ 100 cm³.

A chloride content in the solution causes interference (AgCl is formed); its effect is eliminated by evaporation with H_2SO_4, or minor amounts may be bound with $HgSO_4$. Other disturbing effects can be taken into account by absorbance measurement after reduction of the MnO_4^-, e.g. addition of $NaNO_2$ or NaN_3. Accordingly, Cu^{2+}, Ni^{2+}, Co^{2+} and Cr^{3+} in quantities 200–300 times that of the Mn^{2+} can be tolerated.

Other chemicals too may be used for the oxidation of Mn^{2+}, e.g. KIO_4, PbO_2, $NaBiO_3$ or sodium perxenate [2]. KIO_4 is applied widely. With this oxidant, there is no evolution of oxygen, the Ag^+ catalyst is not required, and the developed colour does not vary over a long period. Here too heating is necessary to develop the colour, formation of which is slower, and here too the medium must be free from Cl^-.

Detailed guidance on how to avoid errors that may arise during the formation of MnO_4^- is to be found in the publications by Gottschalk [1], Strickland and Spicer [3], Nydahl [4] and Lavrukhina and Yukina [5]. In our view, to be able to be confident in the results demands systematic controls with an internal standard method.

Rapid determination of manganese content of steels after oxidation with periodate [6]

Analytical procedure: 0.1–0.2 g sample is dissolved in 5 cm³ conc. HCl+ 5 cm³ 1:4 HNO_3 by heating. The solution is filtered if necessary, transferred to a 100 cm³ volumetric flask, and made up to the mark with water. An aliquot containing 0.01–0.4 mg Mn is taken in a beaker, 7 cm³ 70% potassium pyrophosphate and 5 cm³ 50% ammonium acetate are added, and the solution is diluted to 30 cm³ with water. The pH is adjusted to 8.4 (by the addition of HNO_3 or KOH). The solution is transferred to a 50 cm³ volumetric flask, 2 cm³ 5% sodium metaperiodate (dissolved in 1:3 HNO_3) is added, the mixture is gently boiled and cooled, and the volume is made up to the mark with water (the resulting pH is 7.3–7.6). The absorbance of the solution is measured in 1–5 cm cells at 525 nm. The result is calculated from a calibration curve prepared in a similar way.

The method is suitable not only for steel analysis, but also for determination in aluminium and copper-based alloys. In this case, after the photometry the $KMnO_4$ is reduced by Na_2SO_3 addition, and the original absorbance is corrected after measurement of the absorbance of the reduced solution.

Determination of manganese traces in plants [7]

Analytical procedure: 20 g plant sample is ignited for 8 hours at 600°C. If the residue is black, it is moistened with distilled water, dried, and further ignited. The residue is moistened with distilled water, and evaporated with 10 cm³ 20% HCl on a water-bath. 10 cm³ 1:2 HNO_3 and 1–2 drops 30% H_2O_2 are added, the mixture is heated and filtered, the residue is washed, and the solution is made up to 100 cm³. A 5 cm³ aliquot of the stock solution is pipetted into a 25 cm³ volumetric flask, 2.5 cm³ of a mixed reagent is added, and the solution is well mixed (mixed reagent: 95 g $HgSO_4$ or 103 g $Hg(NO_3)_2$ is dissolved in 60 cm³ 2:1 HNO_3; 300 cm³ conc. H_3PO_4 and 1 g $AgNO_3$ are added, and the volume is made up to 1 dm³.) 1 g $(NH_4)_2S_2O_8$ is added and the volume is made up nearly to the mark. The solution is thermostated at 90–95°C for 30 min, cooled, and made up to the mark. Its absorbance is measured at 525 nm.

(B) DETERMINATION OF MANGANESE WITH FORMALDOXIME

Formaldoxime (CH_2=N—OH) reacts with Mn^{2+} to form a colourless complex; on the action of atmospheric oxygen, in alkaline medium this is converted to the reddish-brown compound $[Mn(CH_2NO)_6]^{2-}$. The oxidation process is fast, and the resulting colour is stable for more than 16 hours.

Formaldoxime similarly forms coloured complexes with Ce^{3+}, Cu^{2+}, Fe^{2+}, V^{4+}, Ni^{2+} and Co^{2+}. However, these can in part be decomposed by heating, and in part masked with CN^- [8]. Metal ions which give precipitates in the alkaline medium used for the determination (0.04–0.05 M NaOH) can be kept in solution with appropriate complex formers (tartrate, etc.).

246

Fast method for determination of manganese in iron-containing waters [9]

Analytical procedure: A 20 cm^3 aliquot of the water sample is taken, 1 cm^3 formaldoxime reagent and 1 cm^3 1:1 NH$_4$OH are added, and 2 min later 1 cm^3 0.1 M EDTA and 2 cm^3 10% NH$_2$OH.HCl are also added. (Formaldoxime reagent: 8 g NH$_2$OH.HCl is dissolved in 100 cm^3 water, 4 cm^3 37% formalin is added, and the volume is made up to 200 cm^3.) After 10 min, when the iron–formaldoxime complex has already decomposed, the absorbance is measured at 450 nm. The calibration curve covers the range 0–60 μg Mn^{2+}. The measurement is not disturbed by Fe^{2+} and Fe^{3+} contents up to 25 mg/dm^3.

The process can also be automated [10].

(C) DETERMINATION OF MANGANESE WITH TRIETHANOLAMINE AND *o*-TOLIDINE

In sensitive determinations of manganese in water, oxine concentration and separation are introduced. Triethanolamine (TEA), pyrophosphate and *o*-tolidine are added to the solution. The colour of the resulting quinone-diimine is measured in 1 cm or 20 cm cells. The disturbing effect of vanadium is eliminated via a blank.

Determination of manganese in water after preliminary concentration [11

Analytical procedure: The pH of the water to be analyzed is adjusted to 10 with NaOH solution, and 5 cm^3 5% oxine is added. The precipitate is separated off by filtration or centrifugation, and dissolved in dilute HCl, and the pH is adjusted to 2–3.

The sample is examined with regard to whether it contains oxidants: 1 cm^3 H$_3$PO$_4$ and 1 cm^3 *o*-tolidine (0.1 g *o*-tolidine is dissolved in 1 cm^3 12 M HCl, and the volume is diluted to 100 cm^3) are added to 1 cm^3 solution. If the solution becomes yellow, a strong oxidant (and vanadium) is present. The effect of a little oxidant can be counteracted by the addition of Fe(II).

An aliquot (up to 10 cm^3) is taken such that the Mn content is 0.1–1.6

247

$\mu g/cm^3$ (or 2–45 ng/cm^3 if the 20 cm cell is used). 1 cm^3 TEA, 1 cm^3 9 M NaOH and 1 cm^3 0.1 M Fe(II) are added. The flask is shaken for 30 s stoppered, and then for 30 s open. 5 cm^3 saturated $Na_4P_2O_7 \cdot 10H_2O$ solution (12 g/100 cm^3) and 5 cm^3 1:1 H_2SO_4 are added, followed after 5 min by 1 cm^3 o-tolidine solution. The volume is made up to 25 cm^3, and the absorbance is measured at 440 nm.

In the presence of vanadium, a blank solution is prepared which is used as comparison. 1 cm^3 TEA, 1 cm^3 1:1 H_2SO_4, 1 cm^3 Fe(II) solution, 1 cm^3 H_3PO_4 and 5 cm^3 pyrophosphate solution are added to the 25 cm^3 volumetric flask, and the sample is well mixed. After 5 min, 1 cm^3 o-tolidine is added, and the solution is made up to the mark.

The calibration plot is produced for the concentrations 0.2, 0.8 and 1.6 μg Mn/cm^3 (or 2, 4, 16 and 32 ng Mn/cm^3 in the case of the 20 cm cell).

References

[1] Gottschalk, G.: Z. Anal. Chem. **212**, 303 (1965).
[2] Bane, R. W.: Analyst **90**, 756 (1965).
[3] Strickland, J. D. H., Spicer, G.: Anal. Chim. Acta **3**, 517 (1949).
[4] Nydahl, F.: Anal. Chim. Acta **3**, 144 (1949).
[5] Lavrukhina, A. K., Yukina, L. V.: Analiticheskaya khimiya margantsa. Nauka, Moscow 1974.
[6] Lazareva, V. I., Lazarev, A. I., Harlamov, I. P.: Zav. Lab. **45**, 193 (1979).
[7] Vazhenina, I. G.: Metody opredeleniya mikroelementov v pochvakh, rasteniyakh vodakh. Kolos, Moscow 1974.
[8] Marczenko, Z.: Spectrophotometric Determination of Elements. Ellis–Horwood, London 1976.
[9] Goto, K., Komatsu, T., Furukawa, T.: Anal. Chim. Acta **27**, 331 (1962).
[10] Henriksen, A.: Analyst **91**, 647 (1966).
[11] Flaschka, H. A., Hornstein, J. V.: Anal. Chim. Acta **100**, 469 (1978).

6.24. Mercury

Mercury belongs in column II/2 of the periodic system. In its aqueous solutions it is to be found in the monovalent and divalent forms. In mercury (I) compounds, the mercury atoms occur in a dimerized state (—Hg—Hg—), so that each Hg atom here too forms two covalent bonds. Mercury(I) compounds are labile in comparison to mercury(II) compounds, for the cation Hg_2^{2+} readily decomposes by disproportionation. As a consequence of the oxidation steps included in the operations, in determinations the form Hg^{2+} is encountered.

The primary role in photometric determinations is played by chelate-forming reagents containing both S and N donor atoms, e.g. dithizone and its derivatives. N and O-containing ligands, e.g. diphenylcarbazone, are of only secondary importance.

Of the reagents suitable for the photometric determination of mercury, dithizone is of outstanding practical value. Some examples are given below of the application of dithizone.

CONDITIONS OF DISSOLUTION AND SEPARATION OF MERCURY [1, 2]

Although the determination of mercury with dithizone is disturbed by few ions, the conditions of the colour development may be made completely selective by means of various procedures. The advantage stemming from the volatility of mercury may be utilized by igniting the sample and condensing the mercury on a gold plate [3], or the mercury in coals may be separated out by brief boiling with Na_2S [4]. If separation is not performed, complex formers are applied. Any silver present is masked with thiocyanate, and other metals (Cu, Pb, Zn, Ni, Bi) with EDTA. The average occurrence of mercury in rocks is of the order of $10^{-6}\%$; it is strongly enriched in arsenic and antimony minerals. Mercury may enter soil and water not only as a consequence of industrial pollution, but also through the use of insecticides, etc.

The dissolution of mercury compounds can be achieved via an acidic–oxidative procedure in a closed space. Evaporation in an open vessel is well known to be accompanied by loss; experiments with isotopically labelled mercury have shown this loss to attain even 5–40%. The acid mixtures used for dissolution always contain an oxidant (HNO_3, $KMnO_4$, KNO_3). The evaporation of mercury vapour is prevented with a reflux condenser, with ground-glass joints. If $KMnO_4$ is used, the excess is decomposed with H_2O_2 or NH_2OH. Before use, the dithizone is to be purified by washing with a little NH_4OH [2].

Determination of mercury in rocks (according to [5] and the authors' modification)

Analytical procedure: 0.5–5 g of the sample is taken in a boiling-flask fitted via ground-glass joints with a condenser. 25–35 cm^3 conc. H_2SO_4 and 1–2 g powdered $KMnO_4$ are added, and the flask is shaken until dissolution has occurred. The mixture is heated until fumes of SO_3 appear, and then for a further 10–15 min. After cooling, the contents of the flask are diluted to ca. 80 cm^3; the precipitated MnO_2 and the excess $KMnO_4$ are reduced by the dropwise addition of 30% H_2O_2. The mixture is boiled, cooled, and made up to 100 cm^3.

An aliquot corresponding to at most 20 μg Hg^{2+} is taken from the stock solution, the pH is raised to 3 with 30% sodium acetate, and 10 cm^3 0.1 M EDTA and 5 cm^3 20% KSCN are added. The volume of the solution to be extracted should be at most 50–60 cm^3. It is extracted with small portions of 0.001% dithizone in CCl_4. The combined organic phase is washed with dilute NH_4OH (1 cm^3 conc. NH_4OH/dm^3), and its volume is made up to 10 cm^3 with CCl_4. The absorbance of the yellow solution is measured against pure CCl_4 at 485 nm. The calibration plot covers the interval 0.5–20 μg Hg^{2+}.

Determination of mercury in water [6, 7]

The determination is disturbed by the colour of organic matter in the water. Accordingly, after the addition of acetate buffer, EDTA and thiocyanate, the water is extracted with small portions of $CHCl_3$ until it is colourless. If a larger quantity of organic matter is present, the water is placed in a boiling-flask fitted with a condenser, and 1 cm^3 conc. H_2SO_4 is added, followed by saturated $KMnO_4$ solution until the colour of the latter persists. The mixture is boiled for 15 min; if it is decolorized, further $KMnO_4$ is added. The slight excess of $KMnO_4$ is decolorized with hydroxylamine or hydrazine sulphate. 10 cm^3 buffer, 10 cm^3 0.1 M EDTA and 10 cm^3 0.1 M KCSN are added, and the volume is made up to 100 cm^3. (Buffer solution: 57 cm^3 glacial acetic acid + 82 g sodium acetate, made up to 1 dm^3.) The mixture is shaken for 2 min with 25 cm^3 0.005% dithizone in $CHCl_3$. The organic phase is washed with complex formers in 25 cm^3 distilled water, then with 50 cm^3 5% NH_4OH, and photometry is performed.

250

Determination of mercury content of cadmium–mercury–tellurium semiconductors with dithizone in the presence of N-cetylpyridine [8]

The determination of mercury with dithizone can be made more sensitive and selective in the presence of surface-active substances, such as N-cetylpyridine chloride.

Analytical procedure: 30 mg sample is dissolved in a 1:3 mixture of HCl and HNO_3, the pH of the solution is adjusted to 9–10 with NH_4OH, and the mixture is diluted to 50 cm^3 with water. Depending on the mercury content, a 0.25–2.5 cm^3 aliquot is pipetted into a 25 cm^3 volumetric flask, 10 cm^3 buffer of pH 12.9 (prepared from a mixture of KOH and acetic acid) and 1 cm^3 saturated EDTA solution (to mask Cd) are added, followed by 0.5 cm^3 36% N-cetylpyridine chloride, 0.5 cm^3 1×10^{-3} M dithizone (in 0.5 M KOH), and buffer up to the mark. After 2 min, the absorbance of the solution is measured at 555 nm, against water. In this way, the limit of determination is of the order of 10^{-1} % Hg.

References

[1] Gladishev, V. P., Levitskaya, S. A., Filippova, L. M.: *Analiticheskaya khimiya rtuti.* Nauka, Moscow 1974.
[2] Iwantscheff, G.: *Das Dithizon und seine Anwendung in der Mikro- und Spurenanalyse.* Chemie, Weinheim 1972.
[3] Wimberley, J. W.: *Anal. Chim. Acta* **76**, 337 (1975).
[4] Vasilevskaya, A. E., Shcherbakov, V. P., Karakozova, E. V.: *Zh. Anal. Khim.* **19**, 1200 (1964).
[5] Knipovich, Yu. N., Morachevskii, Yu. V.: *Analiz mineralnovo syrya.* GKhJ, Leningrad 1959.
[6] Lure, Yu. Yu.: *Unifitsirovannie metody analiza vod.* 2nd. Ed. Khimiya, Moscow 1973.
[7] Lure, Yu. Yu., Rybnikova, A. I.: *Khimicheskii analiz proizvodstvennykh stochnykh vod.* Khimiya, Moscow 1974.
[8] Albota, L. A.: *Zh. Anal. Khim.* **36**, 270 (1981).

6.25. Molybdenum and tungsten

Molybdenum and tungsten belong in column VI/2 of the periodic system. Their hexavalent forms are the most frequent, but these can fairly readily be reduced to the pentavalent forms. Their tetravalent, trivalent and divalent forms are also known.

Both molybdenum and tungsten display a tendency to complex formation. Their hexavalent forms characteristically form iso- and heteropolyacids. In alkaline medium, the hexavalent forms remain in solution as molybdate and tungstate (MoO_4^{2-} and WO_4^{2-}), whereas the pentavalent forms are precipitated as hydroxides.

PHOTOMETRIC REAGENTS FOR MOLYBDENUM AND TUNGSTEN

There is no selective photometric reagent for molybdenum and tungsten. The most frequently applied reagents are thiocyanate and dithiol. The complexes can be extracted with organic solvents. The literature deals very comprehensively with the determination of both molybdenum and tungsten with these reagents in various ores, rocks, alloys, biological substances, etc. [1–7].

The data reported in the literature are often contradictory; the media applied in the determinations frequently differ. All this indicates that the analysis of molybdenum and tungsten is not problem-free. The determination of tungsten with thiocyanate is treated by Fogg et al. [8], the analysis of tungsten by Busev et al. [9].

Besides the two reagents of fundamental importance, other reagents are known for the determination of molybdenum: fluorone derivatives (primarily phenylfluorone), certain di- and triphenol derivatives, tiron, phenylhydrazine and thioglycollic acid. Besides the fluorones, mention may be made of rhodamine B for the determination of tungsten. However, the practical significance of these additional reagents is not sufficient to warrant a detailed account of their properties.

Conditions of determination of molybdenum
and tungsten with thiocyanate

In H_2SO_4 or HCl medium (optimally 1–2 M H_2SO_4 or 2–4 M HCl), Mo^{5+} forms complexes with the thiocyanate ion. The coordination structures and compositions of these have not yet been fully elucidated. The composition of the complex on which the determination is based is probably $MoO(SNC)_3$. The molybdenum can conveniently be reduced with $SnCl_2$, thiourea, ascorbic acid, etc. Strong reductants such as $TiCl_3$ must be avoided, for they may overreduce the molybdenum, thereby causing a decrease in the intensity.

The colour intensity of the thiocyanate complex of molybdenum depends on the acidity, the thiocyanate concentration, and the duration of standing. In the presence of Fe^{3+} there is an increase in the intensity and the stability of the complex, since the resulting redox potential inhibits reduction of the molybdenum to the trivalent state. The sequence of addition of the reagents is important: the thiocyanate is added to the acidic solution, followed by the Fe^{3+}, and finally the $SnCl_2$.

The molybdenum-thiocyanate complex can be extracted with organic solvents (butanol, amyl alcohol, amyl acetate, ether, chloroform, cyclo-hexanone, etc.). In this way, the sensitivity and stability are increased.

The determination is disturbed by larger amounts of Ti, V, Pt, Cr, Re and W. Tungsten can be masked with citric acid or tartaric acid; the other ions must either be separated or corrected for. The intensity is lowered by the presence of Si.

In HCl or H_2SO_4 medium at high acidity (optimally 4–10 M HCl or 2–5 M H_2SO_4) W^{5+} forms a complex with the thiocyanate ion. Here too, stepwise complex formation means that several complexes are involved. It must be noted that, in the strongly HCl medium recommended in the literature, in a short time the thiocyanate undergoes decomposition to give the yellow isothiopercyanic acid, which disturbs the determination of tungsten. It is therefore preferable to use a H_2SO_4 medium. $SnCl_2$ or $TiCl_3$ is suitable as reductant. Disturbing ions: Mo, V, As, Sb, Cr, Cu, Pt, Nb and Si. The thiocyanate complex of tungsten can be extracted with the organic solvents listed for molybdenum. The sensitivity of the determination can be increased if the mixed complex formed in the presence of an organic cation (diphenylguanidine, crystal violet, etc.) is extracted [10]; by this means, ε_{max} can attain a value of 3.2×10^5.

TABLE 6.19.

Conditions of determination of molybdenum and tungsten with thiocyanate

Element	Acidity	SCN^-	Extractant	λ_{max}, nm	$\varepsilon_{max} \times 10^{-3}$
Mo	2 M H_2SO_4	0.25 M	n-amyl acetate	475	19.2
W	3 M H_2SO_4+HCl	0.50 M	n-amyl acetate	407	13.1

For both molybdenum and tungsten, the intensity of the thiocyanate complex is a function of many parameters (acidity, thiocyanate concentration, extractant, etc.); the conditions judged to be best in the procedure used in our laboratory are given in Table 6.19.

Determination of molybdenum and tungsten with dithiol

In HCl or H_2SO_4 medium, dithiol forms complexes with Mo^{5+} and W^{5+}. Molybdenum is reduced by the reagent (at an acidity of 2–4 M HCl or 1–2 M H_2SO_4), and complex formation is complete within a few minutes at room temperature. The tungsten complex is formed only after fairly long heating on a water-bath, in strongly acidic medium in the presence of a strong reductant, or in a weakly acidic medium without a reductant [1–7]. Alkyl acetates, CCl_4, $CHCl_3$, benzene, etc. can be used as extractant. Interfering elements: Fe, Cu, Ag, Ge, Te, Hg, Au and Pt; and Mo or W in the determination of the other. Nevertheless, it is possible to determine molybdenum and tungsten when both are present, because of the difference in the conditions of the reduction. The main data on the dithiol complexes of molybdenum and tungsten are listed in Table 6.20.

TABLE 6.20.

Main data on dithiol complexes of molybdenum and tungsten

Element	Acidity	Reductant	Extractant	λ_{max}, nm	$\varepsilon_{\lambda max} \times 10^{-3}$
Mo	1.8 M H_2SO_4	—	$CHCl_3$	680	21
W	1.8 M H_2SO_4	$TiCl_3$	$CHCl_3$	645	24
	0.15 M or H_2SO_4	—	$CHCl_3$		

DETERMINATION OF MOLYBDENUM AND TUNGSTEN IN ROCKS AND ORES
(procedure of the authors)

Digestion: The literature recommends alkaline fusion with Na_2O_2, pyrosulphate digestion, evaporation with HF and subsequent acidic dissolution, and possibly digestion with carbonate. The alkaline digestion

254

has the drawback that, after aqueous leaching-out, not only molybdenum and tungsten pass into the filtrate, but also silicon, which interferes in the determination. In the pyrosulphate digestion, much platinum dissolves up, and this too disturbs the subsequent determination. In our own experience, molybdenum and tungsten can both be dissolved up quantitatively: following digestion with HF, the residue is dissolved out of the platinum crucible with HCl, HNO_3 is added and the solution is boiled, oxides of nitrogen are removed by evaporation, and the residue is redissolved in HCl.

Analytical procedure: 1 g sample is weighed into a platinum dish, it is moistened with 2 drops 1:1 H_2SO_4 and 0.5 cm³ HNO_3, 10 cm³ HF is added, and the mixture is evaporated to dryness. For samples with high silica contents, the HF evaporation is repeated. The residue is dissolved in ca. 5 cm³ conc. HCl and washed into a beaker, 5 cm³ conc. HNO_3 is added, and the mixture is evaporated. The residue is dissolved in a little conc. HCl, the solution is diluted to ca. 20 cm³ with water and neutralized with NaOH, and 5 cm³ 5 M NaOH is added in excess. The solution is boiled and filtered, the residue is washed with water, and the filtrate is made up to 50 cm³ in a volumetric flask. The resulting solution is suitable for the analysis of both molybdenum and tungsten.

The conditions of determination of molybdenum and tungsten with thiocyanate and dithiol will next be described.

Determination of molybdenum with thiocyanate

At most 10 cm³ ($\leqq 75$ μg Mo) of the stock solution is taken in a separating-funnel (for smaller aliquots the volume is made up to 10 cm³ with 1.0 M NaOH), and 1.5 cm³ water, 5 cm³ 30% citric acid, 6 cm³ 1:1 H_2SO_4 1.5 cm³ 40% KSCN, 0.2 cm³ Fe^{3+} (10 mg Fe^{3+}/cm³) and a little solid ascorbic acid are added. The solution is left to stand for 15 min, and then extracted for 1 min with 10 cm³ *n*-amyl acetate. The aqueous phase is poured off, and the organic phase is washed with 8 cm³ washing liquid (washing liquid: 3 cm³ water+3 cm³ 30% citric acid+1 cm³ 1:1 H_2SO_4+ 1 cm³ 1:1 HCl containing 15% $SnCl_2.2H_2O$). The organic phase is run off into a dry vessel and, after complete clarification, the absorbance is measured at 475 nm, against a solution prepared as above, but containing 10 cm³ 1 M NaOH instead of the tested solution.

The calibration plot is produced with 2–15 μg Mo (in 4–5 cm cells) or 15–75 μg Mo (in 1 cm cells) taken under the conditions described for the analysis. A molybdenum stock solution is prepared from ammonium molybdate in 1 M NaOH. The absorbance of 10 μg Mo in a 4 cm cell is 0.79. The sensitivity of the method is 3 ppm Mo.

Determination of molybdenum with dithiol

At most 5 cm^3 of the solution to be analyzed (a smaller aliquot is made up to 5 cm^3 with 1 M NaOH) is taken in a 100 cm^3 shaking-funnel, and 2 cm^3 30% citric acid, 2.3 cm^3 1:1 H_2SO_4 and 1 cm^3 dithiol solution are added (dithiol solution: 1 g reagent is dissolved in 145 cm^3 water containing 10 g NaOH, and 5 cm^3 thioglycollic acid is added; the solution is kept in a dark bottle in a refrigerator; it must be prepared freshly weekly). After a few min, the aqueous solution is extracted for 1 min with 10 cm^3 $CHCl_3$. The organic phase is run off into a dry vessel, allowed to clarify, and the absorbance is measured in 1–5 cm cells at 675 nm, against a solution prepared as above, but containing 5 cm^3 1 M NaOH instead of the sample solution.

The calibration plot is produced with 1–50 μg Mo under the conditions described. The absorbance of 10 μg Mo in a 4 cm cell is 0.87. The sensitivity of the method is 5 ppm.

Determination of tungsten with thiocyanate

Since the absorbance maximum for the thiocyanate complex of tungsten is at 407 nm, the number of interfering ions is much higher than in the case of molybdenum. Our investigations reveal that in the analysis of rock samples with low tungsten contents the absorption spectrum of the extracted thiocyanate complex is distorted; instead of an absorbance maximum, only an inflection indicates the presence of tungsten, and the relevant calculations lead to false results. (Attention is not drawn to this in the literature.) At higher tungsten contents (>100 ppm) the lower sample weight means that this error does not occur, and thus the method can be applied.

If the maximum sensitivity of the method (5–10 ppm) is to be made use of, the tungsten and molybdenum must first be separated from the disturbing components with α-benzoin-oxime [3].

However, an equivalent quantity of molybdenum also disturbs the determination, and in this case it is not advisable to use the thiocyanate method.

A 2 cm³ aliquot of the stock solution is pipetted into a separating-funnel, and 11.5 cm³ water+8.5 cm³ 1:1 H_2SO_4+3 cm³ 40% KSCN+5 cm³ 15% $SnCl_2 \cdot 2H_2O$ in 1:1 HCl are added. After a waiting period of 15 min, the solution is extracted with 10 cm³ n-amyl acetate. The organic phase is washed with a solution having the same composition as the aqueous phase, but not containing thiocyanate. After complete separation, the absorbance of the organic phase is measured in 1–5 cm cells at 407 nm, against the corresponding solution obtained with the reagents, but without tungsten.

The calibration plot is recorded for 2–100 μg W, by the same method. The absorbance of 10 μg W in a 4 cm cell is 0.30.

Determination of tungsten with dithiol

An aliquot of at most 5 cm³ of the stock solution (if necessary, the volume is made up to 5 cm³ with 1 M NaOH) is pipetted into a separating-funnel, 2.3 cm³ 1:1 H_2SO_4 and 1 cm³ dithiol are added, and the volume is made up to 10 cm³ with water.

For removal of molybdenum, the solution is immediately extracted with 10 cm³ $CHCl_3$. The aqueous phase is transferred to a test-tube, 1 cm³ dithiol and 2 cm³ 15% $TiCl_3$ are added, and the mixture is heated for 30 min on a water-bath. After cooling, it is transferred to a separating-funnel and extracted for 1 min with 10 cm³ $CHCl_3$. After the organic phase has become clear, its absorbance is measured in 1–5 cm cells at 645 nm. The absorbance of 10 μg W in a 4 cm cell is 0.52. The sensitivity of the method is 10 ppm W.

DETERMINATION OF MOLYBDENUM AND TUNGSTEN WITH OTHER REAGENTS

Determination of molybdenum in steel with 1-nitroso-2-naphthol [11]

Analytical procedure: 0.5 g steel is dissolved in a mixture of 1 : 1 HCl and conc. HNO_3. The solution is washed into a 50 cm^3 volumetric flask and the volume is made up to the mark with water. A 6 cm^3 aliquot (20–200 μg Mo) is pipetted into a separating-funnel, the HCl concentration is adjusted to 2–3 M, 5 cm^3 1×10^{-3} M 1-nitroso-2-naphthol is added, and the solution is extracted for 10 min with 10 cm^3 of a 1 : 1 mixture of $CHCl_3$ and isoamyl alcohol. The excess reagent is re-extracted into the aqueous phase by shaking with 10 cm^3 0.1 M NaOH. The absorbance of the organic phase is measured at 380 nm, against the pure mixture of $CHCl_3$ and isoamyl alcohol. The calibration plot is recorded in an analogous manner. With this method, $\sim 10^{-1}\%$ Mo can be determined in steel. It has the advantage that preliminary separation need not be introduced. The error of the method does not exceed $\pm 10\%$ relative.

Determination of tungsten in molybdenum concentrates with pyrocatechol violet in the presence of H_2O_2 [12]

The method is preceded by the acidic digestion, and then the alkaline separation of the concentrate. Outline of analysis: 1–2 cm^3 of the alkaline filtrate (0.5–3 μg W) is pipetted into a 50 cm^3 volumetric flask, H_2O_2 is added, the acidity is adjusted so that the final pH will be 1, 5 cm^3 1% pyrocatechol violet is added, and the volume is made up to the mark with water. The absorbance of the solution is measured in 1 cm cells at 580 nm, and compared with a calibration plot. This method permits the determination of 2–4% W in the presence of 60% Mo. (The concentrate also contains trace amounts of Fe, Cu and As.)

References

[1] Popov, N. P., Stolyarova, I. A.: *Khimicheskii analiz porod i mineralov.* Nedra, Moscow 1974.
[2] Topping, J. J.: *Talanta* **25**, 61 (1978).

258

[3] Stepanova, N. A., Yakushina, G. A.: *Zh. Anal. Khim.* **17**, 858 (1962).
[4] Quinn, B. F., Brooks, R. R.: *Anal. Chim. Acta* **74**, 75 (1975).
[5] Chan, K. M., Riley, J. P.: *Anal. Chim. Acta* **39**, 103 (1967).
[6] Quinn, B. F., Brooks, R. R.: *Anal. Chim. Acta* **58**, 301 (1972).
[7] Dragomirecky, A., Mayer, V., Michal, J., Rericka, K.: *Photometrische Analyse anorganischer Roh- und Werkstoffe.* VEB Deutscher Verlag für Grundstoffindustrie, Leipzig 1968.
[8] Fogg, A. G., Marriott, D. R., Thorburn-Burns, D.: *Analyst* **95**, 848 (1970).
[9] Busev, A. I., Ivanov, V. M., Skolova, T. A.: *Analiticheskaya khimiya volframa.* Nauka, Moscow 1976.
[10] Ivanova, I. F.: *Zh. Anal. Khim.* **30**, 1395 (1975).
[11] Peshkova, V. M., Ivanova, E. K., Memon, S.: *Zh. Anal. Khim.* **35**, 486 (1980).
[12] Lebedeva, L. I., Baisberg, A. S.: *Zav. Lab.* **47**, 21 (1981).

6.26. Nickel

Nickel, a typical transition metal, is to be found in group VIII/2 of the periodic system. The complex-forming properties of nickel resemble primarily those of Co^{2+} and Cu^{2+}, and in part those of Fe^{2+} and the platinum metals. In solution it is generally divalent, but in certain of its stable complexes it is present as Ni^{3+} or Ni^{4+}.

TABLE 6.21.

More important photometric reagents for nickel

Reagent	λ_{max} nm	$\varepsilon \times 10^{-3}$	Medium	Interferences	Notes	Refs
Dimethyl-glyoxime	450	14	Oxidized, NH_4OH	Many elements	Complex formers, precipitation separation	[5, 6, 10]
2,2'-Furyldioxime	436	18	pH 7.5–9.0	Much Cu, Co	$Na_2S_2O_3$, EDTA, masking agents	[11, 12]
Bis(4-Na-tetrazolylazo-5)-ethyl acetate	490	27	pH 4.0	Much Fe, Al, Cu, Zn, Cd, Co	NaF, EDTA, masking agents	[6, 13]
TAN-3,6-S	596	26	pH 8–10	Cu, Zn	Tartaric acid, complex former	[3, 4]
Diantipyryl-methane-di-thiooxalate	505	3.2	2.5 M HCl	Fe, Mo, Co, Cu, V	$Na_2S_2O_3$, NH_4F, masking agents	[14]

It displays tendencies to complex formation with ligands containing S, N and O donor atoms. Of these, dithizone and the dithiocarbamates are primarily of use for extraction separation, while thiocyanate is mainly of value for separation or determination in the form of mixed ligand complexes, e.g. thiocyanate+pyridine. Of the photometric reagents, the α-dioximes are of outstanding importance; this is particularly so for dimethylglyoxime (diacetyldioxime; DMGO), which has long been of great importance in complex-chemical research [1]. The general spectrophotometric use of the oximes has been reviewed (from 1953 on) by Singh *et al.* [2]. Certain hydroxyazo compounds [3, 4] and thiocarboxylic acids are of increasing importance. The more important photometric reagents for nickel are listed in Table 6.21.

Besides the above-mentioned extraction possibilities for the separation of nickel, other procedures of significance are precipitation in the presence of a metal hydroxide carrier, e.g. $Mg(OH)_2$, and separation with NH_4OH via the formation of soluble ammine complexes of Ni(II). Because of their relatively low stability, its halide complexes cannot be separated advantageously by anion-exchange. The nickel complexes of the dioximes are suitable for separation purposes too.

Determination of nickel traces in rocks with dimethylglyoxime [5]

Besides nickel, this reagent forms colours with Cu^{2+}, Co^{2+}, Pd^{2+} and Pt^{2+}, and interference also results from ions which yield hydroxide precipitates at the pH of the determination. The effects of the disturbing components may be eliminated by separation and by the use of complex formers, e.g. citrate, tartrate [6]. By these comparatively simple means, the possibility of selective measurement can be ensured.

An effective separation prior to photometry is provided by extraction of the Ni–DMGO complex into an organic solvent, e.g. $CHCl_3$, and then acidic re-extraction of the Ni. The resulting solution is made basic with NH_4OH, DMGO and an oxidant (persulphate, bromine water, iodine) are added, and the absorbance of the colour produced is measured. The complex formed has the composition Ni : L=1 : 3; its structure has not been elucidated.

Analytical procedure: For pretreatment of the sample, the operations

260

described in connection with zinc (Section 6.43) must be performed. A 100 cm³ stock solution is obtained after digestion with HF and separation with NH_4OH-NH_4Cl. A 5–20 cm³ aliquot of this (at most 50 μg Ni) is taken and diluted to 50 cm³, 2 cm³ 1% DMGO in alcohol is added, and after 2 min the mixture is extracted with 2×5 cm³ $CHCl_3$. The Ni is re-extracted with 2×5 cm³ 0.5 M HCl, and 1 cm³ 33% $(NH_4)_2S_2O_8$ and 2 cm³ 1% DMGO in 5% KOH are added to the acidic solution. After 10 min, the volume is made up to 25 cm³ and the absorbance of the reddish-brown solution is measured at 450 nm. The calibration plot is recorded in 5 cm cells up to 10 μg Ni^{2+}/25 cm³.

Microdetermination of nickel in natural waters [7]

The water to be analyzed is mixed with a resin saturated with PAN (see Appendix), and the absorbance of the resin phase is determined. Other elements too (Cr, Fe, Co, Zn, Cu, Bi, Cd) can be determined successfully with this procedure.

Preparation of sample: The water is filtered through a membrane filter, and is acidified by the addition of 10 cm³ conc. HCl/dm³. It is stored in a polyethylene vessel.

Analytical procedure: 24 cm³ 0.1% PAN in methanol is made up to 100 cm³ and mixed for 1 hour with 30 g 100–200 mesh Dowex 50 W-X2 resin. The resin is then filtered off, washed with water, air-dried and stored in the cold. 0.5 g of this resin is added to 200 cm³ of the water sample (0.1–2 μg Ni) containing 1 cm³ 0.1 M sodium pyrophosphate, the pH is adjusted to 6, and the mixture is stirred for 20 min. The resin is filtered off and stirred for 10 min with a masking solution (0.1 M thioglycollic acid and 10^{-3} M EDTA, adjusted to pH 7.8 with NH_4OH). The absorbance of the Ni–PAN resin complex is determined in 1 mm cells at 566 nm. The absorbance is also measured at 700 nm, this value being used as a blank correction. A double-beam recording spectrophotometer is used for the determination, with the pretreated resin in the comparative light-path. The calibration plot covers the range 0.1–12 μg Ni/dm³. The Co content too may be determined by photometry on the resin at 628 nm.

Determination of nickel in steel and bronze [8]

Principle of method: 6-Br-BTAK (see Appendix) forms a complex with nickel in 0.5–2 M NH_4OH medium; this complex is insoluble in water, but it dissolves in $CHCl_3$ (and other organic solvents). The reagent is synthesized by the diazotization of 6-bromobenzothiazole with *p*-cresol [9]. The determination is not disturbed by Zn, Cd, Cr, Cu, Mn, V, Pb, W, U, Pd and Th. Iron can be masked with potassium sodium tartrate (Seignette salt). The absorbance of the complex is maximum at 620 nm ($\varepsilon=4.2\times10^4$).

Analytical procedure: 0.1 g sample is dissolved in conc. HNO_3 and 2 M HCl with heating. The solution is evaporated to low volume, and then made up to 100 cm^3. A 1–2 cm^3 aliquot (0.05–6 µg Ni) is pipetted into a separating-funnel, and 2 cm^3 0.001 M 6-Br-BTAK in $CHCl_3$, 0.4 cm^3 40% Seignette salt and 8 cm^3 $CHCl_3$ are added. The volume of the aqueous phase is made up to 10 cm^3 with 1 M NH_4OH, and the extraction is performed for 5 min. The absorbance of the organic phase is measured at 620 nm, against an extract of the reagent. The calibration plot is produced by the above procedure, with 0.05–6 µg Ni.

The method is also suitable for the analysis of U–Pb–based alloys.

References

[1] Chugaev, L. A.: *Z. anorg. allgem. Chem.* **46**, 114 (1905).
[2] Singh, R. B., Garg, B. S., Singh, R. P.: *Talanta* **26**, 425 (1979).
[3] Busev, A. I., Zholondovskaya, T. N., Krysina, L. S., Barinova, O. A.: *Zh. Anal. Khim.* **29**, 1758 (1974).
[4] Marchak, T. V., Brykina, G. D., Belyavskaya, T. A.: *Zh. Anal. Khim.* **36**, 513 (1981).
[5] Görbicz, M., Upor, E.: *Acta Chim. Acad. Sci. Hung.* **66**, 373 (1970).
[6] Peshkova, V. M., Savostina, V. M.: *Analaticheskaya khimiya nikelya.* Nauka, Moscow 1966.
[7] Yosimura, K., Toshimitsu, Y., Ohashi, S.: *Talanta* **27**, 693 (1980).
[8] Gusev, S. I., Zhvakina, M. V., Kozhevnikova, I. A., Maltseva, L. S.: *Zh. Anal. Khim.* **33**, 952 (1978).
[9] Gusev, S. I., Zhvakina, M. V., Kozhevnikova, I. A.: *Zh. Anal. Khim.* **26**, 859 (1971).
[10] Pakhomova, K. S., Volkova, L. P., Gorshkov, V. V.: *Zh. Anal. Khim.* **19**, 1085 (1964).
[11] Budanova, L. M., Volodarskaya, R. S., Kasaev, N. A.: *Analiz aluminievykh i magnievykh splavov.* Metallurgiya, Moscow 1966.
[12] Nowlan, G. A.: *Geol. Surv. Profess. Pap.* 700-B, 177 (1970).
[13] Frumina, N. S., Goryunova, N. N., Mustafin, I. S.: *Zh. Anal. Khim.* **22**, 1523 (1967).
[14] Pilipenko, A. T., Maslei, N. N., Skorokhod, E. G.: *Zh. Anal. Khim.* **23**, 227 (1968).

6.27. Niobium

Niobium is situated in column V/2 of the periodic system. In its compounds, it displays similar behaviour to the compounds of V and Ta, elements from the same column. It is generally pentavalent, but occasionally trivalent or tetravalent. Nb_2O_5 is amphoteric in character, but the acidic nature tends to predominate. In aqueous solution, the niobates are readily hydrolyzed; they do not dissolve in NaOH. Nb^{5+} is likewise readily hydrolyzed in mineral acids, and it is therefore kept in solution withs complex formers (fluoride, oxalate, tartrate, H_2O_2, etc.).

Niobium forms coloured complexes with many organic and several inorganic compounds; the spectrophotometric determination of niobium may be based on these complexes. Organic ligands complexing niobium generally contain not only O, but also N and S donor atoms. The reagents are not sufficiently selective, and preliminary separation must therefore be

TABLE 6.22.

More important spectrophotometric reagents for niobium

Reagent	Medium	λ_{max}, nm complex	λ_{max}, nm reagent	$\varepsilon \times 10^3$	Interferences	Refs
KSCN	Extraction with methyl isobutyl ketone, 2.9 M HCl	385	320	37*	Mo, W, Ti, V, Re, Pt	[1]
Bromopyrogallol red	pH 5.8 tartaric acid, complexone III	610	—	53	Al, Th, U	[1]
PAR, in presence of tartrate and acetate	pH 6	555	415	38.7	U, V, PO_4^{3-}	[1]
Lumogallion	0.25–1 M H_2SO_4	515	410	16.8	Ti, Ta, oxalate, citrate	[1, 4]
2,4-Sulpho-chlorophenol S	1–3 M HCl complexone III, tartaric acid	650	560	30	Zr, Hf	[1, 7, 8]
Oxine	2% tartaric acid	385	—	48	Many elements	[1]
Phenylfluorone	0.8% H_3PO_4	520	—	37	Much Ta, Ti, Sb, Ge	[1]

* Datum of the authors.

263

applied (Ta and Ti interfere strongly). Table 6.22 lists spectrophotometric reagents which give niobium complexes with $\varepsilon > 15 \times 10^3$. The spectrophotometry of niobium has been reviewed by Gibalo [1] and Elinson [2].

(A) NIOBIUM DETERMINATION BY THIOCYANATE EXTRACTION

Since the thiocyanate complex of niobium is not sufficiently stable, it gives a coloured compound only in the presence of an organic solvent (acetone) suppressing dissociation, or when extracted into the organic phase. This is one of the most widespread methods for the determination of niobium. Extraction of the niobium–thiocyanate complexes has been investigated in detail by Mari [3]. Depending on the extractant, the absorbance maxima for the complexes lie in the interval 360–400 nm. The occurrence of stepwise complex formation means that care must be taken to maintain the prescribed thiocyanate concentration. A positive error may be caused by the formation of isothiopercyanic acid, with absorbance maximum at 370 nm. This can be avoided if the acid concentration is not too high (< 4 M HCl) [1,2,4].

Certain demands must be met when the choice of extractant is considered: it should not be volatile, it should extract the niobium–thiocyanate complex well, and the complex formed should have a reasonably large molar absorbance. Our studies indicate that, as regards the generally recommended solvents, ether is not suitable because of its high volatility, tributyl phosphate gives rise to a compound with low molar absorbance, and ethyl acetate exerts a considerable extraction effect on titanium. The most suitable solvent for extraction seems to be methyl isobutyl ketone [5].

Determination of niobium content of rocks and ores
(tungsten ore, bauxite, etc.) [5]

The choice of the optimum digestion procedure is of great importance from the aspect of the subsequent analysis. The digestion with pyrosulphate in a platinum dish, that is recommended by various authors, is not suitable, for platinum passes into solution and shows up with a brown colour in the thiocyanate extraction. The presence of pyrosulphate is not advantageous either, as it exerts a masking effect, thereby inhibiting the extraction of niobium.

Fusion with alkali is a favourable digestion procedure, since a number of interfering metals (W, Mo, V, U) can be separated at the same time. After aqueous leaching-out, these are to be found in the solution, while niobium remains in the hydroxide precipitate. Accordingly, the niobium must subsequently be separated only from titanium.

The optimum conditions for formation and extraction of the niobium–thiocyanate complex (>98% Nb extraction) are ensured if the HCl concentration of the aqueous phase is adjusted to 3.1 M, and the thiocyanate concentration to 0.29 M. At higher acidities the reproducibility deteriorates (formation of isothiopercyanic acid), while at lower acidities the efficiency of the extraction decreases. At higher thiocyanate concentrations the disturbing effect of titanium increases, while at lower thiocyanate concentrations the niobium–thiocyanate complex is not extracted so well.

The disturbing effect of titanium can be avoided if the organic phase is washed once or twice with a washing liquid not containing thiocyanate. The titanium–thiocyanate complex, which is otherwise poorly extracted $(D_{Ti} < 0.1)$, then re-enters the aqueous phase. Thus, a quantity of titanium several thousand times larger than that of the niobium does not interfere in the determination. The disturbing effect of Fe^{3+} can be eliminated with $SnCl_2$ reduction.

Analytical procedure: 0.1–1 g sample is fused with 8–10 g NaOH in a silver or nickel dish. The melt is leached out with hot water and filtered, and the solid is well washed with 1% NaOH. The precipitate is washed back into the beaker and is dissolved up at 80°C by the addition of 10 cm³ 15% tartaric acid (the temperature should not go higher, for the niobium may undergo hydrolysis). The solution is transferred to a 50 cm³ volumetric flask, and the volume is made up to the mark with water. An aliquot of 8 cm³ is pipetted into a 50 cm³ separating-funnel, and 4.0 cm³ KSCN (220 g/dm³), 6 cm³ 15% $SnCl_2.2H_2O$ in 4.6 M HCl, and 12 cm³ 1 : 1 HCl are added. The mixture is extracted for 30 s with 5 cm³ methyl isobutyl ketone, and the organic phase is washed with 15 cm³ washing liquid (4 cm 3% tartaric acid+2 cm³ water+3 cm³ 15% $SnCl_2$+6 cm³ 1 : 1 HCl). In the event of a titanium content higher than 1%, the washing is repeated. The organic phase is run into a dry 10 cm³ volumetric flask and diluted with 5 cm³ methyl isobutyl ketone. In this way an optically pure solution is obtained. The absorbance is measured in 1–5 cm cells at 385 nm.

265

The calibration plot is produced with 0.5–25 μg Nb_2O_5 solution, the operations being started from the extraction. Photometry is performed in 5 cm cells in the range 0.5–5 μg, and in 1 cm cells in the range 5–25 μg. The absorbance of 10 μg Nb_2O_5 in a 1 cm cell is 0.28.

Preparation of Nb stock solution: 0.0125 g Nb_2O_5 is fused with 0.5 g $K_2S_2O_7$ for 0.5–1 min at 600–700°C in a platinum dish. The melt is leached out with 3% tartaric acid at 70–80°C, and the solution is made up to 250 cm³ with 3% tartaric acid (50 μg Nb_2O_5/cm³). By dilution, a stock solution containing 5 μg Nb_2O_5/cm³ is obtained. The solution must be prepared freshly at weekly intervals.

The sensitivity of the method is 3 ppm, with an error of ±10%.

Determination of niobium content of steels [6]

In the course of the determination the niobium content of the steel is extracted in the form of the niobium–thiocyanate complex with butyl acetate. HF is added, and the niobium is re-extracted into the aqueous phase. Following H_3BO_3 addition, the thiocyanate complex is again extracted. By this means, an effective separation is achieved; the presence of Mo, Ti, V and W in 50-fold amounts compared to the Nb can be tolerated.

Analytical procedure: 0.2 g sample (0.01% Nb content) is dissolved in a mixture of 10 cm³ HCl and a few drops HNO_3, and the mixture is evaporated to dryness. The evaporation is repeated with 3×5 cm³ HCl. 10 cm³ 10% tartaric acid, 10 cm³ conc. HCl, 2 cm³ $SnCl_2$ solution (50 g $SnCl_2.2H_2O$ is dissolved in 50 cm³ conc. HCl and the volume is made up to 100 cm³), and 5 cm³ 30% KSCN solution are added to the residue. The mixture is transferred to a separating-funnel and extracted for 1 min with 10 cm³ butyl acetate. After separation, the organic phase is shaken with a re-extracting mixture (20 cm³ 6 M HCl+2 cm³ $SnCl_2$ solution+0.8 cm³ 1 : 3 HF). The aqueous phase is washed with 5 cm³ butyl acetate, is then transferred to another separating-funnel, together with 0.5 cm³ niobium-free iron solution (10 mg Fe^{3+}/cm³ in 1.2 M HCl), 5 cm³ 30% KSCN solution and 0.5 g H_3BO_3, and is shaken for 1 min with 10 cm³ butyl acetate. The absorbance of the organic phase is measured at 385 and 470 nm. At 470 nm the disturbing effect of cobalt is measured, and its quantity is corrected for in the calculation of the Nb content. The calibration plot is recorded in the interval 0–30 μg Nb. The sensitivity of the method is 0.01% Nb.

Determination of niobium content of rocks with sulphochlorophenol S [7]

After acidic digestion, the rock sample is fused with pyrosulphate, and its Nb content is extracted from H_2SO_4 medium with methyl isobutyl ketone (MIBK) and re-extracted with water. The complex formed with the reagent is extracted with amyl alcohol.

Analytical procedure: 0.25 g sample (1–40 ppm Nb) is treated with a mixture of 5 cm³ HNO_3 and 10 cm³ HF in a platinum dish, and the contents are evaporated to dryness. The residue is fused with 2 g potassium pyrosulphate. The melt is dissolved in 50 cm³ acid mixture (6 M H_2SO_4 + 2 M HF), and the solution is extracted with 25 cm³ MIBK in a teflon separating-funnel. The organic phase is washed with 20 cm³ acid mixture. The Nb is re-extracted from the organic phase with 2×5 cm³ water. 1 cm³ conc. H_2SO_4 and 0.2 cm³ 1% KCl are added, and the solution is evaporated to dryness in a teflon beaker. The residue is dissolved in 5 cm³ 6 M HCl, and 1.6 cm³ 5% tartaric acid in 4 M HCl, 15 cm³ distilled water and 0.1 cm³ thioglycollic acid are added. The mixture is kept at 60°C for 15 min, 0.3 cm³ 0.3 M EDTA and 0.2 cm³ 0.1% sulphochlorophenol S solution are added, and heating is continued for a further 15 min. 0.5 cm³ 10% diphenylguanidine in conc. HCl is added, and the Nb complex is extracted in a separating-funnel with 5 cm³ amyl alcohol. 1 cm³ ethanol is added to the organic phase, and the absorbance is measured in 1 or 5 cm cells at 656 nm, against a comparison taken throughout the process. The calibration plot covers the interval 0.25–10 µg Nb. The accuracy of the method is ±15%.

References

[1] Gibalo, J. M.: *Analiticheskaya khimiya niobiya i tantala.* Nauka, Moscow 1967.
[2] Elinson, S. V.: *Spektrofotometriya niobiya i tantala.* Atomizdat, Moscow 1973.
[3] Mari, E. A.: *Anal. Chim. Acta* **29**, 303 (1963).
[4] Shcherbov, D. P., *Issledovaniya v oblasti khimicheskikh i fizicheskikh metodov analiza mineralnovo syrya.* ONTI Kaz. IMS, Alma-Ata 1971.
[5] Mohai, M., Upor, E.: *Magy. Kém. Folyóirat* **72**, 394 (1966).
[6] Iyer, C. S. P., Kamath, V. A.: *Talanta* **27**, 537 (1980).
[7] Childress, A. E., Greenland, L. P.: *Anal. Chim. Acta* **116**, 185 (1980).
[8] Savvin, S. B.: *Organicheskie reagenty gruppy arzenazo III.* Atomizdat, Moscow 1971.

6.28. Nitrogen compounds

Nitrogen belongs in column V/1 of the periodic system. Its oxidation number varies between -3 and $+5$. The determination of nitrogen compounds (nitrate, nitrite, ammonia, cyanide) in foodstuffs, soil and drinking-water is of importance, as is the determination of nitrides in metallurgy.

Some of the procedures here are among the oldest methods in photometry (Messler, 1856; Berthelot, 1859; Griess and Ilosvay, 1879). They are characterized by the fact that the determinations are based not on simple complex formation, but on participation in the synthesis of some organic compound.

The most widespread groups of methods for the determination of nitrate are as follows [1, 2]:

1. Nitration of phenols, and measurement of the absorbance of the nitrophenols formed.
2. Oxidation of organic compounds, and utilization of the coloured oxidation product for determination.
3. Reduction, followed by determination of NO_2^- or NH_4^+.

Nitrate absorbs in the UV, and this too can be used for its determination [3], though this absorbance is rather usually regarded as a cause of interference in other determinations in the UV.

Nitrite is primarily determined by the diazotization of primary aromatic amines, and by photometry of the azo dyes formed by coupling with phenols or amines.

In the determination of ammonia (or NH_4^+), use is made of the Nessler or the indophenol method, similarly the participation of NH_3 in the formation of a coloured compound.

The spectrophotometric determination of cyanide is based on the Kőnig reaction. A developed variant of this is the use of the sodium isonicotinate–sodium barbiturate reagent [4].

Data on reagents that are the most important or are used in determinations described here are given in Table 6.23.

Detailed surveys of the analysis of nitrogen compounds are provided by the monograph of Strenli and Averell [5], and by books dealing with the determination of non-metallic elements [1,2]. Sawicki et al. [6] evaluate 52 different methods proposed for the determination of NO_2^-.

TABLE 6.23.

Features of some methods used for determination of nitrate, nitrite, ammonia and cyanide

Basis of determination	λ_{max}, nm	ε_{max}	Interferences	Refs
NO_3^-.aq	302	7.2	NO_2^- and other compounds absorbing in the UV	[3]
2,6-Xylenol + NO_3^- $\xrightarrow{H_2SO_4}$ 4-nitro-2,6-xylenol	310	2.1×10^3	NO_2^-, Cl^-, organic compounds absorbing in the UV	[7]
Brucine + NO_3^- → oxidation	410	1.5×10^3	NO_2^-	[1]
Salicylic acid + NO_3^- → nitrosalicylic acid	410		Fe^{2+}	[9]
Sulphanilic acid + NO_2^- + α-naphthylamine → diazo product	520	3.3×10^4	Fe^{3+}, Cu^{2+}, strong oxidants or reductants	[6]
$2 HgI_4^- + 2 NH_3 \rightarrow$ $NH_2Hg_2I_3$	375	2.1×10^3	Ions yielding a precipitate in alkaline medium	[1]
Phenol + ClO^- + NH_3 → indophenol (in n-BuOH)	680	5.3×10^3	Ions yielding a precipitate in alkaline medium	[1]
CN^- + barbituric acid + chloramine T + pyridine → polymethyne dye	580	1.4×10^5	NCS^-, reductants, oxidants	[14]

Distillation (possibly the microdiffusion variant) is suitable for the separation of NH_3 from interfering components. Precipitation, extraction, ion-exchange and masking of the disturbing metal ions may be considered in the determination of NO_3^- and NO_2^-.

Determination of nitrate content of soils with 2,6-dimethylphenol [7]

Analytical procedure: 25–50 g air-dried soil sample is shaken mechanically for 1 hour with 125 cm³ 1% KCl solution. After filtration, the disturbing chloride is precipitated with Ag_2SO_4 solution and filtered off. A 10 cm³ filtrate aliquot is taken in a 50 cm³ volumetric flask, a little solid sulphanilic acid is added to bind possible nitrite traces, then under cooling 30 cm³ conc. H_2SO_4 and 5 cm³ 2,6-dimethylphenol (12.2 g in 1 dm³ glacial acetic

269

acid) are added, and the solution is made up to the mark. The absorbance is measured at 310 nm. The calibration plot is produced with 0.1–2.5 mg NO_3^-.

Determination of nitrate and nitrite contents of river water [8]

The NO_2^- to be found in the water diazotizes added p-aminoacetophenone, which is then coupled with m-phenylenediamine. The dye formed (2,4-di-amino-4'-acetylazobenzene) is subjected to photometry. The NO_3^- is reduced in a Cd–Cu column and determined as NO_2^-.

Analytical procedure: The water is filtered through a membrane filter (0.45 μm). A 25 cm^3 portion is pipetted into a 50 cm^3 volumetric flask, and 10–20 cm^3 distilled water and 0.5 cm^3 p-aminoacetophenone (0.5 g dissolved in 5 cm^3 conc. HCl, then diluted to 50 cm^3) are added. The pH of the solution should be 2.0–2.5. After 5 min, 0.5 cm^3 m-phenylenediamine (1.34 g m-phenylenediammonium chloride is dissolved in 0.08 cm^3 conc. HCl and made up to 50 cm^3) is added, and the flask is filled up to the mark. After 30 min, the absorbance of the solution is measured in 10 cm cells at 460 nm. The molar absorbance of the complex is 4.7×10^4.

In the determination of the nitrate content of the water, 5 cm^3 buffer (0.1 M NH$_4$Cl, 0.1 M NH$_4$OH, pH 9.3) is added to 10 cm^3 of the sample water and the NO_3^- is reduced to NO_2^- on a Cd–Cu column (grain size 20–60 mesh, column cross-section 0.8 cm^2, length 14 cm, flow rate 0.048 cm^3/s, pH 9.5–10.5, reduction efficiency 97%). The column is washed with 5 cm^3 5-fold diluted buffer solution and with distilled water. The further steps in the determination are as described for nitrite.

The determination is not disturbed by the components of average river-water. The calibration plot covers the range 1–10 μg NO_2^-/50 cm^3.

Determination of ammonium ion content of waste water [9]

The ammonium ion content of water can be determined directly by means of the Nessler reagent (10 g HgI$_2$ + 5 g KI + 20 g NaOH, in 100 cm^3 water) in the presence of Seignette salt (potassium sodium tartrate) as masking agent. The ammonium ion gives a yellowish-brown colour with an alkaline solution of dipotassium tetraiodomercurate. In some cases, the ammonium ion content of the water sample is separated in order to avoid interfering effects. Here, 20 cm³ 40% NaOH or (in the presence of urea) 20 cm³ buffer solution (14.3 g K_2HPO_4+68.8 g KH_2PO_4, in 1 dm³, pH 7.4) is added to 500 cm³ water sample. The solution is distilled into 20 cm³ 0.01 M H_2SO_4 solution, and the volume is made up to 250 cm³. 1 cm³ Nessler reagent is added to an aliquot of the sample (less than 50 cm³), the solution is made up to 50 cm³, and the absorbance is measured at 425 nm. The calibration plot is obtained for 2–100 μg NH_4^+/50 cm³.

Determination of nitrogen dioxide in air [10, 11]

A known volume of air is bubbled through Saltzmann reagent; this solution absorbs NO_2, which gives a pink-coloured azo dye with the reagent.

Analytical procedure: 10 cm³ Saltzmann reagent (5 g sulphanilic acid+400 cm³ water+50 cm³ glacial acetic acid+0.05 g N-(1-naphthyl)-ethylenediamine dihydrochloride+10 cm³ acetone, in 1 dm³) is taken in a bubbling-vessel, and sufficient air is bubbled at a rate of 0.5–1 dm³/min through the solution so as to give a well visible coloration. If necessary, the volume is made up to 10 cm³. After 15 min, the absorbance of the solution is measured at 545 nm.

The SO_2 content of the air lowers the value; this can be avoided by adding H_2O_2 to the solution.

The calibration plot is recorded with solutions containing 0.2–2 μg NO_2^-/10 cm³.

If the sampling process requires a long time, the determination is modified, in that the air is bubbled through 0.1 M triethanolamine solution, which is subsequently mixed with Saltzmann reagent.

If the NO content of the air is to be determined, a MnO_2 oxidizing column is inserted before the sorption vessel.

In the joint determination of nitrogen oxides, the air is bubbled through NaOH solution, and the absorbance of the red diazotized sulphanilic acid–α-naphthylamine product (Griess–Ilosvay reagent) formed after acidification is measured.

271

In the presence of γ-picoline (see Appendix), barbituric acid and chloramine T, cyanide gives a violet-blue product.

Analytical procedure: 5 cm^3 buffer (13.6 g $KH_2PO_4 + 0.28$ g Na_2HPO_4, in 1 dm^3) and 0.25 cm^3 1% chloramine T are added to 10 cm^3 water sample, followed after 1-2 min by 3 cm^3 reagent (3 g barbituric acid + 15 cm^3 γ-picoline + 3 cm^3 conc. HCl, in 50 cm^3). After 5 min, the absorbance is measured at 605 nm at 25°C.

The reagent is also suitable for the determination of thiocyanate.

The calibration plot covers the interval 0.5-5 μg CN^-. The sensitivity of the method is 50 μg CN^-/dm^3.

The sensitivity can be increased if the cyanide is first concentrated [13]. The course of the concentration is as follows: 500 cm^3 water is extracted with 25 cm^3 0.25 M tributyltin (TBT) hydroxide in trichloroethylene, and the CN^- is re-extracted with 2 M NaOH.

10 g $(NH_4)_2SO_4$, 4 g Na_2CO_3 and 1 g EDTA are added to 500 cm^3 of the water to be analyzed. After dissolution, 25 cm^3 0.25 M TBT (8 cm^3 TBT chloride is dissolved in 92 cm^3 trichloroethylene and mixed with 50 cm^3 2 M NaOH; the organic phase is used) is added, and the mixture is shaken mechanically for 40-60 min. The organic phase is transferred to a small separating-funnel, and the cyanide is re-extracted with 3.5 cm^3 2 M NaOH for 30 min. The CN^- content of the resulting solution is determined by the above photometric method.

References

[1] Babko, A. K., Pilipenko, A. T.: *Fotometricheskii analiz. Metody opredeleniya nemetallov.* Khimiya, Moscow 1974.
[2] Boltz, D. F.: *Colorimetric Determination of Nonmetals.* Wiley, New York 1978.
[3] Wetters, J. H., Kenneth, L. U.: *Anal. Chem.* **42**, 335 (1970).
[4] Nagashima, Sh.: *Anal. Chim. Acta* **99**, 1979 (1978).
[5] Strenli, C. A., Averell, P. R.: *Analytical Chemistry of Nitrogen and its Compounds.* Interscience, New York 1968.
[6] Sawicki, E., Stonley, T., Pfoff, J., D'Amico, A.: *Talanta* **10**, 641 (1963).
[7] Hajós, P., Inczédy, J.: *Hung. Sci. Instr.* **34**, 25 (1975).
[8] Okada, M., Miyata, H., Toei, K.: *Analyst* **104**, 1195 (1979).

[9] Chovanecz, T.: *Ipari vízvizsgálatok (Industrial water examinations)*. Műszaki Könyvkiadó, Budapest 1977.
[10] Jacobs, M. B., Hochheiser, S.: *Anal. Chem.* **30**, 426 (1958).
[11] Várkonyi, T., Cziczó, T.: *A levegőminőség vizsgálata (Air-quality investigations)*. Műszaki Könyvkiadó, Budapest 1980.
[12] Nagashima, Sh.: *Anal. Chim. Acta* **91**, 303 (1977).
[13] Wronski, M.: *Talanta* **28**, 255 (1981).
[14] Lure, Yu. Yu. (ed.): *Unifitsirovannye metody analiza vod*. Khimiya, Moscow 1973.

6.29. Noble metals (silver, gold, platinum metals)

The analytical properties of the noble metals and the interdependence of the analytical tasks justify the discussion of these metals together.

Silver and gold are to be found in column I/2 and the platinum metals in column VIII/2 of the periodic system. All of these metals occur in several valency states; the most characteristic ions in solution are Ag^+, Au^{2+}, Au^{3+}, Rh^{3+}, Pd^{2+}, Ir^{3+}, Ir^{4+}, Pt^{2+} and Pt^{4+}. $Os(VIII)$ complexes are characteristic for osmium. In behaviour, these latter are primarily similar to Fe and Mn; Ag^+ is similar to Tl^+; Au^+ to Ag^+ and Hg^+; Pd^{2+} to Ag^+; Rh^{3+} and Ir^{3+} to Co^{3+}; and Pt^{2+} to Au^{2+}, Ni^{2+} and Co^{2+}.

The noble metals display strong complex-forming properties in virtually all their oxidation states. They form complexes with ligands containing S, O and N donor atoms. The complexes are generally of fairly high stability, and in separations and determinations this permits relatively selective complex formation even with complex formers which otherwise react with many ions. The interfering elements are primarily other noble metals, because of their frequent co-occurrence. This does not generally mean a major problem for silver in higher concentrations. The most common analytical tasks arise in the investigation of sulphide ores, ultrabasic rocks, metallurgical side-products and various industrial products.

Since these metals occur in a scattered manner, partly in the elemental state, in rocks, it is often necessary to take an unusually large sample (5–20 g) in order to attain analytical results truly representative of the material. A large role is played in their separation by the elemental state, e.g. the preparation of metallic beads by ignition.

Possibilities for their extraction separation from solution are provided by complex formers also suitable as photometric reagents, and by halide complexes. The latter may be used for anion-exchange separation too.

A detailed guide to the determination of the noble metals is given by a number of relevant books [1–5a] and reviews; as concerns the latter, Beamish [6] deals with the separation and determination of the metals, Faye and Moloughney [7] with their concentration by firing, Lisitsina et al. [8] with the separation of silver, Stolyarov [9] with the photometric reagents for silver, Shkrobot [10] with the photometric reagents for gold and Puddephatt [10a] with complexes of gold. Coombes and Chow [11] give a critical account of the determination of platinum in ores. Khvostova and Golovnya [11a] have published a review of the methods of chemical digestion of platinum-containing ores and rocks. Kalinyin [12] discusses the présent state of the analysis of the platinum metals, and Savvin [13] surveys the possibilities for the photometric determination of the noble metals.

Myasoedova et al. [13a, 13b] deal with the separation and enrichment of noble metals on chelate sorbents.

Dithizone is suitable for the determination of all noble metals. Additionally, p-benzylidene-rhodanine and its azo derivatives, the Michler thioketone (see Appendix) and the mixed complexes of Ag^{2+} (o-phenanthroline + bromopyrogallol red, etc.) are of use in the analysis of silver. Primarily the ion-pairs formed with triphenylmethane derivatives (crystal violet, brilliant green, etc.) and also the rhodanine derivatives, may come into consideration for the photometric determination of gold. The photometric reagents for the platinum metals are of essentially the same type as those for gold, but measurement of the absorbance of the colloidal metals is also suitable for the solution of certain analytical tasks.

With regard to the noble metals, the photometric methods for palladium are the most widespread. This can be attributed to the good complex-forming properties of palladium, and to its high kinetic activity towards organic reagents containing various functional groups [13].

Determination of silver traces in galena [14]

After acidic dissolution, the solution is extracted with dithizone in $CHCl_3$. Ag, Cu and Hg are extracted, but not Pb. The organic phase is shaken with HCl + NaCl solution, which extracts the Ag, but leaves the Cu and Hg in the organic phase. The absorbance of silver dithizonate is measured.

274

Analytical procedure: 20–60 mg ore (Ag content 5–45 μg) is evaporated with 2 cm³ conc. $HClO_4$, and the residue is dissolved in 50 cm³ 1 : 99 HNO_3. The solution is shaken with 10 cm³ 0.01% dithizone in $CHCl_3$. The silver is re-extracted from the organic phase with 3 cm³ 5% NaCl in 0.015 M HCl. The aqueous phase is washed with 2×5 cm³ $CHCl_3$. The pH of the aqueous phase is adjusted to 4.6 with sodium acetate solution (1.6 g/dm³), the volume is made up to 30 cm³, and the silver is extracted with 10 cm³ 0.01% dithizone in $CHCl_3$. Photometry is performed at 610 nm. The calibration plot is produced with 5–50 μg Ag.

If it is desired to increase the sensitivity, a dithizone excess is avoided in the extraction and the photometry is carried out at 460 nm, where the absorbance is maximum (the green dithizone does not absorb at 610 nm, but the absorbance of silver dithizonate is much lower there).

Determination of gold in soil and rocks with brillant green [15]

Analytical procedure: 20 g sample is treated with 100 cm³ acid mixture (400 cm³ conc. HCl+100 cm³ conc. HNO_3 in 1,000 cm³), and then evaporated to dryness. The residue is heated for 15 min with 100 cm³ 2 M HCl, the mixture is filtered, and the filtrate is made up to 200 cm³ with 2 M HCl. An appropriate aliquot is diluted with 2 M HCl to 50 cm³, 1 cm³ 4% $CuSO_4$, 2 cm³ 0.2% $TeCl_4$ and 10 cm³ 20% $SnCl_2$ in 2 M HCl are added, and the mixture is boiled. It is kept hot for 30 min, and the tellurium containing the gold is filtered off, washed with 2 M HCl and dissolved in 1 cm³ hot aqua regia. The solution is evaporated to dryness, and the residue is dissolved with mild heating in 2.5 cm³ 2 M HCl. The solution is diluted to 10 cm³ with water, 1.0 cm³ 0.05% brilliant green in alcohol is added, and the mixture is shaken for 30 s with 5 cm³ toluene. The organic phase is washed with 10 cm³ 0.5 M HCl+0.05 cm³ 1% brilliant green, and the colour is either subjected to photometry at 650 nm or is compared with the colours of a standard series prepared with 0.05–2.0 μg Au. This procedure permits the determination of 0.1 ppm Au.

Determination of a small quantity (0.01 ppm) of gold in metallic copper [16]

Analytical procedure: 1 g sample is boiled for 10 min in 4 cm³ HNO_3+3 cm³ HCl, then 20 cm³ 20% NH_4OH is added, and the mixture is diluted to 50 cm³ with water. 2 cm³ metallic mercury is added to an aliquot containing 0.1–6 μg Au to precipitate the gold, and the mixture is stirred magnetically for several min. The Au-containing mercury is filtered off, washed with water, dried with filter paper, and heated in a stream of N_2 in a quartz tube at 350°C to drive off the mercury. 2 drops conc. HNO_3, 1 drop water and 3 drops aqua regia are added to dissolve the residual gold, and the solution is evaporated to dryness with HCl several times on a water-bath to remove HNO_3. 2.5 cm³ 0.12 M HCl and 5.0 cm³ 4% NaCl are added, the solution is heated for 5–10 min on a water-bath and then cooled, and 0.5 cm³ 1% NaF and 0.5 cm³ 0.003% 5-(p-dimethylaminobenzylidene)-rhodanine in alcohol are added. The absorbance is measured at 562 nm, after 1 min.

Determination of palladium in alloys, with an azo derivative of oxine [17]

Of the azo derivatives of oxine, the most suitable for the determination of palladium is 5-(2-hydroxy-3,5-dinitrophenylazo)-N-oxide-8-hydroxy-quinoline (reagent I). Details of the synthesis of this reagent are given by Nemodruk *et al.* [18].

Reagent I is stable from an acidity of 3 M (HNO_3, $HClO_4$) or 1.5 M (H_2SO_4) up to pH 10. With palladium, it forms a stable complex, insoluble in water, but well soluble in acetone: $\lambda_{max} = 530$ nm, $\varepsilon = 2.1 \times 10^4$. The method is suitable for the determination of the palladium content of nickel, magnesium–nickel and magnesium–silver-based alloys, in which the palladium content attains 0.01–2%.

Analytical procedure: 0.1–1 g alloy is dissolved in 12–25 cm³ 1 : 1 HNO_3, the solution is boiled to eliminate oxides of nitrogen, and it is then evaporated to dryness. The residue is dissolved in 3 M HNO_3, and made up to 50 cm³ in a volumetric flask with 3 M HNO_3. An aliquot of 1–5 cm³ (7–65 μg Pd) is transferred to a 25 cm³ volumetric flask, and sufficient conc. HNO_3, is added for the final acidity to be 3 M. The volume is made up to 12.5 cm³ with water, 10 cm^3 2.5$\times$10^{-4} M reagent I in acetone is added, and the volume is made up to the mark with acetone. The solution is kept at 60–65° C (water-bath) for 20–25 min and, after cooling, the volume is made up to the mark with acetone. The absorbance of the solution is measured in 1–2 cm cells, against a blank taken throughout the process. The calibration plot is produced with 3–70 μg Pd, in the above manner. The error of the analysis does not exceed $\pm$7%.

References

[1] Beamish, F. E.: *The Analytical Chemistry of the Noble Metals.* Pergamon, Oxford 1966.

[2] Grinzburg, S. I., Ezerskaya, I. V., Fedorenko, N. V., Shlenskaya, V. I., Bel'skii, N. K.: *Analytical Chemistry of Platinum Metals.* Halsted–Wiley, New York 1975.

[3] Analytical Chemistry of the Platinum-Group Metals. Symposium, Johannesburg 1972.

[4] Busev, A. J., Ivanov, V. M.: *Analiticheskaya khimiya zolota.* Khimiya, Moscow 1973.

[5] Pyatnitskii, I. V., Sukhan, V. V.: *Analiticheskaya khimiya serebra.* Nauka, Moscow 1971.

[5a] Basargin, N. N., Rozovskiy, Yu. G.: *Novye organichekie reagenty v analize blagorodnykh metallov.* Metallurgiya, Moscow, 1982.

[6] Beamish, F. E.: *Minerals Sci. Eng.* **4**, 3 (1972).

[7] Faye, G. H., Moloughney, P. E.: *Talanta* **19**, 269 (1972).

[8] Lisitsina, D. B., Shcherbov, D. P., Talatynova, N. A.: *Issledovaniya v oblasti khimicheskikh i fizicheskikh metodov analiza mineralnovo syrya.* ONTI Kaz. IMS, Alma-Ata 1973.

[9] Stolyarov, K. P.: *Instrumentalye i khimicheskie metody analiza.* Leningrad University, 1973.

[10] Shkrobot, E. P.: *Analiz rud tsvetnykh metallov i produktov ikh pererabotki.* (Collection of articles.) Metallurgiya, Moscow 1971.

[10a] Puddephatt, R. J.: *The Chemistry of Gold.* Elsevier, Amsterdam, 1978.

[11] Coombes, R. J., Chow, A.: *Talanta* **26**, 991 (1979).

[11a] Khvostova, V. P., Golovnya, S. B.: *Zav. Lab.* **48**, 7, 3 (1982).

[12] Kalinin, S. K.: *Zh. Anal. Khim.* **35**, 2226 (1980).

[13] Savvin, S. B.: *Zh. Anal. Khim.* **35**, 1818 (1980).

[13a] Myasoedova, G. V., Antokolskaya, I. M., Emets, L. V., Danilova, E. Ya, Naumenko, E. A., Danilova, F. I., Fedotova, I. A., Seraya, V. I., Mushchiy, R. Ya., Volf, L. A., Savvin, S. B.: *Zh. Anal. Khim.* **37**, 1574 (1982).

[13b] Myasoedova, G. V., Antokolskaya, I. M., Danilova, F. I., Fedotova, I. A., Grinkov, V. P., Ustinova, N. V., Naumenko, E. A., Danilova, E. Ya., Savvin, S. B.: *Zh. Anal. Khim.* **37**, 1578 (1982).

[14] Bounsall, E. J., McBryde, W. A. E.: *Canad. J. Chem.* **38**, 1488 (1960).
[15] Stanton, R. E., McDonald, A. J.: *Analyst* **89**, 767 (1964).
[16] Mizuitke, A.: *Talanta* **9**, 948 (1962).
[17] Kleimenova, O. K., Nemodruk, A. A., Gibalo, I. M.: *Zh. Anal. Khim.* **35**, 2170 (1980).
[18] Nemodruk, A. A., Gibalo, I. M., Kleimenova, O. K.: *Zh. Anal. Khim.* **32**, 457 (1977).

6.30. Phosphorus

Phosphorus belongs in column V/1 of the periodic system. In its chemical properties it is more similar to arsenic than to nitrogen. It is an element of physiological significance, and thus its analysis is important from many aspects. In its compounds it is either trivalent or pentavalent. When it occurs in the lower oxidation state, e.g. in phosphides, an oxidant, e.g. $KMnO_4$, must be used in the dissolution stage to yield PO_4^{3-}. Polyphosphates possibly present can be transformed to PO_4^{3-} by acidic hydrolysis.

PO_4^{3-} gives precipitates with very many metal ions. This can be used for its concentration and separation, but in many cases it makes the determination more difficult.

As in the case of silica, only heteropolyacids can come into consideration for the photometric determination of phosphate. The two fundamental possibilities are the measurement of the absorbance of the yellow phosphomolybdic acid, $H_3[P(Mo_3O_{10})_4]$, and that of its reduction product, phosphomolybdenum blue. Below, therefore, we shall discuss these in some detail on the basis of monographs and reviews dealing with the analysis of phosphorus compounds [1–6].

The optimum acidity for the formation of phosphomolybdenum yellow is 0.12 M. In such a medium, the otherwise interfering silicomolybdic acid is not formed (if complex formation is carried out at pH 1.5, which is favourable for formation of the latter compound, and the acid concentration is subsequently increased to 2 M, the phosphate complex decomposes, and thus silica can be determined in the presence of PO_4^{3-}). The maximum absorbance of the complex lies at 310 nm ($\varepsilon_{310} = 1.2 \times 10^4$). The molybdate too absorbs in the UV, and hence the phosphorus can be determined only at higher wavelengths (≥ 345 nm) and with a much lower sensitivity. The extraction of phosphomolybdenum yellow (with alcohols or esters) helps;

278

the reagent is not extracted, so that measurements may be made at the maximum in the absorbance curve.

The sensitivity of the method can be improved in two further ways. One possibility is to form the complex also containing vanadium ($P_2O_5 \cdot V_2O_5 \cdot 22MoO_3 \cdot nH_2O$), and the other is the extraction of phosphomolybdenum yellow with basic dyes (crystal violet, safranine, etc.). By this means the sensitivity is raised to $\varepsilon_{max} = 3-8 \times 10^4$.

A more widespread procedure for the determination of phosphorus, but one which is burdened with many possibilities of error, is the photometry of molybdenum blue, formed by the reduction of phosphomolybdenum yellow.

The composition of molybdenum blue has not been elucidated; it contains Mo(V) and Mo(VI) in various proportions, depending on the conditions of reduction, and thus its spectrum too varies. The use of a strong reductant and high acidity increases the possibility of reduction of the reagent, and hence causes a positive error. Milder reductants (1-amino-2-naphthol-4-sulphonic acid, hydroquinone, ascorbic acid, etc.) are better. In contrast with phosphomolybdenum yellow, the dye is a colloid; accordingly, its absorbance depends strongly on the temperature and ionic strength too.

The determination can similarly be combined with extraction. Since the maximum absorbance is in the near infrared (780–830 nm), fewer interferences need be reckoned with.

Both determinations may be disturbed by As^{3+} and Ge^{4+}, which likewise form heteropolyacids. If they are present in larger amounts, they may be removed by extraction or distillation of their chlorides. The disturbing effects of Fe^{3+}, Cu^{2+} and other coloured ions may generally be eliminated by reduction, and in some cases by separation. The effects of ions which form precipitates with phosphate even in the acidic medium, or which yield involved complexes in the presence of the reagent, can be inhibited by separation with NaOH (e.g. Ln^{3+}, Zr^{4+}), by masking with fluoride (Nb, Ti), or by some other special method. For certain samples, this is a question that is very difficult to solve.

279

Determination of phosphorus in rocks (based on [13] and the investigations of the authors)

The phosphorus content of around 0.1% in silicate rocks exists in the form of apatite grains that can be practically completely dissolved out with HNO_3. In the interest of a more exact analysis, total digestion is strived for. This is achieved by fusion with $NaOH + Na_2O_2$, or by dissolution with $HF + HClO_4$, and fusion of the insoluble residue with Na_2CO_3. The phosphorus content is determined in the form of phosphomolybdenum blue, by photometry.

Course of the analysis: 0.200 g sample in a Pt dish is evaporated to dryness with a mixture of 10 cm^3 conc. $HF + 3$ cm^3 conc. $HClO_4$. The residue is again evaporated to dryness with 1 cm^3 saturated H_3BO_3. After dissolution in 5 cm^3 1 : 1 HNO_3 and 20 cm^3 distilled water, the mixture is filtered, and the precipitate is washed 4–5 times with distilled water. The filter paper is ashed, and the residue is digested with 1 g Na_2CO_3. The melt is leached with hot water, and the solution is acidified with a little HNO_3 and mixed with the first filtrate. The volume is made up to 50 cm^3.

30 cm^3 distilled water and 8 cm^3 reagent are added to a 1 cm^3 aliquot of the stock solution, and the timing is begun. The volume of the solution is made up to 50 cm^3, and photometry is performed in 5 cm cells, at 882 nm, after 10 minutes. The composition of the reagent: a mixture of 125 cm^3 2.5 M H_2SO_4, 37.5 cm^3 4% NH_4-molybdate, 12.5 cm^3 K-antimonyl tartrate (1 mg Sb/cm^3) and 75 cm^3 freshly prepared ascorbic acid (1.32 g in 75 cm^3 distilled water).

The calibration plot is prepared in the interval 5–80 μg P_2O_5/50 cm^3.

Notes:

(a) The colour of the solution continuously becomes stronger. The waiting period of 10 minutes must be adhered to exactly.

(b) In the present case, better photometric sensitivity can be ensured not by a larger aliquot, but by a greater cell length.

Determination of phosphorus traces in soil [7]

Analytical procedure: 0.5–1 g material is fused in a platinum dish with ca. 5 g anhydrous Na_2CO_3 or $NaKCO_3$. After leaching-out with water, acidification, and dehydration of silica, a 250 cm³ stock solution is prepared. 10 cm³ (or less) of this is diluted and heated to boiling. The iron is reduced in the hot with aluminium foil until the solution is decolorized. The excess acid is neutralized with 10% NH_4OH using α- or β-dinitrophenol as indicator. The volume is made up to 90–95 cm³, and 4 cm³ H_2SO_4–molybdate solution is added (280 cm³ conc. H_2SO_4 is added to 500 cm³ water, and the solution is cooled; in another vessel, 25 g ammonium molybdate is dissolved in 200 cm³, and the solution is heated to 60°C, cooled, and filtered if opalescent; the molybdate solution is added to the diluted H_2SO_4, and the volume is made up to 1 dm³). Reduction is next carried out with 0.3 cm³ $SnCl_2$ solution, and the volume is made up to 100 cm³ ($SnCl_2$: 2.5 g $SnCl_2$. $2H_2O$ is dissolved in 24 cm³ conc. HCl, and the solution is made up to 100 cm³). After 10 min, the absorbance is measured at 725 nm ($\varepsilon = 1.5 \times 10^4$). The calibration plot covers the interval 10–200 μg P_2O_5/100 cm³.

Determination of dissolved phosphate in natural water (sea water) [8]

In the experience of the authors, the reproducibility of ascorbic acid reduction is better than that in the $SnCl_2$ method. The colour formation is relatively slow, but is accelerated catalytically by potassium antimonyl tartrate.

For the determination, 8 cm³ freshly prepared mixed reagent is added to 40 cm³ of the water to be examined, and the volume is made up to 50 cm³ (mixed reagent: 37.5 cm³ 4% ammonium molybdate, 75 cm³ 0.1 M ascorbic acid, 0.037 g potassium antimonyl tartrate and 12.5 cm³ distilled water are added to 125 cm³ 2.5 M H_2SO_4; the solution is used on the same day). After 10 min, the absorbance is measured at 882 nm. The calibration plot is recorded for 10–70 μg P, in 1–2 cm cells.

Determination of phosphorus in iron ore [9]

Iron ore containing ca. 300 μg P is fused with a mixture of 0.5 g Na_2CO_3 and 2 g Na_2O_2 in a graphite crucible. The melt is dissolved out with water, 10 cm³ 70% $HClO_4$ is added, and the solution is boiled, cooled and made up to 50 cm³. 5 cm³ water, 1 cm³ 70% $HClO_4$ and 100 cm³ freshly prepared

281

15% Na_2SO_3 solution are added to a 10 cm³ aliquot, and the mixture is boiled. 20 cm³ of a molybdate–hydrazine mixture is added (55 cm³ water+25 cm³ 2% $(NH_4)_6Mo_7O_{24}.4H_2O$ in 2 M H_2SO_4+10 cm³ freshly prepared 0.15% hydrazine sulphate, made up to 100 cm³ with water). The solution is heated on a boiling water-bath for 10 min, cooled, and made up to 50 cm³. The absorbance is measured at 725 nm. The calibration plot is recorded in the range 25–150 µg P/50 cm³, as described above.

Determination of phosphorus traces in semiconductor silicon [10]

The silicon is removed by fuming with HF in the presence of Au(III) as catalyst. The phosphorus is measured spectrophotometrically.

Analytical procedure: The surface of the sample is cleaned with a washing liquid (trichloroethylene+acetone+methanol) vibrated with ultrasound, and it is then etched with a 1 : 1 mixture of $HF+HNO_3$. The material is finely ground in an agate mortar. 10–100 mg is taken in a teflon digestion vessel, and 3 drops 1% $HAuCl_4$, 5 cm³ 6 M HF and 2 cm³ 36% H_2O_2 are added. To make the digestion complete, the covered vessel is placed in a glycerol-bath at 110°C for 45 min, and then evaporated, 5 cm³ 1 M HCl, 5 cm³ 2% ammonium molybdate and 5 cm³ *n*-butyl acetate are added to the residue. The solution is extracted for 2 min, and the absorbance is measured in 10 mm quartz cells at 315 nm. The calibration plot is produced with 1–10 µg P.

Extraction–photometric determination of phosphorus in WO_3 [11]

Principle of method: The W(VI) ions formed on alkaline digestion react with the phosphorus to yield tungstophosphoric acid. The absorbance of tungstophosphoric acid coincides with that of the simultaneously formed isopolytungstate ions ($\lambda_{max}=265$ nm). However, the tungstophosphoric acid can readily be separated from isopolytungstate by extraction with oxygen-containing organic solvents [12]. Thus, the phosphorus in the WO_3 can be determined without preliminary separation. The determination is not disturbed by a 20-fold amount of Si.

Analytical procedure: 2 g WO_3 is dissolved in 5 cm³ 4 M NaOH with the

282

addition of a few drops of bromine water. The excess bromine is removed by heating, the remaining bromate and bromite are decomposed by the addition of 30% H_2O_2, and the excess H_2O_2 is removed by boiling. The remaining solution is made up to 25 cm^3 in a volumetric flask with water. If the solution is turbid, it must be filtered. A 1.6 cm^3 aliquot of the solution (1–20 μg P content) is pipetted into a 25 cm^3 volumetric flask, 2.5 cm^3 0.25 M H_2SO_4 is added, the mixture is diluted to ca. 15 cm^3, its pH is adjusted to 1.6 (checked on a pH-meter), and the volume is made up to 25 cm^3 with water. The resulting solution is extracted for 2 min with 5 cm^3 n-amyl alcohol, and the organic phase is then centrifuged for 10 min. 2.5 cm^3 ethanol and 1.5 cm^3 distilled water are added to 1 cm^3 extract, and the absorbance of the solution is measured in 1 cm quartz cells at 265 nm, against a mixture of 1 cm^3 n-amyl alcohol + 2.5 cm^3 ethanol + 1.5 cm^3 water.

Production of calibration plot: 2 g WO_3 (P content $< 1 \times 10^{-4}$ %) is dissolved in alkali, as described above. 1.6 cm^3 of the stock solution is transferred to each of five 25 cm^3 volumetric flasks, and solutions with P contents in the range 1–20 μg are added. The subsequent procedure is as above.

The sensitivity of the method is 8×10^{-4} % P.

References

[1] Boltz, D. F.: Colorimetric Determination of Nonmetals. Wiley, New York 1978.
[2] Skararavskii, Yu. F., Lynchak, K. A., Chernogorenko, V. B.: Zav. Lab. 47, No. 1, 3 (1981).
[3] Kazanskii, L. P., Torchenkova, E. A., Spitsyn, V. I.: Uspekhi Khim. 43, 1137 (1974).
[4] Babko, A. K., Pilipenko, A. T.: Fotometricheskii analiz. Metody opredeleniya nemetallov. Khimiya, Moscow 1974.
[5] Lyalikov, Yu. S.: Analiticheskaya khimiya fosfora. Nauka, Moscow 1974.
[6] Fedorov, A. A.: Novye metody opredeleniya fosfora. Metallurgiya, Moscow 1965.
[7] Arinushkina, E. V.: Rukovodstvo po khimicheskomu analizu pochv. 2nd. Ed. Moscow University, 1970.
[8] Goryushina, V. G., Esenina, N. V., Snesarev, K. A.: Zh. Anal. Khim. 25, 1610 (1970).
[9] Bhargava, Om. P., Gmitro, M., Hines, G. W.: Talanta 27, 263 (1980).
[10] Buldini, P. L., Sandrini, D.: Anal. Chim. Acta 98, 401 (1978).
[11] Chkanikova, O. K., Krivtsova, V. Yu., Dorokhova, E. N.: Zh. Anal. Khim. 34, 1207 (1979).
[12] Wünsch, J., Umland, F. Z.: Z. Anal. Chem. 250, 248 (1970).
[13] Murphy, J., Riley, Y. P.: Anal. Chim. Acta 27, 31 (1962).

6.31. Rhenium

Rhenium is situated in group VII/2 of the periodic system. Similarly to manganese, several oxidation states are known. It can readily be oxidized to the perrhenate anion, ReO_4^-; this is very stable in solution.

The average rhenium content of the Earth's crust is $1 \times 10^{-7}\%$, so that it is a very rare element. It forms no mineral of its own, but is generally to be found in molybdenite and platinum ores, and occasionally in molybdenum-containing chalcopyrites.

The books by Lebedev [1] and Borisova and Ermakov [2] provide excellent accounts of the possibilities of digesting, separating and determining rhenium. Photometric reagents for rhenium are listed in Table 6.24.

TABLE 6.24.

More important photometric reagents for rhenium

Reagent	Medium	Extractant	λ_{max} nm	$\varepsilon \times 10^{-3}$	Interferences	Refs
SCN$^-$	5.4 M HCl + SnCl$_2$	n-butanol	425	23*	Mo, W	[1]
Diethyldithiophosphoric acid	1–3 M HCl + SnCl$_2$	Benzene	436	0.6	Mo	[1]
Thiourea	1.5–6 M HCl + SnCl$_2$	—	390	10.5	Cd, Bi, Sb, Hg, Se, Te, As, and 50-fold Mo and W	[1,3]
8-Mercaptoquinoline	9–10 M HCl	CHCl$_3$	438	8.5	As, Au, Pd, Pt, NO$_3^-$, oxidants	[1,3]
Dimethylglyoxime	1 M HCl + SnCl$_2$	—	440	6.9	Mo, heavy metals	[1,3]
α-Furyldioxime	HCl + SnCl$_2$ + acetone + alcohol	—	330	40.5	Mo, Cu, Pd, Cr, Co, Mn, Ni, Fe, Pt, Ti, W, V	[1,3]
4-Methyl-1,2-cyclohexane-dionedioxime	HCl + reductant	CHCl$_3$	436	69	Mo	[3]

* Datum of the authors

These reagents are not selective for rhenium; in almost every case the determination is disturbed by molybdenum. Since rhenium generally occurs only in molybdenum-rich products, separation must definitely be achieved. The best-known means of determining rhenium is in the thiocyanate form. The probable composition of the complex is $[ReO_2(SCN)_4]^{2-}$, though certain authors consider that the rhenium in the complex is pentavalent: $[ReO_2(SCN)_4]^{3-}$.

Determination of rhenium in the form of the thiocyanate complex,
in molybdenite, molybdenum ores and concentrates, pyrite and chalcopyrite
(based on [1, 4] and the modification by the authors)

Analytical procedure: At most 3 g sample is mixed with 0.1 g $KMnO_4$ and 2–4 g CaO in a porcelain crucible. This is placed in a cold furnace, and the temperature is slowly raised to 650°C and maintained at this level for 2 hours. After cooling, the rhenium is dissolved out of the crucible with 50–60 cm³ water. The poorly soluble calcium molybdate remains in the precipitate, and only a small quantity (50–200 μg) of molybdenum passes into solution. The solution containing the precipitate is boiled, left to stand for 2 hours, and filtered. The precipitate is washed with distilled water, and the filtrate is evaporated to a volume of ca. 30 cm³. The calcium is precipitated with 10 cm³ saturated $(NH_4)_2CO_3$ solution, the mixture is boiled for 1–2 min, then filtered, and the precipitate is washed with water. The filtrate is made up to the mark in a 50 cm³ volumetric flask.

At most 15 cm³ is pipetted from the solution into a separating-funnel, the alkalinity is adjusted to 5 M by the addition of 15 cm³ 10 M NaOH, and the rhenium is extracted for 1 min with 15 cm³ methyl ethyl ketone (separation from the residual molybdenum). After 20 min, the aqueous phase is poured off, and the organic phase is washed with 2×5 cm³ 5 M NaOH (with interim waiting for complete separation of the phases). 15 cm³ $CHCl_3$ is next added to the organic phase, and the rhenium is extracted back into the aqueous phase with 2×7 cm³ water, for 1 min on each occasion. 16 cm³ conc. HCl, 4 cm³ KSCN (220 g/dm³) and 4 cm³ 1 : 1 HCl containing 35% $SnCl_2$ are added to the combined aqueous phases (upper layer). After 5–10 min, the thiocyanate complex formed is extracted for 30 s with 14 cm³ *n*-butanol previously saturated with water. The alcohol phase is transferred to a dry

25 cm^3 volumetric flask, the contents are made up to the mark with n-butanol, and photometry is performed at 425 nm in 1–5 cm cells, depending on the rhenium content. The comparison solution is the n-butanol extract of 14 cm^3 water + 16 cm^3 conc. HCl + 4 cm^3 KSCN + 4 cm^3 SnCl$_2$.

The calibration plot is produced with 2–50 μg Re, starting from the methyl ethyl ketone extraction from 5 M NaOH, as described in the ore analysis.

The absorbance of 10 μg Re at 425 nm, measured in a 5 cm cell, is 0.25.

Assuming a sample weight of 3 g, the sensitivity of the method is 2 ppm Re; the error at low concentrations is ±20%, while at Re contents above 100 ppm it is ±5%.

Determination of rhenium in alloys [5]

Principle of method: In conc. H$_2$SO$_4$ medium in the presence of a KBr or HBr excess, Re(VII) is transformed to Re(VI) and ReOBr$_4$ is formed. This compound has two absorbance maxima: at 556 and 653 nm. Beer's law is obeyed for Re contents of 1.7–5×10^{-4} M. The molar absorbances are $\varepsilon_{556} = 2.2 \times 10^3$, and $\varepsilon_{653} = 2.56 \times 10^3$.

Analytical procedure: 0.05–0.5 g alloy is dissolved in 30% H$_2$O$_2$ with gentle heating, the insoluble residue is filtered off, and the filtrate is evaporated to 1 cm^3. The volume is made up to 25 cm^3 with conc. H$_2$SO$_4$. An aliquot of 0.1–0.3 cm^3 is taken, and diluted to 5 cm^3 with conc. H$_2$SO$_4$. 0.01 g KBr is added, and after 20 min the absorbance of the solution is measured at 556 or 653 nm. A calibration plot is prepared analogously.

The method is suitable for determination of the Re contents (20–40%) of Re–Mo, Re–W and Re–W–Nb–Zr alloys.

References

[1] Lebedev, K. B.: *Renyii.* Metallurgizdat, Moscow 1963.
[2] Borisova, L. B., Ermakov, A. N.: *Analiticheskaya khimiya reniya.* Nauka, Moscow 1974.
[3] Tarayan, B. M., Vartanyan, S. V.: *Zav. Lab.* **11,** 1301 (1964).
[4] Pensionerova, V. M., Paukova, V. E.: *Metody khimicheskovo analiza mineralnovo syrya.* Nedra, Moscow 1965.
[5] Borisova, L. V., Plastinina, E. I., Ermakov, A. N., Marov, I. N., Zhukov, V. V.: *Zh. Anal. Khim.* **33,** 1982 (1978).

6.32. Scandium

Scandium is situated in column III/2 of the periodic system. In its compounds it is always trivalent. Sc^{3+} is readily hydrolyzed; its hydroxide does not dissolve in alkali. In its analytical properties it resembles primarily the lanthanides, but in part aluminium.

There is no selective reagent for Sc^{3+}. The most frequently applied reagents are eriochrome cyanine R [1], xylenol orange [2], fluorone derivatives [3] and reagents of the arsenazo type [4, 5]. With these reagents the complexes are formed in the pH interval 2–7 [5, 6].

Of the complexes mentioned, scandium–xylenol orange has the highest molar absorbance (29×10^3), but it is unfavourable as regards its stability and selectivity. Since arsenazo III is the most selective, and its complex is the most stable, the determination can be carried out even at pH 2. This is very advantageous, for at higher pH (~ 3) the hydrolysis of Sc^{3+} begins and it is necessary to keep it in solution by means of complex formers with general masking effects.

In 0.003 M HCl medium the composition of the Sc^{3+}–arsenazo III complex is $M : L = 1 : 1$; $\lambda_{max} = 670$ nm, where $\varepsilon = 13.5 \times 10^3$.

The photometric determination of scandium is dealt with in the review article by Eremin and Katochkina [6].

Determination with arsenazo III of scandium content of rocks and ores (tungsten, molybdenum, titanium and uranium ores, apatites) and uranium-plant digestion solutions [7]

The following method is a modification of our earlier method [8]. Since the final evaluation is carried out by photometry on the Sc-arsenazo III complex, it is necessary first to remove the disturbing elements:

(a) hydroxide separation (Al, Si, U, V, W, Mo pass into solution);

(b) extraction of Sc from strongly acidic (2–5 M) medium with di-2-ethyl-hexylphosphoric acid, then washing of the organic phase with a $H_2SO_4 +$ $+ HCl + $ oxalic acid mixture (this yields separation from U, Th, Ln, Fe, Zr, Hf, Ti, Ca, Mg, etc.);

(c) antagonistic re-extraction: Sc can be re-extracted quantitatively with

2–3 M HCl in the presence of a moderately high alcohol (*n*-octanol, nonanol); the re-extract contains practically no interfering components.

Analytical procedure: 1–3 g sample is weighed into a nickel dish, moistened with 1–2 cm³ alcohol, and fused with ca. 15 g NaOH, a little Na_2O_2 too being added towards the end of the digestion. The melt is leached out with hot water, the mixture is boiled and filtered, and the precipitate is washed with hot water and washed back into the initial beaker. It is dissolved in HCl, and 20–30 cm³ conc. HCl is added in excess. The solution is boiled, cooled and diluted to 200 cm³ (its acidity should be 2–3 M). The solution is poured into a 300 cm³ separating-funnel, and the Sc is extracted for 1 min with 10 cm³ 0.02 M di-2-ethylhexylphosphoric acid in petroleum ether ($D_{Sc} > 1000$). The aqueous phase is run off, and the organic phase is washed with 2×10 cm³ acid mixture (0.5 M H_2SO_4 + 5 M HCl + 0.1 M $H_2C_2O_4$). To remove the oxalic acid traces, which would later disturb the photometric determination of scandium, the organic phase is washed with a further 10 cm³ 5 M HCl. 10 cm³ *n*-octanol is next added to the organic phase, and the scandium is re-extracted with 2×10 cm³ 2 M HCl (for 1 min each). The aqueous phase is run into a 50 cm³ volumetric flask, and the volume is made up to the mark with water. A 20 cm³ aliquot is pipetted into a 50 cm³ volumetric flask, the solution is neutralized with 1 M NaOH in the presence of 1 drop phenolphthalein, it is just re-acidified with 1 M HCl, and 0.5 cm³ 1 M HCl is added in excess. 0.5 cm³ 3% ascorbic acid (to reduce any Fe^{3+} present) and 1 cm³ 0.1% arsenazo III are added, the volume is made up to the mark with water, and the absorbance is measured in 1–5 cm cells at 670 nm.

Comparison solution: 20 cm³ 2 M HCl is neutralized, 0.5 cm³ 0.5 M HCl and 1 cm³ 0.1% arsenazo III are added, and the mixture is diluted to 50 cm³.

When digestion solutions are to be analyzed, at most 200 cm³ of the solution is pipetted into a separating-funnel (the volume is made up to 200 cm³ if necessary), 20–30 cm³ is added, and the subsequent procedure is as described above.

The calibration plot is produced with 5–100 µg Sc, the operations being started from the extraction.

The method is suitable for the determination of 3 ppm Sc in a solid sample, and of 50 µg Sc/dm³ in solution. For contents above 10 ppm or 100 µg/dm³, the error in the method is ±10 rel.%.

Notes: In the analysis of tungsten and molybdenum ores, it is advisable to repeat the alkaline separation. In the analysis of titanium ores (rutile, ilmenite), digestion with pyrosulphate, dissolution with acid, and precipitation with NH_4OH must be applied.

Other methods of determination

Romantseva *et al.* [9] propose extraction–chromatographic separation for determination of the scandium content of materials containing large amounts of zirconium and rare earth metals. Tributyl phosphate (TBP) is used as extractant, with hydrophobized silica gel as extractant support (1.7 g silica gel and 1 cm³ TBP, grain size 100 μm). The acidity of the solution subjected to sorption: 5 M perchloric acid. Desorption is effected with water. In this way, scandium can be separated quantitatively from 1000–5000-fold amounts of Zr, Fe, Al, Y, Yb, Ti, Cr and Cu. The scandium is determined photometrically with xylenol orange.

Bykhovtsova and Tserkovnitskaya [10] have developed an extraction–photometric method for the determination of scandium in rocks. Separation from the interfering elements is carried out with a mixture of di-2-ethylhexylphosphoric acid, mono-2-ethylhexylphosphoric acid, trialkylamine and TBP from a 6 M HNO_3 medium. After the removal of Zr, Th and U, the Sc is extracted with a mixture of 0.1 M di-2-ethylhexylphosphoric acid + 0.4 M mono-2-ethylhexylphosphoric acid in benzene. The scandium can be determined in the organic phase with chlorphosphonazo III, in alcoholic medium. The sensitivity of the method is $2 \times 10^{-4}\%$.

References

[1] Taitiro, F.: *Japan Analyst* **12**, 399 (1963).
[2] Koukova, G. V.: *Zh. Anal. Khim.* **19**, 73 (1964).
[3] MacDonald, J.: *Anal. Chim. Acta* **28**, 264 (1963).
[4] Alimarin, I. P.: *Vestn. Moskovskogo universiteta,* No. 4, 67 (1963).
[5] Savvin, S. B.: *Arsenazo III.* Atomizdat, Moscow 1966.
[6] Eremin, Yu. G., Katochkina, V. S.: *Zav. Lab.* **12**, 1425 (1969).
[7] Csővári, M., Mohai, M.: Paper presented at 16th Transdanubian Analytical Conference, Győr, Hungary, 1981.
[8] Upor, E., Szalay, J., Klesch, K.: *Magy. Kém. Folyóirat* **74**, 438 (1968).
[9] Romantseva, T. I., Shmanenkova, G. I., Kakhaeva, T. V., Kolenkova, M. A., Sazhina, V. A.: *Zh. Anal. Khim.,* **36**, 1529 (1981).
[10] Bykhovtsova, T. T., Tserkovnitskaya, I. A.: *Zh. Anal. Khim.* **37**, 624 (1982).

6.33. Selenium

Selenium is situated in column VI/1 of the periodic system. It may be divalent, tetravalent and hexavalent. In aqueous solution it is generally present as the anion SeO_3^{2-}. It can readily be reduced from its compounds, to give red metallic selenium.

The book by Nazaranko and Ermakov [1] deals comprehensively with the analysis of selenium. Additionally, a good survey of the separation methods is given by Bock and Jakob [2], while Shcherbov et al. [3] compare the photometric reagents. Boltz [4] and Babko and Pilipenko [5] likewise provide detailed information on the photometric determination.

The organic reagents recommended most frequently for the photometric determination of selenium are listed in Table 6.25.

TABLE 6.25.

More important photometric reagents for selenium

Reagent	Medium	Extractant	λ_{max} nm	$\varepsilon \times 10^{-3}$	Interferences	Refs
3,3'-Diamino-benzidine	pH 4.5–7	$CHCl_3$	420	17	Au^{3+}, Bi^{3+}, Sn^{2+}, Ti^{3+}, Ti^{4+}, W^{6+}, Fe^{3+}, Cu^{2+}, Zr^{4+}, oxidants	[1,3]
2,3-Diamino-naphthalene	pH 2	Cyclohexane, toluene	380	23.8	Oxidants and many elements	[1,3]
o-Phenylene-diamine	pH 2.5	$CHCl_3$, toluene	335	17.8	Fe^{3+}, Sn^{4+}, I^-, V^{5+}, Bi^{3+}, Hg^{2+}	[1,3]
4,5-Diamino-6-thiopyrimidine	pH 1.5–2.5		380	19.2	Fe^{2+}, Fe^{3+}, oxidants	[3]
1,1'-Dianthrimide	96% H_2SO_4		480 585		B, Ge, Te, Br, F	[3]
2-Mercapto-benzimidazole	HCl	Butanol+ $CHCl_3$	335	10.4	Te and many elements	[3]
Dithizone	6–7 M HCl, 5.5–6 M H_2SO_4	CCl_4	420	73.5	Te and many elements	[1,3,13]

290

Those with the most favourable properties are the o-diamines. These are comparatively selective for selenium and they do not form coloured complexes with tellurium, which is advantageous in the case of certain samples, for separation from tellurium is unnecessary. The most widespread photometric methods are those based on the absorbance of the Se(IV) complex with 3,3'-diaminobenzidine (piazoselenol) [4, 6–9], which is formed in the pH interval 1–3. The complex can be extracted with organic solvents at pH 4.5–7. Aqueous solutions of the reagent are very labile; they are readily oxidized on the action of light. The reagent solution must therefore be prepared freshly daily, and stored in a dark bottle.

Dithizone is a very sensitive reagent for selenium. The determination is disturbed by many ions, but this can be avoided by means of preliminary separation.

Another widespread procedure for selenium determination is the photometric measurement of a solution of colloidal selenium [4, 9–12]. The principle of the determination is that the selenium is reduced to the elemental state, and the absorbance of the colloidal selenium is measured in acetic acid medium in the presence of a gelatin protective colloid, at 400–470 nm, with a spectrophotometer or a photocolorimeter (the spectrum does not contain a sharp maximum). The sensitivity of the selenium determination can be increased by the addition of "sensitizers" (Sb^{3+}, Cu^{2+}, Bi^{3+}) [10, 12].

Separation of selenium from disturbing components. No matter which photometric method is employed, it is useful to reduce the selenium to the metal, for it can then readily be separated from many interfering ions. Since the normal redox potential of the Se(IV)-Se(O) system is $+0.74$ V, this reduction may be performed with tin(II) chloride, hydrazine, sulphur dioxide, hydroxylamine or sodium hypophosphite.

The investigations by the authors indicate that 100 μg Se can be precipitated without loss from 100 cm^3 6 м HCl medium with the same volume of 25% $SnCl_2$. After standing on a water-bath for several hours, the precipitate can be filtered off well. For lower selenium contents, 200 μg As is added as carrier; this is sufficient for the loss-free precipitation of 1–2 μg Se.

The selenium precipitate may contain Te, Au, platinum metals, Cu, Sb, Fe, Sn and As as impurities. However, in the average case the quantities of these elements in rocks, and passing into the precipitate, do not disturb photometry in the colloidal or piazoselenol forms.

During the digestion process, care must be taken not only to ensure quantitative dissolution of the selenium, but also to prevent the formation of volatile selenium compounds [2]. Digestion with HF is therefore not suitable (50–90% Se loss). Boiling with HCl too is accompanied by a loss. This may be avoided if the digestion is performed with an oxidizing medium. No loss occurs if the sample is boiled for several hours with conc. HNO_3 or a 3:1 mixture of conc. HNO_3+conc. HCl and then evaporated; by this means, the selenium is digested totally even in the case of pyrite and sulphide ores.

Analytical procedure: Depending on the expected selenium content, 1–5 g rock sample is evaporated almost to dryness with 30 cm³ conc. HNO_3 on a sand-bath (selenium is lost if the mixture is evaporated fully to dryness). In the case of pyrite or sulphide ores, the sample is digested with a 3:1 mixture of HNO_3+HCl, and evaporated. The residue is taken up in a little hot water, the solution is filtered off, and the solid is washed with a little hot water. If possible, the volume of the filtrate should not exceed 20 cm³.

For the precipitation of metallic selenium, an equal volume of conc. HCl and 200 µg As^{3+} carrier are added, the mixture is heated, and the Se is reduced with 25% $SnCl_2$ in 1:1 HCl. The amount of $SnCl_2$ needed for the reduction varies between 20 and 40 cm³, depending on the composition of the sample and the quantity of residual HNO_3. With some practice, this can be established from the decolorization of the solution and the appearance of the red Se precipitate. To promote quantitative precipitation, the solution is kept on a water-bath for several hours and then left to stand until the following day. The precipitate is filtered off, and washed free of reductant with hot water (checked with 0.0002 M $KMnO_4$). The precipitate is dissolved in a small volume of a 10:1 mixture of 1:1 HCl and conc. HNO_3. Depending on the selenium content of the resulting solution, it is determined either in the colloidal form or as piazoselenol.

Measurement of absorbance of colloidal selenium (25–500 µg Se). The solution obtained by dissolution of the selenium precipitate is transferred to a 50 cm³ volumetric flask, and approximately neutralized with NH_4OH. The selenium is reduced with 2 cm³ 25% $SnCl_2$ in 1:1 HCl in the presence of 5 cm³ conc. HCl. To keep the colloidal selenium in solution, 0.5 cm³ freshly prepared 0.2% gelatin solution and 3 cm³ 5% sodium tartrate solution are added. Sufficient solid NH_2OH. HCl is added for stannic acid just to begin to precipitate, then 2–3 cm³ glacial acetic acid is added, the volume is made up to the mark, and the absorbance of the solution is measured on a photometer with an appropriate colour filter or on a spectrophotometer, in 2–5 cm cells at 400–470 nm. The solution of the reagents is used as comparison. The determination is disturbed by tellurium. In rock samples, however, the Clark value for tellurium is generally two orders of magnitude lower than that for selenium, so that in practice there is no interference. (If the Se : Te $\leqq 5$, the determination with diaminobenzidine is recommended.) The calibration plot is recorded with 25–200 μg or 100–500 μg Se, in 5 or 2 cm cells. For a sample weight of 5 g, the sensitivity of the method is 4×10^{-4} %.

Measurement of absorbance of piazoselenol. A little crystalline NH_2OH. . HCl is added to a solution containing 2–50 μg Se, to prevent the oxidizing action of NO_3^-, and the pH is then adjusted to 1–3 with NH_4OH (indicator paper). 0.5 cm³ freshly prepared 1% 3,3'-diaminobenzidine hydrochloride is added, followed 20 min later (when the colour of the complex has developed by 2 cm³ acetate buffer of pH 5.5 (57 cm³ glacial acetic acid + 82 g sodium acetate, in 1 dm³ water). The volume of the solution should not be more than 30–40 cm³. The piazoselenol is extracted with 10 cm³ $CHCl_3$ in 2–3 portions, and the absorbance is measured in 1–5 cm cells at 420 nm, against $CHCl_3$ that has been shaken with the reagent solutions. The calibration plot is produced analogously, with 2–10 μg or 10–50 μg Se, in 5 or 1 cm cells. The absorbance of 10 μg Se in 10 cm³ $CHCl_3$ in a 1 cm cell is 0.10. For a sample weight of 5 g, the sensitivity of the method is 1×10^{-5} %.

Notes: Because of the tendency of 3,3'-diaminobenzidine to decompose, care must be taken that the reagent is always freshly prepared and is stored in a dark bottle. Although the violet-coloured oxidation product is not extracted into $CHCl_3$, its formation leads to a decrease in the reagent concentration, and hence to incomplete development of the complex, with a resulting low colour intensity.

Determination of selenium with dithizone [3]

Selenium (and tellurium) is separated from disturbing substances in the customary way, by reduction to metallic selenium. The precipitate is filtered off, washed free of reductant with hot water, and dissolved in 5.5 M H_2SO_4.

An aliquot containing 0.2–10 μg Se is pipetted into a separating-funnel, its volume is made up to 25 cm³ with 5.5 M H_2SO_4, and 15 cm³ freshly prepared 0.002% dithizone in CCl_4 is added. The mixture is shaken for 30 s, and the organic phase is run off into another separating-funnel and washed with 20 cm³ 25% ferrocyanide solution (shaking for 2 min). For removal of the excess dithizone, the organic phase is shaken with 30 cm³ 1 : 100 NH_4OH. After separation of the phases, the absorbance of the organic solution is measured in 1–3 cm cells at 420 nm. The calibration plot is recorded in a similar manner.

If the sample to be analyzed also contains tellurium ($\cong 10$ μg Te in the aliquot taken), the tellurium content of the same stock solution is determined with butylrhodamine B or ethylrhodamine B, and the result for selenium is corrected accordingly.

References

[1] Nazarenko, I. I., Ermakov, A. N.: *Analiticheskaya khimiya selena i tellura.* Nauka, Moscow 1971.
[2] Bock, R., Jakob, D.: *Z. Anal. Chem.* **200**, 81 (1964).
[3] Shcherbov, D. P., Ivankova, A. I., Gladysheva, G. P.: *Zav. Lab.* **33**, 683 (1967).
[4] Boltz, D. F.: *Colorimetric Determination of Nonmetals.* Wiley, New York 1978.
[5] Babko, A. K., Pilipenko, A. T.: *Fotometricheskii analiz. Metody opredeleniya nemetallov.* Khimiya, Moscow 1974.
[6] Roquebert, J.: *Bull. Soc. Pharmac. Bordeaux* **101**, 29 (1962).
[7] Cheng, K. L.: *Anal. Chem.* **28**, 1738 (1956).
[8] Barcza, L.: *Z. Anal. Chem.* **199**, 10 (1963).
[9] Szalay, J., Upor, E., Klesch, K.: *Magy. Kém. Folyóirat* **74**, 507 (1968).
[10] Vinogradov, A. P.: *Metody opredeleniya i analiza redkikh elementov.* AN SSSR, Moscow 1961.
[11] Blum, I. A.: *Khimicheskie, fizikokhimicheskie i spektralnye metody issledovaniya rud redkikh i rasseyannikh elementov.* Gosgeoltekhizdat, Moscow 1961.
[12] Tarayan, V. M., Arstamyan, Zh. M.: AN Arm. SSR, **15**, 415 (1962).
[13] Campbell, A. D., Yahaya, A. H.: *Anal. Chim. Acta* **119**, 171 (1980).

6.34. Silicon

Silicon is situated in column IV/1 of the periodic system. It is a non-metallic element. In its compounds it is tetravalent. In its aqueous solutions it is present as the silicate anion, SiO_3^{2-}; this is very susceptible to polymerization.

Detailed accounts of the photometric determination of silicon are to be found in the publications by Babko and Pilipenko [1], Mishlyaeva and Krasnoshchekov [2], Boltz [3] and Kato [4].

Only one reagent, molybdate, is known for the photometric determination of silicic acid; this forms a heteropolyacid. The determination is disturbed by anions (phosphate, arsenate, germanate, vanadate) which form similar heteropolyacids. Zr, Sn, Ti and W also interfere, by yielding precipitates with the reagent. Fluoride is added to eliminate the interfering effects in some cases, and to promote depolymerization. Since an excess of fluoride likewise causes a disturbance, the unbound fluoride is rendered ineffective by the addition of boric acid or Al^{3+}.

Two isomeric forms of the silicomolybdate heteropolyacid can be used for the photometric determination: the α- and the β-modification. The kinetics of formation of these two modifications have been studied by Truesdale *et al.* [5]. In aqueous solutions, the β-form transforms spontaneously to the α-modification. The β-form is stabilized by organic solvents. The formation of the two modifications depends on the acidity of the starting solution: the α-modification is formed in the pH interval 2.5–4, and the β-modification in the pH interval 1–1.8. At the customary photometric wavelength, 400 nm, the absorbance of the β-form is substantially higher (about twice) than that of the α-modification. On photometry in aqueous solution, the colour is produced in the presence of buffer, e.g. monochloroacetic acid, at pH 3.0–3.7, when $\lambda_{max} = 390$ nm, with $\varepsilon = 1.7 \times 10^3$. The sensitivity of the determination may be increased if the heteropolyacid formed at pH 1.2 is extracted from more strongly acidic medium (2.2 M HCl) with an organic solvent, e.g. isoamyl alcohol, and silicomolybdenum blue is formed by shaking with a reductant, e.g. a mixture of Mohr salt and ascorbic acid. Photometry is then carried out at 800 nm, where $\varepsilon = 23.3 \times 10^3$.

For colour development, the monomeric form of silicic acid is required. This can be attained by the use of a freshly digested solution, or by depolymerization, e.g. by HF addition in the case of an older solution. The disturbance due primarily to phosphate may be prevented by the addition of oxalic acid, citric acid or tartaric acid to the aqueous solution. At higher phosphate contents, the phosphoheteropolyacid is removed by extraction with ethyl acetate or a mixture of benzene and isobutanol at pH 1.2. Any arsenic possibly present is co-extracted with the phosphate. The colour intensities of the phosphorus and arsenic complexes are otherwise weaker: from this respect, 1 mg Si is equivalent to 6.9 mg P or 187 mg As. The disturbing effect of phosphate may also be diminished by increasing the acidity of the solution after formation of the silicomolybdate complex. The heteropolyacid of phosphorus then decomposes, whereas that of silicon remains stable even in 1.5–1.75 M H_2SO_4.

Silicomolybdenum yellow is usually reduced to silicomolybdenum blue with Fe^{2+}, tin(II) oxalate, tin(II) chloride, hydrazine sulphate, metol, ascorbic acid, sodium sulphite, 1-amino-2-naphthol-4-sulphonic acid or combinations of these. Butanol and isoamyl alcohol are convenient extractants. The maximum absorbance of the compound formed varies in the range 720–805 nm, depending on whether the α- or β-isomer is obtained in the aqueous solution, and on the reductant [1, 2, 6–8a].

Photometric determination of silica in rocks [9]

In the analysis of rocks, an exact dehydration–gravimetric determination necessitates the photometric measurement of the silicic acid remaining in the filtrate. TiO_2 in the rock does not interfere up to 5%, nor P_2O_5 up to 10%.

Analytical procedure: 1 g silicate rock sample is fused with 5 g anhydrous Na_2CO_3 in a platinum dish. After aqueous leaching-out, the solution is acidified with conc. HCl in a covered porcelain dish, and 10 cm^3 conc. HCl is added in excess. The solution is evaporated to dryness on a water-bath. 10 cm^3 conc. HCl and 50 cm^3 hot water are added to the residue, the mixture is filtered on a Whatman No. 41 filter paper, and the precipitate is washed 2–3 times with dilute HCl solution and then with hot water. The filtrate is made up to the mark in a 200 cm^3 volumetric flask.

A 5 cm^3 aliquot of the stock solution is pipetted into a 100 cm^3 volumetric flask, and 10 cm^3 water and 1 cm^3 ammonium molybdate (10 g dissolved in 100 cm^3 1 M NH_4OH) are added. After a standing period of 10 min, 5 cm^3 oxalic acid (10 g/100 cm^3 water) is added, followed without delay by 2 cm^3 reductant solution (0.15 g 1-amino-2-naphthol-4-sulphonic acid + 0.7 g Na_2SO_3 + 9 g sodium metabisulphite, dissolved in 100 cm^3 water). The contents of the flask are made up to the mark, and after 30–60 min the absorbance is measured in 2 cm cells at 650 nm. The calibration plot is produced analogously, with 0–200 μg SiO_2.

The silica on the filter paper is ashed and fumed in the normal way. The SiO_2 content of the sample is given by the sum of the gravimetric and photometric results.

Determination of silica in micro amounts of mineral fractions [10]

The Fe, Ti, Th, Zr, Ca and F contents of the mineral grain to be analyzed may cause a loss of up to even 50% in the course of the process. After digestion by means of fusion, therefore, the silicic acid is depolymerized by heating with a molybdate solution in H_2SO_4, and the determination then follows.

Prior to the digestion, borax and Na_2CO_3 are mixed in a ratio of 1 : 1.7 and fused in a platinum vessel. Bidistilled water is added to the still hot melt, which is then dried and powdered.

5–20 mg sample is fused with ca. 0.1 g digestion mixture. The melt is dissolved by the addition of 40 cm³ 0.1 M H_2SO_4 and 5 cm³ 5% ammonium molybdate, with heating for 30 min to achieve depolymerization. The volume of the solution is then made up to 50 cm³.

A 5–20 cm³ aliquot of the stock solution is taken, the volume is made up to 30 cm³ with 0.1 M H_2SO_4, and 5 cm³ 5% ammonium molybdate is added. After 15 min, 5 cm³ 4 M H_2SO_4 and 4 cm³ 0.5% freshly prepared $SnCl_2$ are added, and the volume is made up to 100 cm³. Photometric measurements are made in 2 cm cells at 740 nm. The sensitivity of the method is 0.02 μg Si/cm³, and the relative error of the determination does not exceed 10%.

Determination of silicon traces in metallic iron, by an extraction method [8]

With modification of the digestion and pretreatment operations, the method may be used for the analysis of iron, steel, Cu–Ni alloys, BeO and U_3O_8.

1 g metallic iron turnings with a low Si content is weighed into a 250 cm³ teflon beaker, and 20 cm³ digestion acid mixture (72 cm³ H_2SO_4+91 cm³ HNO_3, in 1 dm³) and 4 cm³ 4% HF are added. The sample is dissolved at

85°C, 20 cm^3 water is added, and the heating is continued for a further 30 min. The digestion solution is poured into 30 cm^3 saturated boric acid, and the volume is made up to 100 cm^3.

A 15 cm^3 aliquot (0–15 μg Si) of the stock solution is taken, its pH is adjusted to 0.9–1.1, 10 cm^3 10% NH$_2$OH.HCl is added, and the mixture is diluted to 40 cm^3. After the addition of 10 cm^3 5% ammonium molybdate, the pH is again adjusted to 1.1, and the solution is left to stand for 10 min. 5 cm^3 20% tartaric acid is added, after 1 min the acidity of the solution is increased with 5 cm^3 4.5 M H$_2$SO$_4$, and the volume is made up to 55 cm^3. Extraction is performed with 10 cm^3 isoamyl alcohol for 1 min, and then 5 cm^3 isoamyl alcohol for 1 min, a separation time of 2 min being allowed. The combined organic phase is left to stand for 5 min, for the water droplets to separate. The organic phase is shaken for 30 s with 4 cm^3 reductant solution (0.25 g Fe(NH$_4$)$_2$(SO$_4$)$_2$.2H$_2$O + 7.5 g ascorbic acid, in 100 cm^3), and the mixture is allowed to stand for 5 min. The organic phase is transferred to a 25 cm^3 volumetric flask, the separating-funnel is rinsed out with a little methanol, which is also added to the flask, and the volume is made up to the mark. The absorbance is measured in 2 cm cells at 800 nm. The limit of the measurement can be extended up to 25 μg Si, when a 1 cm cell is used.

Determination of silica content of natural waters [11]

Principle of method: During the analysis, α-silicomolybdic acid is formed at pH 3.8–4.8; this is reduced with a mixture of Sn(IV) + ascorbic acid + oxalic acid, without reduction of the excess of the molybdate reagent. Photometry is performed at 740 nm.

Analytical procedure: 3 cm^3 buffer (20 g sodium acetate + 58 cm^3 acetic acid, diluted to 500 cm^3) and 5 cm^3 4.4% ammonium molybdate are added to 25 cm^3 of the water sample. After 15 min, 5 cm^3 combined reductant solution (4 cm^3 anhydrous SnCl$_4$ + 40 cm^3 distilled water + 2 g ascorbic acid, diluted to 50 cm^3; 40 cm^3 freshly filtered 8.5% oxalic acid is added to 10 cm^3 of this solution) is added, with photometry 20 min later at 740 nm. The calibration plot covers the interval 0–2000 μg Si/dm^3.

References

[1] Babko, A. K., Pilipenko, A. T.: *Fotometricheskii analiz. Metody opredeleniya nemetal-lov*. Khimiya, Moscow 1974.
[2] Mishlyaeva, L. V., Krasnoshchekov, V. V.: *Analiticheskaya khimiya kremniya*. Nauka, Moscow 1972.
[3] Boltz, D. F.: *Colorimetric Determination of Nonmetals*. Wiley, New York 1978.
[4] Kato, K.: *Anal. Chim. Acta* **82**, 401 (1976).
[5] Truesdale, V. W., Smith, P. J., Smith, C. J.: *Analyst* **104**, 897 (1979).
[6] Nemodruk, A. A., Bezrogova, Ye. V.: *Zh. Anal. Khim.* **25**, 1587 (1970).
[7] Ringbom, A., Ahlers, P. E., Siitonen, S.: *Anal. Chim. Acta* **20**, 78 (1959).
[8] Pakalns, P., Flynn, W. W.: *Anal. Chim. Acta* **38**, 403 (1967).
[8a] Kircher, C. C., Crouch, S. R.: *Anal. Chem.* **55**, 248 (1983).
[9] Jeffery, P. G., Wilson, A. D.: *Analyst* **85**, 478 (1960).
[10] Bikhovtsova, T. T.: *Zav. Lab.* **10**, 25 (1974).
[11] Smith, J. D., Milne, P. J.: *Anal. Chim. Acta* **123**, 263 (1981).

6.35. Sulphur compounds

Sulphur is situated in column VI/1 of the periodic system. It is a non-metallic element. In its compounds, sulphur has oxidation states of -2, $+2$, $+4$ and $+6$. In aqueous solution it exists only as an anion. The possibility of a variation in its oxidation state is often utilized in its separation and determination.

Similarly as for other anions, the majority of the methods used for the photometric determination of sulphur compounds are not direct procedures based on complex formation. Most of the methods employed for the determination of H_2S and SO_2 (SO_3^{2-}) utilize the synthesis of some coloured organic compound, while those most widely used for SO_4^{2-} are indirect procedures based on the determination of the excess of a known addition of Ba^{2+}. Some of these procedures are summarized in Table 6.26.

The most suitable means of separating H_2S and SO_2 from interfering components is distillation (generally in a current of inert gas). A simple solution in the determination of sulphate is removal of the disturbing cations by ion-exchange. The analysis of sulphur compounds has been reviewed by Karchmer [1], Boltz [2], Babko and Pilipenko [3] and Busev and Simonova [4].

299

TABLE 6.26.

Possibilities for photometric determination of sulphur compounds

Basis of determination	λ_{max} nm	$\varepsilon \times 10^{-3}$	Interferences	Refs
$S^{2-} + Fe^{3+} + N,N$-dimethyl--p-phenylenediamine $\rightarrow$ methylene blue	670 740	96 34	Heavy metals, oxidants	[13]
$SO_3^{2-} + Hg^{2+} + CH_2O + p$-rosaniline $\rightarrow$ coloured product	560	30	NO_2, H_2S, reductants	[3]
$SO_4^{2-} + Ba^{2+} + $ orthanil K	650	62	Cations (with the exception of alkali metals); anions forming complexes with Ba^{2+}	[14]
$SO_4^{2-} + Ba^{2+} + $ nitchromeazo	640	50	Cations (with the exception of alkali metals); anions forming complexes with Ba^{2+}	[15]
$S_nO_6^{2-} + (n-2)\,CN^- \rightarrow$ $\rightarrow (n-2)\,NCS^-$ $2NCS^- + Fe^{3+} \rightarrow$ $\rightarrow Fe(NCS)_2^+$	460		Ions giving precipitates in alkaline medium; ions forming complexes with NCS^-	[12, 16]
Extraction of $Fe(NCS)_2^+$ with trioctylphosphine oxide (TOPO)		3		[11]

Determination of hydrogen sulphide and soluble sulphides in the form of methylene blue

In acidic medium in the presence of Fe^{3+}, the sulphide ion reacts with N,N-dimethyl-p-phenylenediamine to give methylene blue. The Fe^{3+} does not merely participate as an oxidant in the process; its chloro complex also forms an ion-association with the methylene blue cation [5].

Analytical procedure: Disturbing ions must not be present. 0.5 cm³ 0.5 M NaOH solution is taken in a 25 cm³ volumetric flask, and 0.5 cm³ 0.2% reagent solution (2 g N,N-dimethyl-p-phenylenediamine sulphate dissolved in 100 cm³ 2 : 1 H_2SO_4) and the solution to be examined (S^{2-} content 1–10 μg) are added, followed after a few min by 1.0 cm³ 1% $FeCl_3$ in 1 : 20 HCl. The volume is made up to the mark. The flask is placed in a hot water-bath for 5 min, then cooled, and the absorbance is measured at 670 nm. The calibration plot covers the interval 1–10 µg S^{2-}.

Methylene blue forms an ion-pair with ClO_4^- and can also be measured after the extraction of this into $CHCl_3$ [6].

In the analysis of water samples, separation from disturbing ions can be achieved by coprecipitation of the S^{2-} as ZnS on a $Zn(OH)_2$ carrier [7]. For the determination of sulphide in rocks, the recommended procedure is the evolution of H_2S with HCl, distillation in a stream of N_2, and absorption in zinc acetate solution [8]. It must be noted, however, that numerous sulphide minerals (pyrite, chalcopyrite, etc.) are not soluble in acid, and accordingly this technique is suitable only for purposes of phase analysis. The total sulphide content can be determined most conveniently by heating the sampling at 1,100°C in a stream of O_2, the SO_2 formed being absorbed and determined by iodometric titration.

Determination of SO_2 content of air with p-rosaniline [9]

The SO_2 content of the air is absorbed in a solution of potassium tetrachloromercurate. The complex with *p*-rosaniline (or *p*-fuchsin) and formaldehyde is transformed to a reddish-violet compound, which is subjected to photometry.

Sampling: Air is passed for 1 hour at a rate of 1 dm^3/min through 10 cm^3 absorption solution (10.9 g $HgCl_2 + 66$ mg EDTA + 6 g KCl, in 1 dm^3).

Analytical procedure: The 10 cm^3 absorption liquid is washed with 7 cm^3 water into a flask, and 1 cm^3 0.6% sulphamic acid (NH_2SO_3H) solution is added. This is followed after 10 min by 2 cm^3 0.2% formalin and 5 cm^3 0.03% *p*-rosaniline solution (0.325 g *p*-rosaniline or *p*-fuchsin is mixed with 15 cm^3 ethanol, 400 cm^3 water and 30 cm^3 conc. HCl are added, and the volume is made up to 1 dm^3). After 30 min, photometry is carried out at 540–580 nm. The calibration plot is produced for 2–30 µg SO_2.

Determination of sulphate traces in river water

As a consequence of precipitate formation, sulphate traces lead to a decrease in intensity of the colour of the added reagent 6-(*p*-acetylphenylazo)-2-aminopyrimidine. The change in intensity is measured by photometry [10].

Analytical procedure: 3 cm³ buffer (a mixture of 1 M acetic acid and 1 M sodium acetate, pH 3.8) and 2 cm³ 7×10^{-4} M reagent in ethanol are added to 5 cm³ of the water to be analyzed. The mixture is shaken for 5 min, left to stand for 10 min, and centrifuged for 10 min at 1500 rpm. The absorbance of the supernatant is measured in 0.5 cm cells at 480 nm. The calibration plot (with negative slope) is recorded for 1–10 µg SO_4^{2-}/cm^3. The determination is not disturbed by the components in average water.

Another method is based on the fact that orthanil K and nitchromeazo (see Appendix) form stable complexes with Ba^{2+} in weakly acidic medium [14, 15]. The stabilities of these complexes are much lower than that of $BaSO_4$, and this is utilized for the indirect determination of SO_4^{2-}. Cations interfering in the determination may be removed by cation-exchange.

Analytical procedure: The water sample to be analyzed is passed through a strongly acidic cation-exchange resin in H^+ form. At most 10 cm³ of the cation-freed effluent (SO_4^{2-} content 5–20 µg) is pipetted into a 25 cm³ volumetric flask. 1.0 cm³ 0.05% orthanil K and 0.3 cm³ 0.001 M $BaCl_2$ solution are added, the pH is adjusted to 3–5 with 0.01 M HCl, 15 cm³ ethanol is added, and the volume is made up to the mark. The absorbance is measured at 540 nm.

With nitchromeazo the determination can also be carried out at pH 2.0. For the equilibria to be established, photometry can be performed only after a waiting period of 1.5 hours.

An advantageous technique with both reagents is determination by titration to colour development.

Determination of small amounts of polythionate in solutions from uranium plants [11]

The principle of the determination is that the polythionates can be converted to NCS^- with CN^- in the presence of Cu^{2+} as catalyst [12]. We have found the extraction of $Fe(NCS)_2^+$ with trioctylphosphine oxide (TOPO) to be the most suitable for determination of NCS^-. In cyanolysis in alkaline medium, the interfering ions are removed by precipitation with NaOH (pH 10). At the same time, this eliminates the differences caused by the variable Fe^{3+} concentration in the extent of extraction of NCS^- (and hence in its absorbance). The high concentration (>0.15 M) of SO_4^{2-} present diminishes the extraction of $Fe(NCS)_2^+$, and it is therefore precipitated in the form of $BaSO_4$ after completion of the cyanolysis.

Analytical procedure: 20 cm³ of the solution to be examined is taken in a 50 cm³ volumetric flask. Sufficient 1 M NaOH is added to give a pH of 10, and the volume is made up to the mark. The hydroxide precipitate is filtered off, and 40 cm³ of the filtrate is taken for cyanolysis. 1.0 cm³ KCN (50 g/dm³) and 0.6 cm³ 0.1 M $CuCl_2$ solution are added. After 10 min, 5 M HNO_3 solution is added from a burette until the solution is neutral to phenolphthalein, followed by a 4 cm³ excess. Next, 5 cm³ $BaCl_2$ (100 g/dm³) is added, and the volume is made up to the mark with distilled water. Some minutes later, the $BaSO_4$ precipitate is filtered off, and 25.0 cm³ of the filtrate is pipetted into a separating-funnel. This is shaken for 1–2 min with 10.0 cm³ 0.05 M TOPO in petroleum ether. The aqueous phase is run off, and the amount necessary for photometry is transferred into the cell with a dry pipette from the upper phase containing the Fe^{3+}–NCS^-–TOPO complex. The absorbance is measured at 460 nm, against petroleum ether. The calibration plot is produced with $1–10 \times 10^{-5}$ M NCS^-. The sensitivity of the method is 0.1 mg S/dm^3; the error does not exceed $\pm 10\%$.

Our investigations demonstrate that with other extractants (TBP, isoamyl alcohol, etc.) tested with regard to the extraction of $Fe(NCS)_2^+$ there is no extraction of NCS^- at all under these conditions.

Naturally, the high-stability extracted complex is also suitable for the direct determination of NCS^-.

References

[1] Karchmer, J. H.: *Analytical Chemistry of Sulfur and its Compounds.* Interscience, New York 1969.
[2] Boltz, D. F.: *Colorimetric Determination of Nonmetals.* Wiley, New York 1978.
[3] Babko, A. K., Pilipenko, A. T.: *Fotometricheskii analiz. Metody opredeleniya nemetallov.* Khimiya, Moscow 1974.
[4] Busev, A. I., Simonova, L. N.: *Analiticheskaya khimiya sery.* Nauka, Moscow 1975.
[5] Hoffmann, K., Hamm, R.: *Z. Anal. Chem.* **232,** 167 (1967).
[6] Marczenko, Z., Cholni-Lenarczyk, L.: *Chem. Anal.* **10,** 729 (1965).
[7] Cline, J. D.: *Limnol. and Oceanogr.* **14,** 454 (1969).
[8] van Loon, J. C., Parissis, C. M., Kingston, P. W.: *Anal. Chim. Acta* **40,** 334 (1968).
[9] Várkonyi, T., Cziczó, T.: *A levegőminőség vizsgálata (Air-quality examinations).* Műszaki Könyvkiadó, Budapest 1980.

[10] Toei, K., Miyata, H., Yamawaki, Y.: *Anal. Chim. Acta* **94**, 485 (1977).
[11] Upor, E., Nagy Gy.: *Magy. Kémikusok Lapja* **28**, 340 (1973).
[12] Urban, P. J.: *Z. Anal. Chem.* **179**, 415 (1961).
[13] Mecklenburg, W., Rosenkronzer, F.: *Z. anorg. Chem.* **86**, 143 (1914).
[14] Savvin, S. B., Akimova, T. G., Dedkova, V. P.: *Organicheskie reagenty dlya opredele-niya Ba²⁺ i SO₄²⁻*. Nauka, Moscow 1971.
[15] Basargin, N. N., Menshikova, V. L., Belova, Z. S., Myasischcheva, L. G.: *Zh. Anal. Khim.* **23**, 732 (1968).
[16] Koh, T., Aoki, Y., Iwasaki, I.: *Analyst* **104**, 41 (1979).

6.36. Tellurium

Tellurium is situated in column VI/1 of the periodic system. It is a metalloid, and a very valuable alloying material. In aqueous solution it is stable in the form of the tellurite anion, TeO_3^{2-}.

Tellurium behaves similarly to selenium analytically, and their separation from each other is not an easy task. However, selenium and tellurium do display different properties towards some reagents, and hence they can be separated from each other or determined in the presence of each other.

The analysis of tellurium is dealt with in the publications by Nazarenko and Ermakov [1], Boltz [2] and Babko and Pilipenko [3].

The tellurium in rocks and other products can generally be dissolved out by digestion with an oxidizing acid (HNO_3). Its halides are not so volatile as those of selenium.

The photometric reagents for tellurium are given in Table 6.27.

None of these reagents is selective; they can be applied only after appropriate separation.

Sodium diethyldithiocarbamate reacts with tellurium to yield a complex of composition $Te(DDTC)_4$; this complex can be extracted. Selenium gives a similar colour, but no extraction occurs in the pH interval 8.5–8.7. A possibility therefore arises for the separation of selenium and tellurium. In the presence of complexone III and cyanide, only Fe(III), Cu(II) and Bi(III) interfere [4, 5].

The bromide complex of tellurium reacts with diantipyrylmethane, and the product can be extracted with dichloroethane. This principle can be used to determine tellurium sensitively (10^{-4}–10^{-1} %) and selectively in concentrates and brass after extraction of the interfering ions as xanthates [6].

TABLE 6.27.

More important photometric reagents for tellurium

Reagent	Medium	Extractant	λ_{max} nm	$\varepsilon \times 10^{-3}$	Interferences
Sodium diethyl-dithiocar-bamate	1.5 M H_2SO_4 or 4 M $HClO_4$ or 3 M HCl	CCl_4	340 (428)	3.2	Many elements, but not Se
Bismuthiol II	3.5–6 M HCl	Alcohols, ketones, $CHCl_3$	330	2.8	Many elements
Isobutyl-dithio-pyrylmethane	1.5–1.75 M H_2SO_4 + KBr		358	51.0	Disturbing ions masked
Butylrhodamine B, or ethyl-rhodamine B	5 M H_2SO_4 + saturated boric acid	Benzene + butyl acetate (5 : 1)	565	7.7	Zn, Tl, Au, Hg, Fe^{3+}, Cu^+, Sn, Ag, Bi; but not Se

Butylrhodamine B and ethylrhodamine B have similar analytical properties; both form complexes with bromotellurite. The use of the latter is somewhat more advantageous, for the effects of the interfering ions are less extensive. Selenium does not react with these reagents [7].

Tellurium may be determined by photometry of a solution of colloidal tellurium too, in a similar way to selenium. The determination is disturbed by selenium, and accordingly this can be recommended only for the analysis of such selenium-free samples as tellurium-containing iron alloys.

Separation of tellurium from disturbing ions [7–11]. As described in connection with the determination of selenium, tellurium can be reduced to the elemental state with $SnCl_2$, sodium hypophosphite, SO_2, etc. The addition of As^{3+} is necessary if only a little tellurium is present. The precipitation with $SnCl_2$ and the subsequent dissolution of the precipitate are the same as described for selenium.

Metallic tellurium is precipitated most suitably with sodium hypophosphite from 6 M HCl medium, with Cu^{2+} as catalyst. This operation is usually carried out with boiling, with the same volume of 25–50% hypophosphite solution. Washing and dissolution of the precipitate are as described for selenium.

Whatever final means of measurement is selected, precipitation as metallic

305

tellurium is very advantageous, as this is a simple way to remove the bulk of the interfering ions. Selenium accompanies tellurium quantitatively.

The separation of tellurium by extraction is discussed in the articles by Havezov and Jordanov [12] and Torocheshnikova *et al.* [13].

Determination of tellurium content of iron ore and slag, after separation with tetrabutylethylenediamine [13]

Te(IV) can be extracted from a bromide-containing solution with a solution of tetrabutylethylenediamine (TBEDA) in $CHCl_3$. The Te(IV) in the extract may be determined photometrically via its own absorbance. The extraction means simultaneous separation from tin and lead and, in the presence of NH_4F, from iron too.

Analytical procedure: 0.1–0.5 g of the material to be analyzed is dissolved in 5 cm^3 conc. HBr with heating (1–1.5 hours, refluxing). The insoluble residue is filtered off, the filtrate is transferred to a separating-funnel, the volume is made up to 10 cm^3 with water, 5 g NH_4F is added, and the Te(IV) is extracted for 15 min with 10 cm^3 0.1 M TBEDA in $CHCl_3$. The organic phase is washed with 5 cm^3 washing liquid (3.6 M with respect to HBr, and 1 M with respect to NH_4F), and the absorbance of the organic phase is measured at 455 nm. The calibration plot is produced in the same manner, with 2–50 μg Te.

Note:
The Te(IV) can be re-extracted from the organic phase with dilute (0.01 M) oxalic acid, and then determined more sensitively with some other photometric method (e.g. with butylrhodamine B).

Determination of tellurium content of selenium mud with isobutyl-dithio-pyrylmethane [14]

In H_2SO_4 medium, isobutyl-dithiopyrylmethane (IDTM) forms a yellow complex with tellurium. The effects of disturbing components are eliminated by the addition of complex formers. The preparation and purification of the reagent are described in [15].

Analytical procedure: 1 g selenium mud is measured into a 150 cm^3

conical flask and heated with 50 cm³ conc. HNO₃ for 10 min. 7 cm³ conc. H_2SO_4 is added, and the mixture is evaporated until the appearance of fumes of SO_3. After cooling, 20–30 cm³ distilled water is added, the mixture is filtered and the precipitate is washed with 10–15 cm³ 1 M H_2SO_4. The filtrate is transferred to a 100 cm³ volumetric flask and made up to the mark with water. A 5–10 cm³ aliquot is pipetted into a 50 cm³ volumetric flask, 5 cm³ ethanol is added, and the pH is adjusted to 4 with 10% NaOH. After the addition of 10 cm³ 2 M H_2SO_4, 0.5 cm³ 10% ascorbic acid, 1 cm³ 0.2% tartaric acid, 1 cm³ 2% EDTA and 10 cm³ 0.02% IDTM in 1.5 M H_2SO_4, the volume is made up to the mark with 1 M H_2SO_4. The absorbance of the solution is measured at 358 nm, against the corresponding mixture of the reagents. The calibration plot covers the range 1–10 µg Te, the above procedure being followed. The sensitivity of the method is $1 \times 10^{-3}\%$ Te.

References

[1] Nazarenko, I. I., Ermakov, A. N.: *Analiticheskaya khimiya selena i tellura*. Nauka, Moscow 1971.
[2] Boltz, D. F.: *Colorimetric Determination of Nonmetals*. Wiley, New York 1978.
[3] Babko, A. K., Pilipenko, A. T.: *Fotometricheskii analiz. Metody opredeleniya nemetallov*. Khimiya, Moscow 1974.
[4] Bode, H.: *Z. Anal. Chem.* **144**, 90 (1955).
[5] Pavlova, V. N., Vasileva, N. G., Kashlinskaya, S. E.: *Zav. Lab.* **27**, 965 (1961).
[6] Donaldson, E. M.: *Talanta* **23**, 823 (1976).
[7] Ivankova, A. I., Blum, I. A.: *Zav. Lab.* **27**, 371 (1961).
[8] Blum, I. A., Glazkova, A. F.: *Khimicheskie, fizikokhimicheskoe i spektralnye metody isledovaniya rud redkikh i rasseyanykh elementov*. Gosgeoltekhizdat, Moscow 1961.
[9] Arstamyan, Zh. M., Tarayan, V. M.: *Arm. Khim. Zh.* **19**, 590 (1966).
[10] Vladimirova, V. M., Davidovich, N. K., Kuchmistaya, G. I., Razumova, L. S.: *Zav. Lab.* **29**, 1419 (1963).
[11] Voronkova, M. A.: *Metody khimicheskovo analiza mineralnovo syrya*. **11**, 51 (1968).
[12] Havezov, I., Jordanov, I.: *Talanta* **21**, 1013 (1974).
[13] Torocheshnikova, I. I., Gibalo, I. M., Dmitrienko S. G., Pashekova, N. A.: *Zh. Anal. Khim.* **36**, 478 (1981).
[14] Dolgorev, A. V., Ziborova, Yu. F.: *Zav. Lab.* **46**, 17 (1980).
[15] Dolgorev, A. V.: Otkrytie izobrezeniya promyshlennye obraztsov, tovarnye znaki. No. 20, Moscow, 1976. (Patent No. 515 747, Moscow, 1974.)

6.37. Thallium

Thallium belongs in column III/1 of the periodic system. In solution it occurs in the monovalent and trivalent forms. The high standard redox potential of the Tl^{3+}/Tl^+ system $(+1.25$ V) means that Tl^+ is the more stable in solution; H_2O_2 or strong oxidants $(MnO_4^-$, etc.) are needed for its oxidation to Tl^{3+}. Tl^+ is rather reminiscent of Ag^+ in its behaviour [1]; for instance, its sulphide and hydroxide are poorly soluble. Tl^{3+} remains in solution only in relatively strongly acidic medium (pH < 0.3); it is otherwise hydrolyzed or $Tl(OH)_3$ separates out [2]. Its hydroxide is not amphoteric. From an analytical aspect, its halide complexes are very important. The complexes with oxyacids are of lower stability. As a consequence of its oxidizing properties, Tl^{3+} does not form complexes with strongly reducing ligands, e.g. those containing S donor atoms.

The extraction of the dithizonate or dithiocarbamates of Tl^+, and the precipitation of TlOH or Tl_2S [2, 3] are suitable procedures for the separation of Tl^+. The extraction of the halides TlX_4^- is particularly advantageous; if $TlBr_4^-$ is extracted into ethyl ether from 1 M HBr, it is accompanied only by $AuBr_4^-$ and a little $FeBr_4^-$.

Separation can also be achieved by the extraction of TlX_4^- complexes with basic dyes. Such basic dyes (crystal violet, rhodamine B, malachite green, etc.) are at the same time the most important photometric reagents for thallium. For these complexes, the molar absorbance lies in the range $2-9 \times 10^4$, so that they provide a possibility for sensitive determination. Dithizone and the dithiocarbamates are not merely suitable for separation purposes; only Bi^{3+} and Pb^{2+} interfere in the determination of thallium with dithizone, and only Bi^{3+} in that with dithiocarbamates [4]. Measurement of the absorbance of TlX_4^- can also be used for determination.

A survey of the analysis of thallium is to be found in the book by Korenman [1].

Determination of thallium in rocks and sulphide ores, with crystal violet [6]

$TlCl_4^-$ can be extracted from 0.15–0.20 M HCl into toluene with crystal violet. Under such conditions, only Sb^{5+}, Au^{3+} and Hg^{2+} interfere. They are separated by cementation onto a copper wire.

308

Analytical procedure: A $HF+HNO_3+H_2SO_4$ mixture is recommended for the digestion of rock samples, whereas sulphide ores necessitate aqua regia and subsequent evaporation with HCl. A sample weight of 0.2–1.0 g is taken. To the dry residue remaining after HCl evaporation, 20 cm³ 2.5% HCl is added, the mixture is boiled and filtered, and the filtrate is diluted to 30–35 cm³. For the removal of disturbing ions, a copper spiral is inserted into the solution, which is then boiled for 2–3 min. After cooling, the volume should be 20–25 cm³. If there is relatively extensive deposition (this can be seen from the dark colour of the copper spiral), 0.5 cm³ conc. HCl is added to the solution to increase the acidity, and the deposition of the disturbing ions is repeated with a fresh copper spiral. After removal of the spiral, H_2O_2 is added (a few drops in excess) to oxidize thallium and to dissolve and oxidize precipitated copper(I) compounds. After 30 min, 10 drops 0.2% crystal violet is added, and the thallium is extracted with 2×15 cm³ toluene. If the second extract is colourless, it is rejected. If it is strongly coloured, the analysis is restarted with a smaller sample weight (it is preferable to prepare a stock solution before the extraction, and to work with aliquots of this). The absorbance of the complex in a volume of 50 cm³ is measured at 570 nm, or with a corresponding colour filter. The calibration plot is produced with 1–100 µg Tl, from a solution containing 5 cm³ HCl, 7 cm³ 10% $FeCl_3$, 10 drops H_2O_2 and 10 drops reagent in 50 cm³. The sensitivity of the method is 2×10^{-4} %.

For the analysis of manganese ores [3], the sample is dissolved in $HCl+H_2C_2O_4$, and precipitation of the thallium on a $Fe(DDTC)_3$ carrier is recommended for its concentration. The precipitate is dissolved in $HNO_3+H_2O_2$, and the nitrate is removed by evaporation with H_2SO_4. The thallium is next extracted from 1 M HBr into ethyl ether, the ether is evaporated off, and the thallium is determined with crystal violet in the described manner.

For the determination of $\geqq 200$ µg Tl/dm³ in industrial effluents [7], the same procedure is used after evaporation of the sample (2–20 cm³) with H_2SO_4, dissolution in HCl, and addition of $NaNO_2$ (for oxidation to Tl^{3+}).

Determination of thallium content of 99.9–99.999% pure lead, cadmium, iridium and zinc, by extraction of TlI_4^- [5]

The principle of the method is that $TlBr_4^-$ is extracted with trioctylamine (TOA). The organic phase is shaken with KI to yield TlI_4^-, the absorbance of which is measured.

5 g sample is dissolved in 15 cm³ 1:1 HNO_3 with heating. 1 cm³ H_2O_2 is added to the solution (and also to 20 cm³ 2 M HNO_3 treated in parallel with this as a blank), and the mixture is boiled gently to oxidize the Tl^+ and to decompose the bulk of the H_2O_2 (30 min). 1% $KMnO_4$ is added to decompose the residual H_2O_2, and 0.1% $NaNO_2$ is added to decolorize the excess $KMnO_4$. The solution is made up to 50 cm³, 5.0 cm³ 1 M HBr is added, and the $TlBr_4^-$ is extracted with 5 cm³ 6% TOA in benzene. The organic phase is washed with 2×20 cm³ 1 M HNO_3 and then 20 cm³ 1 M $HClO_4$. It is next shaken with 20 cm³ KI solution, and the absorbance of the TlI_4^- is measured at 400 nm (KI solution: 1.2 g KI is dissolved in 120 cm³ water, and 30 cm³ 0.1% sodium thiosulphate solution and 50 cm³ of acetate buffer of pH 6.0 are added). The calibration plot is produced for the interval 0–50 μg Tl.

References

[1] Korenman, I. M.: *Analiticheskaya khimiya talliya*. Nauka, Moscow 1960.
[2] Marczenko, Z.: *Spectrophotometric Determination of Elements*. Ellis–Horwood, London 1976.
[3] Efremov, G. V., Parshina, N. V.: *Spektralie i khimicheskie metody analiza materialov*. Metallurgiya, Moscow 1964.
[4] Keil, R.: *Z. Anal. Chem.* **258**, 97 (1972).
[5] Tsukahara, Y., Sakakibara, M., Yamamoto, T.: *Anal. Chim. Acta* **83**, 251 (1976).
[6] Popov, N. P., Stolyarova, J. A.: *Khimicheskii analiz gornykh porod i mineralov*. Nedra, Moscow 1974.
[7] Raikova, I. G., *et al: Nauchnye trudy v redkometallicheskoi promysl.* **47**, 233 (1973)

6.38. Thorium

Thorium, a member of the actinides, belongs in column III/2 of the periodic system. Although thorium exists in aqueous solution as a tetravalent cation, it is not very susceptible to hydrolysis, and it is therefore not difficult to keep it in solution. The best-known photometric reagents for its determination are the azo-arsonic acid derivatives. These include thoron [1] SPADNS [2], arsenazo II [3] and arsenazo III [4] (see Appendix). The first three of these reagents are not suitable for sensitive and accurate determination, as a consequence of the low stability and partly the low molar absorbance of the complexes. Very many ions interfere, and accordingly, complicated and lengthy separations are necessary in the analysis.

The properties of arsenazo III are far better than those of any other reagent, and it is therefore particularly recommended for the determination of thorium. The analysis of thorium is surveyed in the monograph by Ryabchikov and Golybraykh [5]. Thorium forms a very stable complex with arsenazo III in 4–10 M HCl. At a reagent concentration of 1×10^{-3} M in 4.8 M HCl medium, $\lambda_{max} = 665$ nm and $\varepsilon = 1.12 \times 10^5$. The composition of the complex is M : L = 1 : 1.

Zirconium forms an arsenazo III complex which is similar in stability to the thorium complex, but it can be masked with oxalic acid. Thus, not even a 100-fold amount of zirconium interferes.

Tetravalent uranium in a ratio of $Th^{4+} : U^{4+} = 1 : 1$ disturbs the determination, but since U^{4+} is formed only on the action of reductants, in practice its presence need not be reckoned with. Hexavalent uranium does not interfere in a ratio of $Th^{4+} : U^{6+} = 1 : 10$. Separation (precipitation with oxalate) allows the presence of even a 50-fold amount of uranium. In the analysis of samples with a higher uranium content than this, extraction (TBP) must be performed.

Lanthanides in a 50-fold amount do not interfere. For the analysis of samples with high lanthanide contents, extraction is carried out with 0.01 M di(n-butyl)-phosphoric acid from 4 M HCl medium; the lanthanides remain in the aqueous phase, while the thorium passes into the organic phase. The thorium may be determined after decomposition of the organic phase with HBr.

Determination of thorium content of rock samples [6]

Depending on the expected thorium content, at most 1 g sample is weighed into a platinum dish. It is moistened with 2–3 drops conc. H_2SO_4, 10 cm³

311

HF is added, and the mixture is evaporated to dryness on a sand-bath. 1 cm^3 saturated boric acid is added to bind the fluoride traces, and the mixture is again evaporated to dryness. The residue is washed into a beaker with ca. 20 cm^3 1:1 HCl, and the mixture is boiled to achieve dissolution. Complete dissolution is promoted by the addition of a few drops of H$_2$O$_2$. The solution is diluted to ca. 80 cm^3 with water and boiled, 5 g solid oxalic acid is added, and the pH is adjusted to 2 with 5 M NaOH (universal indicator paper). If the sample to be analyzed does not contain sufficient calcium (this can be seen from the quantity of the oxalate precipitate), 4 cm^3 5% CaCl$_2$ is added to serve as a carrier. After 30 min, the oxalate precipitate is filtered off on a Whatman No. 41 filter paper, and is washed with a little 1% oxalic acid. The precipitate is washed back into the beaker with 1:1 HCl, dissolved by boiling, transferred into a 50 cm^3 volumetric flask, and made up to the mark with 1:1 HCl.

At most 10 cm^3 of this stock solution is pipetted into a 25 cm^3 volumetric flask, and 4 cm^3 10% oxalic acid is added, together with sufficient conc. HCl for the final HCl concentration to be 4.8–5 M. Next, 1 cm^3 0.1% arsenazo III solution is added, the volume is made up to the mark with water, and the absorbance is measured in 1–5 cm cells at 665 nm. Comparison solution: 10 cm^3 conc. HCl + 4 cm^3 10% oxalic acid + 1 cm^3 0.1% arsenazo III, in 25 cm^3. The calibration plot is recorded with 1–25 μg Th, in 5 cm cells up to 10 μg Th, and in 2 cm cells for 10–25 μg Th. The absorbance of 10 μg Th in a 5 cm cell is 0.97. The sensitivity of the method is 2.5 ppm Th. The error in the determination of low concentrations is ±20%, but for above 20 ppm Th it does not exceed ±5%.

Notes:

1. The rock sample may also be digested by KOH fusion in a nickel or iron crucible over a gas flame. Towards the end of the digestion, ca. 1 g Na$_2$O$_2$ is added, and the fusion is continued for a further 30 s. The melt is leached out with hot water, the solution is boiled, the precipitate is filtered off, washed with hot water, and then washed back into the original beaker, the mixture is acidified, and the oxalate precipitation is carried out at pH 2 as described above.

2. This digestion can also be employed in the analysis of ores of tungsten, molybdenum and chromium, when the main components pass into the filtrate, while the thorium remains in the hydroxide precipitate. If the rock sample to be analyzed does not contain sufficient iron, 50 mg Fe^{3+} is added to the aqueous leaching solution to ensure the quantitative precipitation of the thorium.

312

3. Depending on the nature of the sample to be analyzed, the separation steps may be repeated. For instance, in the analysis of bauxite it is advisable to carry out the oxalate precipitation twice because of the high titanium content, while repeated hydroxide precipitation is recommended in the analysis of tungsten ores.

4. In the determination of the thorium content of uranium ores, the sample is digested and the stock solution is prepared as described in connection with the analysis of rock samples. 10 cm^3 of the 1 : 1 (6 M) HCl solution is pipetted into a 50 cm^3 separating-funnel, 10 cm^3 20% tributyl phosphate (TBP) in petroleum ether is added, and extraction is performed for 1 min. The lower, aqueous phase is run off into another separating-funnel, and the extraction is repeated with 10 cm^3 TBP. The lower phase is run off into a 25 cm^3 volumetric flask, the organic phases are combined and shaken with 5 cm^3 6 M HCl to avoid any slight thorium loss, this aqueous phase too is transferred to the 25 cm^3 volumetric flask, and the volume is made up to the mark with 6 M HCl. In this way uranium can be separated from thorium: $D_U = 15$, and $D_{Th} \leq 0.02$. The thorium is determined on an appropriate aliquot of the stock solution.

Determination of a small quantity of thorium in the presence of much lanthanide (in monazite, apatite, yttrium oxide and lanthanide oxide) [7]

The digestion (with HF or KOH) of the sample to be examined, and the oxalate precipitation, are carried out as above. The filtered-off oxalate precipitate is ignited at 600–700°C, and the oxides are dissolved in HCl and H$_2$O$_2$. The final volume of the solution should be 20 cm^3, and its HCl concentration 4 M. In a separating-funnel, the solution is extracted for 1 min with 20 cm^3 0.01 M di(n-butyl)-phosphoric acid (HDBP) in petroleum ether. The lanthanides remain in the aqueous phase, while the thorium (together with zirconium and scandium) pass into the organic phase. The organic phase is washed with 20 cm^3 0.01 M HDBP that has previously been shaken with 4 M HCl. The organic phase is transferred to a 50 cm^3 beaker, the petroleum ether is evaporated off on a water-bath, 5 cm^3 HBr and 5 cm^3 glacial acetic acid are added to destroy the HDBP, and the mixture is left covered with a watch-glass for 1 hour on a water-bath. For removal of the acetic acid and bromobutane, the watch-glass is taken off and the solution is boiled; this step is continued after the addition of a little conc. HCl. The residue is washed into a 25 cm^3 volumetric flask, and sufficient conc. HCl is added for its concentration to be 6 M after the solution has been made up to the mark. A suitable aliquot of this stock solution is taken for thorium determination by the above procedure.

313

The several $\mu g/dm^3$ thorium and uranium in surface waters can be well sorbed at pH 3 on ion-exchange resins containing iminodiacetic acid as functional group (Dowex-A-1, Chelex-100, Amberlite XE-318, Wofatit MC-50, Ligandex E. etc.). Elution may be performed with 2 M HCl.

Analytical procedure: The pH of 1,000–1,500 cm^3 filtered water is adjusted to 3. For acidic waters this is done with 5 M NaOH, and in other cases with 6 M HCl. 1–1.5 g air-dried Wofatit MC-50 ion-exchange resin in Na$^+$ form is added to the resulting water, and mixing is carried out with compressed air for 3 hours. The resin is then filtered off and washed with distilled water. 10 cm^3 2 M HCl is added to it, the mixture is left to stand for half an hour with periodic mixing, the eluate is next poured off the resin, and the elution is repeated with 10 cm^3 2 M HCl. The eluate is evaporated to ca. 10 cm^3 and transferred to a 25 cm^3 volumetric flask, 10 cm^3 conc. HCl, 2 cm^3 10% oxalic acid and 1 cm^3 0.1% arsenazo III are added, the volume is made up to the mark, and photometry is performed as in [6].

The calibration plot is obtained by the above procedure following the addition of 1–10 μg Th to ion-exchanged water. The sensitivity of the method is 1 μg Th/dm^3.

Korkisch and Krivanecz [8] use Dowex 1×8 anion-exchange resin in citrate form to bind the uranium and thorium content of natural waters, and determine the thorium content with arsenazo III.

In another procedure [9], the oxalate precipitation in the presence of a calcium carrier is followed by the extraction of the thorium content of the natural water with Aliquat 336 (tricaprylmethylammonium chloride). This concentrates the thorium, which is subsequently similarly determined with arsenazo III.

References

[1] Kuznetsov, V. I.: *Dokl. AN SSSR* **31**, 895 (1941).
[2] Banerjee, G.: *Anal. Chim. Acta* **16**, 56 (1957).
[3] Kuznetsov, V. I.: *Zh. Anal. Khim.* **14**, 7 (1959).
[4] Savvin, S. B.: *Arzenazo III.* Atomizdat, Moscow 1966.
[5] Ryabchikov, D. I., Golybraykh, E. K.: *Analiticheskaya khimiya toriya.* AN SSSR, Moscow 1960.

[6] Upor, E., Jurcsik, J., Mohai, M.: *Acta Chim. Acad. Sci. Hung.* **37**, 1 (1963)

[7] Mohai, M., Upor, E.: *Magy. Kém. Folyóirat* **77**, 553 (1971).

[7a] Horváth, Á, Csővári, M., Hohmann, J.: Felszíni vizek urán- és tóriumtartalmának meghatározása előzetes kémiai dúsítás után (Determination of the uranium and thorium content of surface waters after preliminary chemical enrichment). Paper presented at the 17th Transdanubian Analytical Conference, Keszthely, Hungary, 1982.

[8] Korkisch, J., Krivanecz, H.: *Talanta* **23**, 295 (1976).

[9] Cospito, M., Rigali, M.: *Anal. Chim. Acta* **106**, 385 (1979).

6.39. Tin

Tin is situated in column IV/1 of the periodic system. Accordingly, its properties are similar to those of titanium and germanium. Its main similarities to lead (the element below it in the same column) are that its divalent form is stable and it readily gives complexes with S-containing ligands. The standard redox potential of the Sn^{2+}/Sn^{4+} system is $+0.15$ V, i.e. Sn^{2+} is a strong reductant. It is more stable against hydrolysis than is Sn^{4+}, which is hydrolyzed even in strongly acidic medium. This can be utilized for the precipitation of Sn^{4+} in acidic medium, with MnO_2 as carrier. $Sn(OH)_2$ dissolves in solutions more alkaline than pH 13, and $Sn(OH)_4$ in solutions more alkaline than pH 9. This likewise provides a possibility for separation.

Particularly tin(IV), but tin(II) too, forms stable complexes with ligands containing N, S and O donor atoms. One possibility for separation is the extraction of the halide complexes.

Perhaps the most important of the photometric reagents for tin is phenylfluorone [1], and this will be dealt with in some detail. Other trihydroxy compounds are pyridyl-3-fluorone [2] and lumogallion [3]. Of the triphenylmethyl derivatives, pyrocatechol violet [4] is worthy of mention; dithiol [5] and salicylideneamino-2-thiophenol [6] are important sulphur-containing reagents. These latter two reagents form complexes with Sn^{2+}, and just like the diethyldithiocarbamates these are suitable for extraction from 0.5–5 M H_2SO_4.

The spectrophotometry of tin is reviewed in the book by Spivankovskii [7].

315

(A) DETERMINATION OF TIN WITH PHENYLFLUORONE

In acidic medium (≤ 1 M), Sn^{4+} reacts with phenylfluorone to yield a red complex ($\lambda_{max} = 510$ nm), which can be kept in a state suitable for photometric determination by the addition of a protective colloid; this is necessary, as the complex is polymeric. The medium recommended for the determination has pH 1.0–5.0. Depending on the composition of the sample to be examined, and on the medium applied, the molar absorbance varies in the range $\varepsilon = 5$–8×10^4.

The determination is disturbed by multivalent metal ions (As^{3+}, Cr^{3+}, Fe^{3+}, Mo^{5+}, Sb^{3+}, Bi^{3+}, Ti^{4+}), strong oxidants and PO_4^{3-}. In some cases the interference may be eliminated by a change in the valency state (Sb^{5+}, As^{5+}) or by masking (Ti^{4+}, Mo^{5+}), but in general it is not possible to avoid separation.

Table 6.28 lists the digestion and separation procedures suggested for the analysis of various samples, together with the recommended medium for photometry with phenylfluorone.

TABLE 6.28.

Features of some suggested methods for determination of tin with phenylfluorone

Sample examined	Digestion medium	Separation procedure	Medium for photometry	Refs
Rocks	Na_2O_2	Distillation from HBr	0.5 M H_2SO_4	[11]
Rocks	Na_2O_2 after fuming with HF	Sorption on silica gel	pH 2–3	[8]
Steel, various metals	Various acid mixtures	SnI_4 extracted from 4 M H_2SO_4 with benzene		[12]
Brass	Aqua regia	Sorption of $SnCl_6^{2-}$ on anion-exchange resin from 3 M HCl	pH 4.0, acetate buffer	[13]
Nickel	1 : 1 HNO_3	Precipitation of $Sn(OH)_4$ with anthraquinone-α-arsonic acid	pH 3.5, acetate buffer	[14]
Lead	HNO_3 + tartaric acid	Extraction of $SnCl_6^{2-}$ from 6 M HCl with Amberlite LA-2	0.15 M H_2SO_4	[15]

316

Tin(II) and tin(IV) can be sorbed on silica gel from weakly acidic medium, and then eluted with 6 M HCl [7, 9]. The eluate does not contain interfering components, and the tin content, after oxidation, can be determined with phenylfluorone.

Analytical procedure: A mixture of 0.5–2 g sample, a few drops of H_2SO_4 and 10 cm^3 HF is evaporated to dryness in a platinum dish. The residue is transferred to a nickel crucible and fused with a few grams of Na_2O_2. The melt is leached out with hot water, acidified with ca. 40 cm^3 conc. HCl, and diluted to 100 cm^3. 300 cm^3 0.5 M sodium citrate + citric acid buffer of pH 5.5 is added [buffer: 2 g each of EDTA, K_2SO_4, NH_4Cl and $(NH_4)_2SO_4$, in 1 dm^3]; the pH of the solution is checked and, if necessary, is adjusted to 5.5 by the addition of NaOH or citric acid.

Pretreatment of silica gel column for sorption: 100 cm^3 0.5 M sodium citrate + citric acid buffer of pH 5.5 is passed through a column 5 mm in diameter, containing 5 g silica gel with a grain size of 0.2–0.4 mm. The buffered solution to be analyzed is passed through this column at a rate of 2 cm^3/min. The column is then washed with 500 cm^3 distilled water. Elution is performed with 10 cm^3 1:1 HCl; the acid is first retained in the column for 5 min, and then allowed to elute at a very low rate (0.5 cm^3/min). The eluate is run into a 50 cm^3 volumetric flask. The column is then washed with water, the washings are combined with the earlier eluate, and the volume is made up to 50 cm^3. A 5 cm^3 aliquot (1–6 µg Sn) is pipetted into a 50 cm^3 volumetric flask, 2 cm^3 3% H_2O_2, 1 cm^3 0.1% gelatin, 10 cm^3 0.5 M H_2SO_4, 10 cm^3 buffer (225 g sodium acetate + 120 cm^3 acetic acid, made up to 500 cm^3) and 10 cm^3 $7.5 \times 10^{-4}\%$ phenylfluorone in alcohol are added, the mixture is shaken, after 5 min the volume is made up to the mark with water, and the absorbance is measured in 4 cm cells at 510 nm, against a mixture of the same quantities of the reagents. The calibration plot is produced with 5–70 µg Sn, the operations being started from the sorption. The sensitivity of the method is 5–10 ppm Sn.

Determination of tin in ores with PAN, after preliminary separation [10]

In a medium of 0.25–0.75 M H_2SO_4 and 2.5 M NaI in the presence of dimethylformamide (DMFA), Sn(IV) can be extracted selectively with benzene, and can be determined in the organic phase with 1-(2-pyridylazo)-2-naphthol (PAN): $\lambda_{max} = 570$ nm, $\varepsilon_{570} = 51 \times 10^3$.

Analytical procedure: 0.5–1 g ore is digested with 5 g Na_2O_2 in a nickel or iron crucible in a furnace at 600°C (20 min). The melt is leached out with water and then HCl, and the mixture is transferred to a 10 cm³ volumetric flask. Sufficient conc. HCl is added for the final acidity to be 1 M, and the volume is made up to the mark with water. A 1–2 cm³ aliquot is pipetted into a 25 cm³ ground-glass stoppered test-tube, 9 cm³ mixture (0.5 M in H_2SO_4, 2.5 M in NaI, 0.15% in ascorbic acid) and 2 cm³ DMFA are added (together with 100 mg thiourea in the presence of Cu, Bi, Se and Te, and with 100 mg tartaric acid in the presence of W), and the solution is extracted for 4–5 min with 10 cm³ benzene. The solution is transferred to a separating-funnel, and the aqueous phase is poured off. In the presence of a large amount of accompanying ions, the organic phase is washed by careful shaking for 5 s with 2×5 cm³ washing liquid (0.15 M in H_2SO_4, 2.5 M in NaI, 10% in DMFA). 5 cm³ 0.05% PAN in benzene is added to the organic phase, and the covered extract is heated for 5 min at 55–65°C on a water-bath. After cooling, it is filtered through filter paper, and the absorbance of the solution is measured in 20, 10 or 3 mm cells, against a mixture of 10 cm³ benzene and 5 cm³ 0.05% PAN in benzene. The calibration plot is produced via the same procedure, with 5–200 µg Sn(IV). The sensitivity of the method is of the order of 10^{-2} % Sn.

References

[1] Luke, C. J.: *Anal. Chem.* **28**, 76 (1956).
[2] Asmus, E., Kraetsch, J.: *Z. Anal. Chem.* **216**, 391 (1966).
[3] Lukin, A. M., Bozhevolnov, E. A.: *Zh. Anal. Khim.* **15**, 43 (1960).
[4] Newmann, E. I., Jones, P. D.: *Analyst* **91**, 406 (1966).
[5] Onishi, H., Sandell, E. B.: *Anal. Chim. Acta* **14**, 153 (1956).
[6] Gregory, G., Jeffery, P. G.: *Analyst* **92**, 293 (1967).

[7] Spivanovskii, V. B.: *Analiticheskaya khimiya olova*. Nauka, Moscow 1975.
[8] Schreck, M.: Unpublished work.
[9] Jones, I. C. H.: *Analyst* **93**, 214 (1968).
[10] Rakhmatullaev, K., Zakirov, B. G.: *Zh. Anal. Khim.* **35**, 284 (1980).
[11] Onishi, H., Sandell, E. B.: *Geochim. Cosmochim. Acta* **12**, 262 (1957).
[12] Luke, C. L.: *Anal. Chim. Acta* **39**, 304 (1967).
[13] Styapin, V. V., Silaeva, E. V.: *Analiz tsvetnykh metallov i splavov*. Metallurgiya, Moscow 1965.
[14] Yakovlev, P. Ya., Razumova, G. P., Dymova, M. S.: *Novye metody ispytaniya metallov*. Sbornik trudov Ts NII chernoi metallurgii, No. 24, Moscow 1962.
[15] Hofer, A., Landl, B.: *Z. Anal. Chem.* **244**, 103 (1969).

6.40. Titanium

Titanium belongs in column IV/2 of the periodic system. In solution it behaves similarly to zirconium, which is situated in the same column. Tetravalent titanium exists in the form Ti^{4+} only at high acidities. The most stable complexes are formed with OH^-; the monomeric state TiO^{2+} is not even ensured in relatively acidic medium (pH 1). Polynuclear hydroxo complexes may be eliminated with auxiliary complex formers (oxyacids, SO_4^{2-}, EDTA, F^-, etc.). $Ti(OH)_4$ is precipitated even at pH 2–3.

At high acidities, Ti^{3+} too is fairly stable; however, it is not of importance in the photometric determination of titanium.

Primarily reagents containing O donor atoms (phenolic or alcoholic OH groups) are suitable for the photometric determination of titanium, but N-containing functional groups are also good. The bonding between the titanium and the ligand is stabilized by a second O donor atom, e.g. polyphenols, or by simultaneous N and S coordination. One of the most convenient photometric reagents for Ti is a polyphenol, chromotropic acid. To increase the solubility of this reagent in water, one or more sulphonic acid groups are incorporated during the synthesis.

Colorimetry of the yellow complex of titanium with H_2O_2 can be applied well, because of the simplicity of the process. If the solution also contains vanadium (e.g. in bauxites), this interferes by giving a red colour. The vanadium must then be removed by alkaline separation.

The most important reagents are listed in Table 6.29.

More detailed information on the analysis of titanium and the photometric reagents is to be found in the reviews by Busev *et al.* [1] and Sommer [2].

TABLE 6.29.

Features of some more important reagents for titanium

Reagent	λ_{max}, nm	$\varepsilon \times 10^{-3}$	Medium	Interferences	Refs
Tiron	390–410	13	pH 4.3–9.6	Fe^{3+}, Mo^{6+}, Os^{7+}, U^{6+}, V^{4+}, Ce^{4+}, Al^{3+}, Ca^{2+}, Th^{4+}, Hg^{2+}, Sn^{2+}, Sn^{4+}, CrO_4^{2-}, WO_4^{2-}, F^-, oxalate	[11]
H_2O_2	410	0.74	0.75–1.75 M H_2SO_4	V, Mo, U, Nb, Cr, Fe, F^-, oxalate, citrate	[12]
Chromotropic acid	460	17	pH 3–5	Fe^{3+}, much V, Zr	[12]
Diantipyrylmethane	360–380	13	0.3–6 M HCl	Fe^{3+}, V^{5+}, Cr, Ni, Co (F^-, borate, silicate do not interfere)	[12]
Salicylic acid ($CHCl_3$)	380	59	pH 2.3	F^-, oxalate	[8]
Dibromotichromin + diphenylguanidine (n-butanol)	480	10.3	0.1–5 M H^+	Mo^{6+}, W^{6+}, Nb^{5+}	[4]
Thiocyanate–diantipyryl-methane ($CHCl_3$)	420	60	2–3 M HCl	Fe^{3+}, Cu, Co, W, Mo, Ni	[7, 14]
9-(2′,4′-Disulphophenyl)-2,3,7-trihydroxy-6-fluorone	570	108	pH 6	Ge, Sn^{4+}, Sb^{3+}, Zr, Al, Mo	[10]

Possibilities for dissolution of titanium. In rocks and minerals, titanium is generally present in forms that are difficult to digest. In a general case it can be dissolved by means of fusion with NaOH; on aqueous leaching-out, the $Ti(OH)_4$ remains quantitatively in the hydroxide precipitate. This process means the simultaneous separation from numerous ions (Mo, W, V, SiO_2, Al, etc.). If the solution is optically pure after dissolution of the hydroxide precipitate in acid, the digestion is satisfactory. In the event of rocks that are difficult to digest, a reasonable procedure is to evaporate the sample with $HF + H_2SO_4$ (the fluoride excess can be removed by evaporation to dryness after the addition of saturated boric acid), next to digest it with potassium pyrosulphate, and then to dissolve the residue in acid.

320

Determination of titanium in rocks by extraction of TiO(H$_2$O$_2$)$^{2+}$ with HDBP [3]

The complex can be extracted quantitatively with 0.5 M HDBP from a solution containing 0.1 M HCl and 0.4 M H$_2$O$_2$. Separation from the interfering V and Mo is achieved by alkaline precipitation, while Fe^{3+} is extracted with TBP. The method is suitable for the analysis of samples with low titanium contents (30–100 ppm).

Analytical procedure: The sample to be analyzed is fused with NaOH, the melt is leached out with hot water, the precipitate is filtered off and it is then dissolved in sufficient conc. HCl for the final acidity to be $\sim$6 M. The solution is shaken with 2×20 cm^3 TBP in petroleum ether, the residual aqueous phase is evaporated to low volume, and it is neutralized with 1 M NH$_4$OH. 2.5 cm^3 1 M HCl and 1 cm^3 30% H$_2$O$_2$ are added, and the volume is made up to 25 cm^3 with distilled water. The resulting complex is extracted with an identical volume of 0.5 M HDBP in petroleum ether. The necessary quantity of the organic phase is pipetted into a 5 cm cell, and the absorbance is measured at 425 nm. The comparison solution is the extract of the corresponding solution not containing titanium. The calibration plot is obtained with 25–250 μg Ti, starting with the HDBP extraction.

Extraction–photometric determination of titanium in steel with dibromotichromin [4]

Tichromin and dibromotichromin give orange-red complexes with titanium in 0.01–5 M HCl (0.005–2.5 M H$_2$SO$_4$) medium. These complexes can be extracted in the presence of large organic cations such as diphenylguanidine (DPG) or triphenylguanidine (TPG). The most satisfactory solvent is *n*-butanol. Of the masking agents or reductants, tartaric acid, citric acid, oxalic acid and hydroxylamine do not interfere (Mo, W and Nb can be masked). Strong interference is caused by fluoride and ascorbic acid.

Analytical procedure: 0.1 g steel is dissolved in 15 cm^3 1:3 H$_2$SO$_4$, after dissolution a few drops of conc. HNO$_3$ are added, and the solution is boiled to remove oxides of nitrogen. After cooling, the solution is washed into a 100 cm³ volumetric flask and made up to the mark. 2.0 cm³ is pipetted into a separating-funnel, 1.0 cm³ 0.5% dibromotichromin (dissolved in 3% Na$_2$SO$_3$ solution) and 5 cm³ 20% DPG or 5 cm³ 2% TPG are added, and the mixture is extracted with 10 cm³ n-butanol previously saturated with water. A few drops of ethanol are added to the organic phase to ensure optical purity, and the absorbance is measured at 480 nm. The calibration plot is recorded with 0.02–80 μg Ti.

Determination of titanium in iron ore with chromotropic acid [5]

From the titanium in a solution of digested iron ore, the yellowish-red complex with chromotropic acid is produced. The determination is disturbed by Fe(III) and V(V); reduction is carried out with ascorbic acid in the presence of CuSO$_4$ solution.

Analytical procedure: 0.5 g iron ore is digested with 4 g Na$_2$O$_2$ in a zirconium (or nickel) crucible. The melt is leached out with hot water, 10 cm³ conc. HCl is added, the solution is boiled and cooled, and the volume is made up to 100 cm³. A 10 cm³ aliquot is taken, and 10 cm³ 5% ascorbic acid and 1–2 drops 5% CuSO$_4$ are added, followed after 5 min by 15 cm³ 1% chromotropic acid. The pH of the solution is adjusted to 3.5–4.0 with acetate buffer, and the volume is made up to 100 cm³. Photometry is performed 15 min later, in 2 cm cells at 470 nm. The calibration plot covers the interval 10–30 μg Ti.

Determination of titanium in pure molybdenum, tungsten, vanadium and their oxides with diantipyrylmethane [6]

After digestion of the material by fusion, the titanium–H$_2$O$_2$ complex is sorbed on silica gel. The eluted titanium is determined spectrophotometrically with diantipyrylmethane.

Analytical procedure: 0.25 g tungsten sample is converted to the oxide by ignition at 750–800°C in a platinum vessel, and the oxide is fused with 2 g NaKCO$_3$. The melt is dissolved up with 40 cm³ water and 10 cm³ 1:1 H$_2$SO$_4$. The mixture is boiled and cooled, and 5 cm³ 30% H$_2$O$_2$, 1 cm³ 1 M tartaric acid and 10 cm³ 0.5 M acetic acid are added. The pH of the solution is adjusted to ca. 6.5, and the volume is made up to 100 cm³.

322

The same digestion procedure is used for tungsten oxide and molybdenum oxide. Metallic molybdenum and vanadium are dissolved in a mixture of $HNO_3 + H_2SO_4$.

Pretreatment of silica gel column: The grain size is 1–2 mm, and the exchange capacity is 2.2 meq g^{-1} (e.g. Biosil). The silica gel is washed several times with 3 M HCl, and then kept at 150°C for 48 hours for activation. About 2 g silica gel is packed into each column (column height 2 cm). The column is washed with 30 cm^3 washing liquid (50 cm^3 0.5 M acetic acid + 5 cm^3 30% H_2O_2, made up to 1 dm^3, the pH being adjusted to 6.5 with NH_4OH).

The sample is allowed to flow through the column at a rate of 3–5 cm^3/min. The column is washed with 20 cm^3 washing liquid. The titanium-peroxo complex is eluted with 15 cm^3 3 M HCl and decomposed by boiling. 2 cm^3 1% ascorbic acid is added to the solution, followed after a few min by 10 cm^3 3% diantipyrylmethane in 3 M HCl. The volume of the solution is made up to 50 cm^3 and after 20–30 min the absorbance is measured in 1–5 cm cells at 390 nm. The calibration plot is recorded with 10–120 μg Ti.

Determination of titanium in sea water
with diantipyrylmethane-thiocyanate [7]

The titanium is precipitated from sea water with sodium diethyldithio-carbamate, the precipitate is dissolved up, and the titanium is determined with KSCN–diantipyrylmethane. No interference is caused by the presence of 5–20 μg Mo^{6+}, Cu^{2+} and Co^{2+}, and 30 μg Fe^{3+}. The disturbing effect of a larger quantity of Fe^{3+} can be eliminated by the addition of $SnCl_2$.

Analytical procedure: The sea water is acidified to pH 2 with HCl, and 1 cm^3 4% sodium diethyldithiocarbamate is added per dm^3. A 1–6 dm^3 sample volume is taken, the pH is adjusted with NH_4OH to 9.0–9.5, the solution is stirred for 5 min, and it is then filtered through a 0.45 μm membrane filter. The precipitate is dissolved in 10 cm^3 2 M HCl with gentle heating. 5 cm^3 30% KSCN, 3 cm^3 5% diantipyrylmethane in 2 M HCl, 2 cm^3 conc. HCl and 2 cm^3 10% $SnCl_2$ solution are added. The titanium complex is extracted for 2 min with 5 cm^3 $CHCl_3$, and the absorbance is measured at 420 nm, against the extract of the reagents as blank. The result is calculated via the data for sea water to which 0, 0.5 and 1 μg Ti has been added. The sensitivity of the method is 0.1 μg Ti/dm^3, and its relative error is 5%.

323

Some other methods for the determination of titanium

Ramakrishna and Gunawardona [8] report a method for the determination in alloys and slags containing iron, aluminium and titanium, based on extraction with $CHCl_3$ of the salicylate complex of titanium.

For the determination of the titanium content of steels, ores and aluminium alloys, Pilipenko et al. [9] recommend n-furoyl-phenylhydroxylamine, which in strongly acidic medium forms a titanium complex that can be extracted with $CHCl_3$. The sensitivity may be increased by the addition of CNS^-. The method is suitable for the determination of $10^{-5}\%$ titanium.

Nazarenko and Biryuk [10] describe a sensitive method for the determination of titanium in pure germanium and silicon with disulpho-phenyl-fluorone.

References

[1] Busev, A. I., Tiptsova, V. G., Ivanov, V. M.: *Rukovodstvo po analiticheskoi khimii redkikh elementov*. Khimiya, Moscow 1978.
[2] Sommer, L.: *Talanta* **9**, 439 (1962).
[3] Upor, E., Klesch, K., Szalai, J.: *Magy. Kém. Folyóirat* **82**, 385 (1976).
[4] Basargin, N. N., Yakovlev, P. Ya., Deinikina, R. S.: *Zav. Lab.* **39**, 1043 (1973).
[5] Bhargava, Om. P., Gmitro, M., Hines W. G.: *Talanta* **27**, 263 (1980).
[6] Cziczek Z., Dolezal J., Sulzek Z.: *Anal. Chim. Acta* **100**, 479 (1980).
[7] Yang C. Y., Shin J. S., Yeh Y. C.: *Analyst* **106**, 385 (1981).
[8] Ramakrishna R. S., Gunawardona H. D.: *Talanta* **20**, 21 (1973).
[9] Pilipenko A. T., Shpak, E. A., Boiko, Yu. P.: *Zav. Lab.* **31**, 151 (1965).
[10] Nazarenko, V. A., Biryuk, E. A.: *Zh. Anal. Khim.* **15**, 306 (1960).
[11] Fries, J., Getrost, H.: *Organische Reagenzien für die Spurenanalyse*. Merck, Darmstadt 1975.
[12] Jeffery, P. G.: *Chemical Methods of Rock Analysis*. Pergamon, Oxford 1978.
[13] Polyak, L. Yu.: *Zh. Anal. Khim.* **17**, 206 (1962).
[14] Tananaiko, M. M., Nebilitskaya, S. L.: *Zav. Lab.* **28**, 263 (1962).

6.41. Uranium

Uranium, one of the actinides, is a naturally radioactive element in column III/2 of the periodic system. Its oxidation states 3, 4, 5 and 6 are known, but in solution only the tetravalent and hexavalent forms are stable. The standard redox potential of the UO_2^{2+}/U^{4+} system is $+0.65$ V, and UO_2^{2+} can therefore be reduced to U^{4+} only with relatively strong reductants (metallic zinc, Ti^{3+}, $Fe^{2+}-PO_4^{3-}$, etc.). In its analytical properties, U^{4+} resembles Th^{4+} and Zr^{4+}. The hydroxide of UO_2^{2+} begins to separate out in weakly acidic medium, and in low concentration (<0.001 M) it dissolves up at $pH > 13$. It undergoes complex formation primarily with oxygen-containing ligands (CO_3^{2-}, $P_2O_7^{2-}$, $C_2O_4^{2-}$, oxyacids, β-diketones, etc.), but it also does so with halides in strongly acidic medium.

For its separation, various extraction methods (phosphoric acid esters are important), anion-exchange (sorption of chloro or sulphato complexes) and certain precipitation methods, e.g. separation of uranium from sea-water with titanium dioxide hydrate dodecylsulphate [1], are of use. The precipitation of U^{4+} [UF_4, $U(HPO_4)_2$, etc.] is frequently utilized.

Numerous reagents are used for the photometric and spectrophotometric determination of uranium. The most important are chelate-forming reagents containing N and O, or two adjacent O donor atoms. A number of reviews have been published on the available reagents [2–6].

The more essential data on the best-known reagents and their uranium complexes are given in Table 6.30.

With regard to its analytical properties and field of application, arsenazo III is the most widespread of the reagents. A similarly valuable photometric reagent is chlorphosphonazo III. Its advantage over arsenazo III is that the masking effects of complex-formers (fluoride, oxalate, citrate, tartrate) are less strongly manifested [4, 6, 7].

The uranium in rocks does not generally occur bound to silicate, its dissolution can usually be achieved if the sample is boiled and evaporated with conc. $HCl + H_2O_2$.

Evaporation with $HF + HClO_4$ may be applied in the case of samples with unknown properties. Fluoride traces are removed by evaporation to dryness after the addition of saturated boric acid. The dry residue is dissolved by boiling with a mixture of conc. $HCl + H_2O_2$.

Determination of uranium content of rocks via the absorbance of the U(IV)-arsenazo III complex [8]

With arsenazo III, uranium that has been reduced to U(IV) with zinc in strongly acidic medium ($\geqq 6$ M HCl) forms a stable complex that can be

TABLE 6.30.

More important reagents used for photometric determination of uranium

U ion	Reagent	Medium	Extractant	λ_{max}, nm	$\varepsilon \times 10^{-3}$	Interferences	Refs
UO_2^{2+}	HTTA (thenoyltrifluoroacetone)	pH 3.5–8	Benzene	410	1.95	Th, Zr, Ag, Cu^{2+}, Fe^{3+}, oxalate, citrate, carbonate, tartrate	[4]
UO_2^{2+}	Thiocyanate	0.1–2 M HCl	32.5% TBP in CHCl$_3$	350	5.3	Fe^{3+}, V, Ti, Bi, Mo, etc.	[4, 6]
UO_2^{2+}	Na-DDTC	pH 2.5–7	Ether, ChCl$_3$, alcohols	350	4.5	Many ions	[2]
UO_2^{2+}	Dibenzoylmethane	pH 5–9	Butyl acetate	400	20	Mo, W, Ti, V, Cr^{3+}, CO$_3^{2-}$, arsenate, phosphate, tartrate, citrate	[2, 4, 6]
UO_2^{2+}	Arsenazo I	pH 4.5–8	—	595	23	Th, Ti, Zr, F$^-$, PO$_4^{3-}$, SO$_4^{2-}$, etc.	[2, 6]
U^{4+}	Arsenazo III	4–10 M HCl	—	650	100	Th, Zr, Ti	[6, 8]
UO_2^{2+}	Arsenazo III	pH 1–3	—	655	53–73	Th, Ti, Fe^{3+}, Ln, Al, Ca, etc.	[6, 9]
UO_2^{2+}	Chlorophosphonazo III	pH 1–3	—	670	73	Th, Zr, Hf, Ce^{4+}, Ti^{4+}, etc.	[2, 6]
UO_2^{2+}	H$_2$O$_2$	10% Na$_2$CO$_3$	—	380–450	0.7	V, Mn, Cr, Fe, organic matter	[2, 12]
UO_2^{2+}	Br-PADAP	pH 7.6	Trioctylphosphine oxide (TOPO)	578	74	Cr, V, PO$_4^{3-}$	[16]

subjected to photometry at 650 nm. The composition of the complex is U(IV):L=1:2.

Interferences. For a rock sample of average composition, the determination is disturbed only by Th^{4+}; Zr^{4+} and Hf^{4+} can be masked with oxalic acid. In the case of high titanium and iron contents (Ti : U>200; Fe : U>2000), the reagent begins to be decomposed because of the low redox potential after the reduction. Up to a ratio U : Ti=1:500, this effect of titanium can be eliminated by the addition of hydroxylamine. At high lanthanide contents (Ln : U>200), the absorbance due to Ln^{3+} is too large to be corrected for.

Analytical procedure: 1 g rock sample is evaporated nearly to dryness with 30 cm³ conc. HCl+3 cm³ H_2O_2, the residue is dissolved in 1:1 HCl, and the solution is diluted to 50 cm³ and filtered. At most 5 cm³ is pipetted into a 50 cm³ beaker, 5 cm³ conc. HCl is added, and the uranium is reduced to U(IV) by the addition of ca. 1 g zinc filings in small portions (duration of reduction: ca. 2 min). The solution is filtered into a 25 cm³ volumetric flask containing 5 cm³ conc. HCl, 4 cm³ 10% oxalic acid, 1 cm³ 10% hydroxylamine and 2 cm³ 0.06% arsenazo III. The volume is made up to the mark with water, and the absorbance is measured in 5 cm cells at 650 nm with a spectrophotometer. The comparison solution: 10 cm³ conc. HCl+ 4 cm³ oxalic acid+2 cm³ 0.06% reagent, in a volume of 25 cm³.

Thorium, titanium and lanthanides are corrected for by repeating the photometry on the solution to be examined, but without reduction of the uranium, and subtracting the latter absorbance. The calibration plot is produced with 2–10 μg U. The sensitivity of the method is 10 ppm U, and its error is ±4%.

Determination of uranium content of rocks and mine waters via the absorbance of the U(VI)–arsenazo III complex [9]

At pH 1–3, UO_2^{2+} forms a stable complex with arsenazo III: $\varepsilon_{max}=57 \times 10^{-3}$. Since the medium for the determination is only weakly acidic, the number of interfering ions is higher here (Th^{4+}, Ti^{4+}, Ln^{3+}, Al^{3+}, Ca^{2+}, etc.), and thus the separation of the uranium is necessary. A suitable procedure for this is extraction from 6 M HCl medium with 30% tributyl phosphate (TBP) in petroleum ether. The partition coefficients: $D_{UO_2^{2+}} \sim 30$; $D_{Th^{4+}}$, $D_{Ti^{4+}} \leqq 0.1$; for the other disturbing ions, D is generally $\leqq 0.01$.

The partition coefficient of U(VI) permits the washing of the organic phase with 6 M HCl; the slight loss of uranium is corrected for via the calibration plot. In this way it is also possible to determine the uranium in samples with high titanium and thorium contents.

In the extraction, care must be taken to maintain the HCl concentration at 6 M: at lower acid concentrations the partition coefficient of U(VI) is smaller, while at higher acid concentrations the extraction of thorium increases.

The uranium can be re-extracted quantitatively in a single shaking with water. The uranium determination is not disturbed by a 50-fold amount of zirconium under these conditions; for higher quantities of zirconium it is better to use the method based on the absorbance of the U(VI)–arsenazo III complex.

Fe^{3+} accompanies the uranium throughout the separation process. Its disturbing effect can be eliminated by reduction with ascorbic acid in analysis of an aliquot containing up to 5 mg Fe^{3+}; for 5–10 mg Fe^{3+}, reduction is first carried out with hydroxylamine, and the residual Fe^{3+} is reduced with ascorbic acid. In the case of higher iron contents than this, the bulk of the iron is removed by extraction from 6 M HCl medium with ether, before the extraction with TBP.

Analytical procedure: A rock sample of at most 3 g is evaporated almost to dryness with 30 cm³ conc. HCl + 3 cm³ H_2O_2, the volume is made up to 50 cm³ with 6 M HCl, and the solution is filtered. An aliquot of 25 cm³ is taken in a separating-funnel, and is extracted with 25 cm³ 30% TBP in petroleum ether for 1 min. The organic phase is washed for 30 s with 25 cm³ 6 M HCl, and the uranium is re-extracted by shaking for 1 min with 25 cm³ water. At most 10 cm³ of the resulting solution (U < 25 µg, Fe < 5 mg) is pipetted into a 25 cm³ volumetric flask, 1 cm³ 3% ascorbic acid is added, and the solution is neutralized with 1 M NaOH in the presence of methyl orange as indicator. 1.2 cm³ 1 M HCl is added, followed immediately before photometry by 1 cm³ 0.1% arsenazo III, the volume is made up to the mark, and the absorbance is measured in 5 cm cells at 650 nm, against a solution containing 1.2 cm³ 1 M HCl and 1 cm³ arsenazo III in a volume of 25 cm³. The calibration curve is produced with 2–20 µg U. The sensitivity of the method is 3 ppm U, and for U contents above 20 ppm the error-is $\pm 5\%$.

Notes: If the quantity of iron in the aliquot taken for analysis is 5–10 mg (this can be seen from the yellow colour of the solution), 1 cm^3 10% NH$_2$OH is added to the solution in a beaker, and reduction of the iron is accelerated by boiling. After cooling, the residual Fe^{3+} is reduced with a few drops of 3% ascorbic acid. pH-adjustment and photometry are as described above.

If the iron content of the solution to be analyzed is high (>10 mg in the solution aliquot), the course of the analysis is modified: from the 6 M HCl solution obtained after the digestion, the bulk of the iron is removed by extraction with an equal volume of diethyl ether, and the uranium is extracted with TBP from the residual aqueous phase in the above manner.

Mine waters containing more than 100 μg U/dm^3 can be analyzed without separation. At most 20 cm^3 solution is taken in a 25 cm^3 volumetric flask, 1.2 cm^3 1 M HCl and 1 cm^3 0.1% arsenazo III are added, the volume is made up to the mark, and the absorbance is measured.

Extraction–photometric determination of the uranium content of rocks with chlorphosphonazo III [9a]

U(VI) forms a stable complex with chlorphosphonazo III. Numerous elements (Th, Zr, lanthanides, Ca, etc.) interfere with the determination, and quantitative separation of the uranium is therefore necessary. U(VI) can be extracted without loss from 4 M HCl medium with 1×10^{-3} M tri-octylamine (TOA). Photometry can be carried out in the organic phase with chlorphosphonazo III reagent after the addition of ethanol.

Analytical procedure: 0.2–2.0 g sample is boiled for 40 minutes with a mixture of 30 cm^3 conc. HCl + 1 cm^3 30% H$_2$O$_2$, followed by evaporation to dryness. The residue is dissolved in 20 cm^3 4 M HCl with mild heating. The solution is transferred to a 25 cm^3 volumetric flask, the volume is made up to the mark with 4 M HCl, and the solution is filtered. A 5–10 cm^3 aliquot is taken in a shaking-funnel, and the Fe(III) is reduced with 0.5 g ascorbic acid (to complete loss of colour). The uranium is then extracted for 2 minutes with 2 cm^3 0.1 M TOA in benzene. The shaking-funnel is washed round with 2×10 cm^3 4 M HCl, and the organic phase is next shaken for 1 minute with 2×10 cm^3 4 M HCl (complete removal of disturbing elements). 2 cm^3 conc. HCl, 8.5 cm^3 ethanol and 2.0 cm^3 0.02% chlorphos-

329

phonazo III in ethanol are added to the separated organic phase, and the optical density of the mixed solution is measured in 1–5 cm cells at 670 nm. A blank taken throughout the process is used as comparison solution. The sensitivity of the method is 1 ppm.

Experience from application of the method [9b]

(a) The demands of geological research necessitate the accurate determination of uranium contents above 0.5 ppm. Such a demand is met by the use of this method. At uranium contents of 0.5–5 ppm the error in the method is ±15 rel.%, while at higher contents it is ±3–5 rel.%.

(b) Application of the method requires careful work. Great attention must be paid to ensuring quantitative washing-out of the disturbing elements from the organic phase and from the walls of the separating funnel.

(c) The applied reagents must be of appropriate quality. For example, if the benzene is not sufficiently pure, two phases are formed in the extract, which makes the results irreproducible.

(d) Since the solution to be measured might otherwise undergo subsequent contamination, it is advisable to run off a few cm^3 of the solution first, and then to run the solution directly from the separating-funnel into the cell.

(e) The use of a recording spectrophotometer is recommended. A shift in the position of the absorption maximum of the U(VI)–chlorphosphonazo III complex is indicative of contamination (most generally Ca). In this case the analysis must be repeated.

(f) The method is also suitable for the analysis of samples with high phosphate and fluoride contents.

(g) When necessary (e.g. in the analysis of coal ash), digestion is performed with a mixture of HF and H_2SO_4. The dry residue is dissolved in 4 M HCl. The subsequent procedure is the same as described above.

Analysis of effluents from uranium plants by photometry of the UO_2^{2+}-diphenylguanidine–arsenazo III mixed ligand complex [10]

Uranium-plant effluents contain much SO_4^{2-}, Ca^{2+}, Mn^{2+} and Mg^{2+}, while their uranium content is low. They can be analyzed only after appropriate separation and concentration. Only the concentration is necessary for natural waters. A suitable procedure for both purposes is extraction from a 0.1 M HCl medium with 0.05 M HDBP, followed by re-extraction of the uranium with 0.2 M NaOH. A 10-fold concentration can be achieved in this way. Although the dibutyl phosphate passing into the aqueous phase does not disturb the subsequent analysis (it has no masking effect at an acidity of 0.05 M), the aqueous phase is not optically pure, and thus the determination cannot be carried out directly.

The uranium–arsenazo III complex is water-soluble because of the hydrophilic groups (SO_3H, AsO_3H_2, OH) in the reagent, and the complex cannot be extracted with organic solvents. In the presence of large hydrophobic cations, e.g. diphenylguanidine, however, it can be extracted with organic solvents (e.g. *n*-butanol) and its absorbance can be measured in the organic phase [11].

Analytical procedure: 100 cm³ of the solution to be examined (0.1 M in HCl) is extracted with 20 cm³ 0.05 M HDBP in petroleum ether. The organic phase is washed with 3×10 cm³ 0.1 M HCl, and the uranium is then re-extracted with 10 cm³ 0.2 M NaOH. The solution is neutralized with 1 M HCl in the presence of methyl orange as indicator. 1 cm³ 1.0 M HCl is added in excess, followed by 0.4 cm³ 0.1% arsenazo III and 2 cm³ 1 M diphenylguanidine (210 g dissolved in 84 cm³ conc. HCl, and the volume made up to 1 dm³), and the contents are made up to 20 dm³ with water. The resulting complex is extracted with 10 cm³ *n*-butanol, and the absorbance of the organic phase is measured in 5 cm cells at 655 nm. As comparison, the extract of the uranium-free solution, but otherwise containing the same reagents, is used. The calibration plot covers the interval 1–10 μg U. The sensitivity of the method is 5 μg U/dm³.

Determination with H_2O_2 of uranium content of solutions obtained by sodium carbonate digestion [12]

In alkali metal hydroxide or carbonate medium, uranium(VI) forms a yellow complex with H_2O_2. The determination is disturbed by the organic matter dissolved out of the rock sample on digestion with sodium carbonate, and also by the product of decomposition of the ion-exchange resin. This makes evaluation uncertain. The quantities of other interfering ions (Cr,

Mn, V, Mo, Cu, Fe) in the examined solution can be neglected. The effect of the organic matter is eliminated in a wet oxidation procedure.

Analytical procedure: 1–5 cm³ 0.1 M $KMnO_4$ is added to 1–20 cm³ carbonate solution, and the mixture is boiled for 20 min. H_2O_2 is added dropwise to the hot solution, to decompose the excess $KMnO_4$. The resulting precipitate is filtered off, and washed with a little hot 0.2 M Na_2CO_3 solution (U loss < 1%). 2 cm³ 3% H_2O_2 is added to the cooled filtrate, the volume is made up to 25 cm³ with 0.2 M Na_2CO_3, and the absorbance is measured at 420 nm. The calibration plot is recorded with 0.2–3 mg U. The absorbance is not influenced by the Na_2CO_3 concentration in the interval 0.1–1 M.

Determination of the uranium content of surface waters via the absorption of the U(IV)–arsenazo III complex [12a]

For the concentration of the uranium content of water we use the Wofatit MC-50 resin recommended in Section 6.38, in the manner described there. After the joint binding of U and Th, elution is performed with 2×10 cm³ 2 M HCl. Since the calcium content of the eluate is relatively high, direct U(VI) determination with arsenazo III is not possible. Accordingly, the much more selective U(IV)–arsenazo III method is used [8]:

The total eluate, or an aliquot of it, is evaporated to ca. 5 cm³. 7 cm³ conc. HCl and a little Fe^{3+} (ca. 0.1 mg) are added, and the uranium is reduced with 1–2 g zinc turnings (for 1–2 minutes after the disappearance of the yellow colour of the iron). The solution is filtered into a 25 cm³ volumetric flask containing 5 cm³ conc. HCl, 2 cm³ 10% oxalic acid and 1 cm³ 0.06% arsenazo III. The volume is made up to the mark with water, and photometry is carried out at 650 nm in 5 cm cells.

If the water also contains thorium, a blank solution is prepared from another aliquot, without reduction, and this is used for comparison.

The sensitivity of the method is 1 μg U/dm³.

The method is more favourable than the fluorimetric method of determining uranium, and yields a more accurate result.

Korkisch and Krivanecz [13, 14] use Dowex 1×8 anion-exchange resin in citrate form to bind the uranium and thorium content of natural waters. The uranium is determined by photometry of the U(IV)–arsenazo III complex.

332

In another procedure [15], the uranium content of natural water is concentrated by extraction from a 3.5 M LiCl+0.2 M HCl medium with Aliquat 336 (a 5% solution of tricaprylmethylammonium chloride in xylene). If the organic phase is shaken with 0.15 M HNO_3, the uranium is transferred to the aqueous phase. The solution is evaporated, the residue is evaporated to dryness several times with HCl, and the uranium is reduced in HCl medium with metallic zinc. Photometry is performed after the addition of arsenazo III. The determination is disturbed only by zirconium (a 50-fold excess can be tolerated).

References

[1] Williams, W. J., Gillam, A. M.: *Analyst* **103**, 1239 (1978).
[2] Ryabchikov, D. I., Senyavin, M. M. (eds.): *Analiticheskaya khimiya urana*. Khimiya, Moscow 1962.
[3] Korkisch, J., Hecht, F.: *Handbuch der analytischen Chemie: Uran*. Springer, New York 1972.
[4] Havel, J., Sommer, L.: *Chromogenic reaction of uranium*. Folia Chemica Universita J. E. Purkine v Brné Vol. 14, No. 12 (1973).
[5] Suten, A., Hodisan, T.: *St. cerc. chim.* **18**, 85 (1970).
[6] Savvin, S. B.: *Arsenazo III*. Atomizdat, Moscow 1966.
[7] Strelow, F. W. E., Van der Walt, T. N.: *Talanta* **26**, 537 (1979).
[8] Mohai, M., Upor, E., Jurcsik, I.: *Magy. Kém. Folyóirat* **71**, 334 (1965).
[9] Mohai, M.: Uranium determination methods based on spectrophotometry of UO_2^{2+}-arsenazo III complex. Paper presented at the Second European Conference on Analytical Chemistry, Budapest 1975.
[9a] Bykhovtsova, T. T., Tserkovnitskaya, I. A.: *Zh. Anal. Khim.* **32**, 745 (1977).
[9b] Mohai, M.: Alacsony urántartalmú kőzetminták klórfoszfonazo III-mal történő meghatározásánál szerzett tapasztalatok (Experience with the determination of low uranium contents in rocks with chlorphosphonazo III). Paper presented at the 18th Transdanubian Analytical Conference, Pécs, Hungary, 1983.
[10] Mohai, M.: *Magy. Kém. Folyóirat* **81**, 164 (1975).
[11] Savvin, S. B.: *Uspekhi Khimii* **32**, 195 (1963).
[12] Upor, E., Görbicz, L., Novák, Gy.: *Magy. Kémikusok Lapja* **21**, 487 (1966).
[12a] Horváth, Á, Csővári, M., Hohmann, J.: Felszíni vizek urán- és tóriumtartalmának meghatározása előzetes kémiai dúsítás után (Determination of the uranium and thorium content of surface waters after preliminary chemical enrichment). Paper presented at the 17th Transdanubian Analytical Conference, Keszthely, Hungary, 1982.
[13] Korkisch, J., Krivanec, H.: *Talanta* **23**, 295 (1976).

[14] Korkisch, J., Krivanec, H.: *Talanta* **23**, 283 (1976).
[15] Barbano, P. G., Rigali, L.: *Anal. Chim. Acta* **96**, 199 (1978).
[16] Johnson, D. A., Florence, T. M.: *Anal. Chim. Acta* **53**, 73 (1971).

6.42. Vanadium

Vanadium is situated in column V/2 of the periodic system. In its aqueous solutions, it exhibits valencies of 2, 3, 4 and 5. Vanadium compounds are readily oxidized and reduced. If the vanadium in the more reduced form is left to stand, it is slowly oxidized to V^{5+}. In some of its reactions, vanadium behaves similarly to Nb and Ta, also in column V/2, but in other cases it rather resembles Mo and W.

Vanadium(IV) and vanadium(V) primarily form stable complexes with ligands containing O donor atoms. With the exception of Nevazol-NO (see Appendix), which reacts with V^{4+}, it is necessary for the vanadium to be in the +5 oxidation state for coloured complexes to be formed. Special mention may be made of the frequently used benzhydroxamic acid derivative, N-benzoylphenylhydroxylamine (BPHA), which is a selective reagent for vanadium(V). The water-insoluble violet complex is formed in strongly acidic medium (2–10 M), and can be extracted well with $CHCl_3$ (ethanol-free). The strongly acidic medium also ensures separation from many interfering elements. The results attained in the analytical chemistry of vanadium are to be found in the excellent review by Svehla and Tölg [1]. The photometric reagents for vanadium are listed in Table 6.31.

Determination of vanadium in rocks, bauxites and coal ash [2–4]

The same procedure is used for the determination of vanadium in rocks, bauxites and coal ash. For vanadium contents higher than 0.03%, vanado-phosphotungstic acid is formed and used for photometry; if the content is less than 0.03%, the vanadium complex of BPHA is similarly utilized.

(a) Vanadophosphotungstic acid $[H_4P(W_2O_5)(V_2O_6)_3]$ is a yellow heteropolyacid. Favourable colour-formation conditions are a W(VI) concentration of 0.01–0.1 M, and a P(V) concentration of 0.5 M. The determination

334

TABLE 6.31.
More important photometric reagents for vanadium

Reagent	λ_{max}, nm	$\varepsilon \times 10^{-3}$	Medium	Interferences	Notes	Refs
N-Benzoyl-N-phenylhydroxylamine (BPHA)	530	4.8	2.9 M HCl	Mo^{6+}, Ti^{4+}, Zr^{4+}, strong oxidants and reductants	Sodium triphosphate masking agent	[3, 7, 8]
N-furoylphenylhydroxylamine	536	5.6	6 M HCl	Ti, Nb, Mo, W	NH_4F addition	[9, 10]
PAR–H_2O_2	540	16.4	pH 0.5	Much Ti, Fe	NH_4F addition	[11]
Nevazol-NO (V^{4+})	565	15	pH 1.8–2.0	Many elements	NaF, thiourea, thioglycollic acid addition	[12]
H_2O_2	290–460	0.3	0.3–3 M H_2SO_4	Ti, Mo, U, W, Nb	HF, H_3PO_4 addition	[13]
Phosphovanadotungstate	365	2	0.5 M H_3PO_4	Many elements	Alkaline separation	[2, 13]
3,5-Br-MEPADAP–H_2O_2	615	54.3	0.5 M H_2SO_4	Cr^{6+}, Fe^{3+}, Ti^{4+}	NH_4F addition	[5]
3,3′-dimethylnaphthidinedisulphonic acid	555		8 M H_3PO_4	Ce^{4+}, Cr^{6+}, NO_3^-, I^-		[14, 15]

is disturbed by Sn, Ti, Zr, Bi and Sb (which give precipitates with phosphate), and by much Ni, Cr, Cu and Co (due to their own colour).

(b) Benzhydroxamic acid and its derivatives, e.g. BPHA, can be used to advantage for the photometric determination of vanadium. The determination is sensitive and selective, and the resulting colour is stable. The acidity of 2–10 M HCl used in the colour development strongly diminishes the disturbing effects. The compound formed can be extracted with organic solvent, e.g. $CHCl_3$. The determination is disturbed by Ti, Mo and Zr.

Analytical procedure: 0.5–1 g sample is fused with a mixture of 10 g $NaOH + 1$ g Na_2O_2 in a nickel crucible. The melt is dissolved out with hot water, and a 250 cm³ stock solution is prepared and filtered.

(a) A 25 cm³ aliquot is taken, and 6 cm³ conc. HNO_3 and 5 cm³ 1 : 2 H_3PO_4 are added. The mixture is boiled, 1 cm³ 15% Na_2WO_4 is added in the hot, and after cooling the volume is made up to 50 cm³. The absorbance is measured 30 min later, in 1 or 5 cm cells at 435 nm. An aliquot without Na_2WO_4 is used as comparison, to eliminate the background colour of the solution itself. The calibration plot covers the interval 50–500 µg V.

(b) An aliquot of 1–50 cm³ (if necessary the volume is diluted to 50 cm³) is pipetted into a separating-funnel containing 10 cm³ 20% NH_4F (this is capable of masking 50 mg Fe, 50 mg Al, and 10 mg Ti). The acidity of the solution is adjusted to be 3–5 M in HCl. Extraction is performed with 25 cm³ 0.1% BPHA in $CHCl_3$ (ethanol-free). The absorbance is measured in 1 or 4 cm cells at 525 nm. The calibration plot is recorded with 5–200 µg V.

Determination of vanadium traces in silicate rock samples, with 3,5-Br-MEPADAP [5]

If V^{5+} is heated with 3,5-Br-MEPADAP in the presence of H_2O_2, an intensely blue triple complex is formed. The coloration is specific for V^{5+} in the presence of fluoride as masking agent. With minor modifications, the analytical procedure given below can also be applied to determine the vanadium content of bitumens, biological substances, ground-water, sea water, etc.

Analytical procedure: 0.2–1 g rock sample with 5 cm³ water, 5 cm³ 1:1 H_2SO_4 and 20 cm³ HF in a platinum vessel is evaporated to low volume. 5 cm³ HF and 3 cm³ HNO_3 are added to the residue, the mixture is evaporated until fumes of SO_3 appear, and the volume is made up to 250 cm³. (Quantitative dissolution must be attained, as the residue may contain an appreciable amount of vanadium.) An aliquot of at most 25 cm³ of the stock solution is taken (V content $\leqq 20$ μg). The solution is oxidized by the dropwise addition of 0.5% $KMnO_4$. 5–10 cm³ 20% NH_4F and 5 cm³ 5 M H_2SO_4 are added, and the volume is made up to ca. 40 cm³. For colour development, 2 cm³ 2×10^{-3} M 3,5-Br-MEPADAP and 3 cm³ 3% H_2O_2 are added, and the volume is made up to 50 cm³. The solution is placed for 5 min on a boiling water-bath, and then cooled. It is shaken for 1 min with 20 cm³ $CHCl_3$ in a separating-funnel. The water droplets are removed by filtration, and the absorbance is measured in 1 cm cells at 615 nm. The calibration plot is recorded with 2–20 μg V.

Determination of vanadium content of steels [6]

In the determination of a small amount (<0.01%) of vanadium, the iron content of the steel interferes. The iron is removed by extraction with tributyl phosphate (TBP). The vanadium is determined with N-benzoyl-N-phenylhydroxylamine (BPHA).

Analytical procedure: 1 g steel sample is dissolved in a 3:1 mixture of HCl + HNO_3, and the solution is evaporated to dryness. The residue is again evaporated to dryness 3–4 times with a little conc. HCl. The salts are dissolved in 50 cm³ 10 M HCl. 0.5 g KCl and 50 cm³ of a 1:1 extractant mixture of TBP + petroleum ether are added, and the phases are shaken together for 1 min. After separation, the organic phase is again shaken for 1 min with 30 cm³ 10 M HCl. The aqueous phases are combined, 4 cm³ 1:1 H_2SO_4 is added, and the volume is evaporated to 20 cm³. A little conc. HNO_3 is added to the solution, which is boiled to decompose the organic matter; this step is repeated 3–4 times. Evaporation is continued until the appearance of fumes of SO_3. The crystallized-out salts are dissolved in 15 cm³ water, and the V(IV) in the solution is oxidized with 0.002 M $KMnO_4$ (the pink colour of the solution should persist for 1 min). Next, 15 cm³ conc. HCl is added, and the solution is extracted with 10 cm³ 0.2%

BPHA in $CHCl_3$. The extract is run through a dry filter paper into a 25 cm^3 volumetric flask, the aqueous phase is again shaken with $CHCl_3$, the organic phases are combined, and the contents of the flask are made up to the mark. The absorbance is measured at 530 nm. The calibration plot covers the interval 10–200 μg V. A blank test is carried out with an artificial mixture with the same iron content as that of the sample.

References

[1] Svehla, G., Tölg, G.: *Talanta* **23,** 755 (1976).
[2] Popov, N. P., Stolyarova, I. A. (eds.): *Khimicheskii analiz gornykh porod i mineralov.* Nedra, Moscow 1974.
[3] Pilkington, E. S., Wilson, W.: *Anal. Chim. Acta* **43,** 461 (1969).
[4] Upor, E., Schreck, M.: Unpublished work.
[5] Kiss, E.: *Anal. Chim. Acta* **77,** 205 (1975).
[6] Onorina, I. A., Fedorova, N. D., Dolgorev, A. V.: *Zav. Lab.* **47,** No. 4, 3 (1981).
[7] Ryan, D. E.: *Analyst* **85,** 569 (1960).
[8] Usha Priyadorshini-Tandon, S. G.: *Anal. Chem.* **33,** 435 (1961).
[9] Pilipenko, A. T., Sereda, I. P., Shpak, E. A.: *Zav. Lab.* **32,** 660 (1966).
[10] Pilipenko, A. T., Shpak, E. A., Kurbatova, G. T.: *Zh. Anal. Khim.* **22,** 1014 (1967).
[11] Bagdasarov, K. N., Akhmedova, H. A., Tataev, O. A.: *Zav. Lab.* **35,** 12 (1969).
[12] Basargin, N. N., Yakovlev, P. Ya., Busev, A. I., Zanina, I. A.: *Zav. Lab.* **35,** 411 (1969).
[13] Dragomirecky, A., Mayer, V., Michal, J., Rericka, K.: *Photometrische Analyse anorganischer Roh- und Werkstoffe.* VEB Deutscher Verlag für Grundstoffind. Leipzig 1968.
[14] Sanke Godva, H., Shakunthala, H.: *Analyst* **103,** 1215 (1978).
[15] Feigl, F., Anger, V.: *Spot Tests in Inorganic Analysis.* Elsevier, Amsterdam 1972.

6.43. Zinc

Zinc belongs in column II/2 of the periodic system. In its analytical properties, it exhibits similarities primarily to Cd^{2+} and Hg^{2+} (from the same column) and to Co^{2+} and Cu^{2+} (from the same period). In solution and in its complexes it is divalent. With ligands containing O donor atoms, it forms complexes of only low stability; however, relatively stable complexes are obtained with ligands containing N and S donor atoms. Its hydroxide is amphoteric; thus, at pH > 13 it may be separated as $Zn(OH)_4^{2-}$

from other metal hydroxides. Its ammine complexes are similarly suitable for its separation. Its halide complexes, e.g. $ZnCl_4^{2-}$, permit separation by extraction or ion-exchange. Its complexes with dithizone and dithiocarbamates may be used not only for photometric determination, but also for the separation of zinc. The reagents for zinc are not selective, and therefore the colour development must be preceded by separation or masking. Zinc may be masked with cyanide, citrate, tartrate, thiocyanate, etc., while it may be released from its complexes by treatment with formaldehyde.

The critical compilation by Ackermann and Kothe [1] compares the properties of some reagents used for the photometric determination of zinc: azo dyes, zincon, xylenol orange and oxine. The best sensitivity and reproducibility were observed for the azo dyes. Detailed data on the analysis of zinc are to be found in the book by Zhivopistsev and Selezneva [2].

The photometric reagents for zinc are listed in Table 6.32. In analytical practice, dithizone is of the greatest importance.

TABLE 6.32.

More important photometric reagents for zinc

Reagent	λ_{max} nm	$\varepsilon \times 10^{-3}$	Medium	Interferences	Notes	Refs
Dithizone	536	96	pH 4.5–6.0	Ni^{2+}, Co^{2+}, Cu^{2+}, Cd^{2+}, Bi^{3+}, Pb^{2+}, Fe^{3+}, V^{5+}, Hg^{2+}	Masking with oxalate, cyanide, thiosulphate; separation	[2, 10, 11]
Zincon	625	24	pH 8.5–9.5	Al, Be, Bi, Cd, Co, Cr, Cu, Fe, Mn, Ni, Ti	Ion-exchange separation; DDTC extraction	[8, 2]
PAN	546		pH 6.6	Fe^{2+}, Co^{2+}, Cu^{2+}, Ni^{2+}, Mn^{2+}, Ga^{3+}, etc.	Extraction, separation, masking	[9, 2]
NAAN	646	39	pH 6.4	Co^{2+}, Ni^{2+}, Cu^{2+}, Mn^{2+}, PO_4^{3-}	Ion-exchange separation	[2, 10]

Determination of zinc in rocks after separation with NH_4OH [3, 4]

The determination of zinc with dithizone is disturbed by many ions. Since these are generally present in low concentrations, their influence can be eliminated by masking (thiourea, thiosulphate, cyanide). If the analytical method is constructed appropriately, there is a possibility for the determination of Cu^{2+}, Ni^{2+} and Co^{2+}, in addition to Zn^{2+}, in the same solution. The ions to be determined are therefore separated in soluble form from those metal hydroxides that are insoluble in this medium (Pb^{2+}, Fe^{3+}, Bi^{3+}, V^{5+}, Al^{3+}, etc.). It has been demonstrated that the separation can be performed without loss if the pH of the solution is ~ 9.5, while the concentration of NH_4OH and NH_4^+ is $\cong 2$ M. In our combined method, after the joint extraction of Zn^{2+}, Co^{2+} and Cu^{2+} with dithizone, the Zn^{2+} is re-extracted with 0.1 M HNO_3; the Co^{2+} and Cu^{2+} remain in the organic phase. Since the Zn^{2+} would otherwise be accompanied by Ni^{2+}, the latter is preliminarily extracted with dimethylglyoxime. The Zn^{2+} can be determined after a further extraction with dithizone.

Analytical procedure: The sample is evaporated to dryness with $HF + HClO_4$ in the presence of a few drops of HNO_3 in a platinum dish. The residue is dissolved in 10–15 cm³ 1:1 HCl. 10 g NH_4Cl and 30 cm³ 1:1 NH_4OH are added, and after a few min the resulting precipitate is filtered off and washed with a little washing liquid (100 g $NH_4Cl + 70$ cm³ conc. NH_4OH, in 1 dm³). The filtrate is made up to 100 cm³. A 5–20 cm³ aliquot is taken from this and diluted to 50 cm³, 2 cm² 1% dimethylglyoxime in alcohol is added, and the Ni^{2+} is extracted with 2×5 cm³ $CHCl_3$. The Zn^{2+}, Cu^{2+} and Co^{2+} are extracted from the aqueous phase with small portions of 0.01% dithizone in $CHCl_3$. The Zn^{2+} is re-extracted by shaking the organic phase with 10 cm³ 0.1 M HNO_3. The aqueous phase is neutralized with NH_4OH, 5 cm³ phosphate–citrate buffer is added (1/15 M Na_2HPO_4 and 0.1 M sodium citrate in a 1:4 mixture, adjusted to pH 5.5, and purified with dithizone before use), and the solution is shaken with small portions of 0.001% dithizone in $CHCl_3$ to extract the Zn^{2+} for determination. The final portion, which is not green, is not added to the other portions; alternatively, if it is added, the dithizone excess is re-extracted with 10 cm³ 0.1% Na_2S (in this way, measurement at two wavelengths is avoided).

Depending on its volume, the extract is made up to 10–25 cm³ with $CHCl_3$, and its absorbance is measured in 1–5 cm cells at 530, against $CHCl_3$ The calibration plot is produced with 0.5–5.0 μg Zn^{2+}.

The method is very sensitive; in principle it is suitable for the determination of even 1 ppm Zn. However, because of the relatively high Clark value for zinc, this is not required; further, the presence of contamination would mean the performance of determinations with specially pure reagents and apparatus. The distilled water and the reagents prepared with it must be carefully purified. Isothermal distillation is recommended for the purification of HCl and NH_4OH. Even then, it may occur that the blank test carried out in parallel with the analysis reveals that the determination must be repeated, with purer conditions being guaranteed.

Determination of zinc in soil [5]

The dried sample is first decomposed in the wet in a Kjeldahl flask, with a mixture of HNO_3, H_2SO_4 and $HClO_4$.

In this method the separation with NH_4OH is omitted; the first dithizone extraction on the stock solution obtained by means of acidic digestion is performed at pH 9–10 in the presence of sodium citrate. The Zn^{2+} is re-extracted with 0.02 M HCl, and the second extraction takes place at pH 5–6, in the presence of $Na_2S_2O_4$ added to bind the Pb^{2+}. The dithizone excess is removed by washing with 0.01 M NH_4OH. The absorbance is measured at 536 nm. The dithizone is applied in CCl_4.

Determination of zinc in natural waters [6, 7]

Prior to the analysis, the water sample is evaporated to dryness, the residue is ignited at 550°C to remove organic matter, and the ignition residue is dissolved in HCl. The Zn-dithizonate is extracted at pH 5 in the presence of KCN and $(NH_4)_2C_2O_4$ as masking agents. The dithizone excess is removed by shaking with a $Na_3PO_4 + Na_2S$ solution of pH 11.

(B) APPLICATION OF SOME OTHER REAGENTS FOR DETERMINATION OF ZINC

Huffmann *et al.* [8] recommend zincon for the determination of zinc in rocks. Since the determination is disturbed by many ions, separation is achieved in several steps. First, $ZnCl_4^{2-}$ is sorbed on an anion-exchange resin from 1.2 M HCl. The column is washed with 1.2 M HCl to remove iron, and the Zn^{2+} is then eluted with 0.01 M HCl. NaDDTC is added at pH 8.5 to precipitate $Zn(DDTC)_2$, which is extracted with $CHCl_3$, and re-extracted with 0.16 M HCl. After the addition of zincon, the pH is adjusted to 9.0 ± 0.5, and the absorbance of the complex is measured at 620 nm. The sensitivity of the method is 10 ppm.

Pohl [9] proposes PAN for the determination of the zinc content of iron ore. The interfering ions are removed in several steps. The Zn(II)–thiocyanate complex is first extracted with methyl isobutyl ketone (separation from iron reduced to Fe^{2+} with $SnCl_2$). The organic phase is re-extracted with 1:1 HCl, the aqueous phase is evaporated to dryness, and the solid is ignited at 280°C (to drive off $GaCl_3$). Tartaric acid, nitroso-R salt and $Na_2S_2O_3$ are added to mask the residual interfering ions, and the Zn–PAN complex is extracted at pH 6.6 with $CHCl_3$. The absorbance is measured at 546 nm. This method allows the determination of 30 ppm Zn.

References

[1] Ackermann, G., Kothe, J.: *Talanta* **26**, 693 (1979).
[2] Zhivopistsev, V. P., Selezneva, E. A.: *Analiticheskaya khimiya tsinka.* Nauka, Moscow 1975.
[3] Upor, E., Nagy, Gy.: *Acta Chim. Acad. Sci. Hung.* **61**, 1 (1969).
[4] Görbicz, M., Upor, E.: *Acta Chim. Acad. Sci. Hung.* **66**, 373 (1970).
[5] Vazhenina, I. G. (ed.): *Metody opredeleniya mikroelementov v pochvakh, rasteniyakh i vodakh.* Kolos, Moscow 1974.
[6] Freier, R. K.: *Wasseranalyse.* de Gruyter, Berlin 1964.
[7] *The Testing of Water.* 5th Ed. Merck, Darmstadt 1965.
[8] Huffmann, C., Lipp, H. H., Rader, L. F.: *Geochim. Cosmochim. Acta* **27**, 209 (1963).
[9] Pohl, H.: *Erzmetall* **16**, 18 (1965).
[10] Kamaeva, L. V., Podchainova, V. N., Fedorova, N. D.: *Zav. Lab.* **37**, 258 (1971).
[11] Iwantscheff, G.: *Das Dithizon und seine Anwendung in der Mikro- und Spurenanalyse.* 2nd. Ed. Verlag Chemie, Weinheim 1972.

6.44. Zirconium

Zirconium is situated in column IV/1 of the periodic system. In its compounds it occurs in the tetravalent form. As concerns its chemical behaviour, it shows close similarities to hafnium. In aqueous solution, it is present as the zirconyl ion, ZrO^{2+}, or as aquo-hydroxo ions, with a tendency to undergo polymerization. It is usually kept in solution with complex-forming anions (SO_4^{2-}, oxyacids) or with a strongly acidic medium. Its hydroxide is a precipitate insoluble in alkali.

Reviews of zirconium are presented in the books by Elinson and Petrov [1] and Mukherji [2].

Various reagents are suggested for zirconium in the literature. However, it is not advisable to use those which form complexes only in weakly acidic medium, due to the tendency of ZrO^{2+} to undergo hydrolysis and polymerization. In general, only azo dyes and oxygen-containing chelate-forming reagents are used for the photometric determination.

The organic reagents that can be most highly recommended for the determination of zirconium are to be seen in Table 6.33 [3].

The Table shows that arsenazo III has excellent properties from the aspects of sensitivity and selectivity; if the average occurrence of the interfering elements in nature is taken as basis, the selectivity is perfectly satis-

TABLE 6.33.
Some photometric reagents for zirconium

Reagent	Medium	λ_{max} nm	$\varepsilon \times 10^{-3}$	Interferences	Refs
Arsenazo III	7–9 M HCl	665	120	Hf, Th, much Ti, Ln, U, SO_4^{2-}	[4]
Xylenol orange	pH 1.5	535	5.2	Hf, Ti, Th, U, Nb, Fe, Sn, Mo, Bi, etc.	[3]
Sulphochlorophenol S	1.2 M HCl	650	75	Hf, Nb	[3]
Alizarin-sulphonic acid sodium salt	0.2 M HCl	520	7.0	Fe^{3+}, F^-, PO_4^{3-}, much SO_4^{2-}	[1]
Pyrocatechol violet	pH 5.0	600	31	Fe, Al, Th, Ti; interference eliminated with EDTA	[1]
Phenylfluorone	pH 2	550	19	F^-, PO_4^{3-}, Ti, Sn, Sb, Ta, Nb	[1]

factory. The absorbance of the ZrO^{2+}–arsenazo III complex does not vary in the concentration range 6.7–9 M HCl; a medium of 7–7.2 M HCl is recommended for the determination. Practically only Th^{4+}, larger amounts of Ti^{4+}, Ln^{3+} and UO_2^{2+} interfere in the determination of ZrO^{2+}. Since ZrO^{2+} can be masked with oxalic acid, whereas these interfering ions are not affected, if photometry is carried out in the absence and in the presence of oxalic acid, even the effects of larger amounts of these elements can be eliminated.

If the sample to be analyzed contains very high contents of Th, Ln, Nb or U, it is preferable to use the following reagents: picramine II, 2,4-nitrosulphophenol S, or its analogue sulphochlorophenol S [4, 5].

Determination of zirconium in rocks, ores and cast iron (based on [4] and the authors' investigations)

Analytical procedure: 0.1–1 g rock or ore sample is digested with 10 g NaOH in a nickel dish. Towards the end of the fusion, a little Na_2O_2 too is added. The melt is leached out with hot water, the mixture is boiled, and it is then filtered. (If the iron content of the sample is low, 50–100 mg Fe^{3+} is added before filtration.) The precipitate is washed with water, then dissolved in 1:1 HCl with boiling, the solution is transferred to a 50 cm³ volumetric flask, and the volume is made up to the mark with 1:1 HCl.

Two 10 cm³ aliquots are pipetted into two 25 cm³ volumetric flasks, and sufficient conc. HCl is added for the final acidity to be 7–7.2 M. 1 cm³ 0.1% arsenazo III is added to one flask, and the volume is made up to the mark with water. 2 cm³ 10% oxalic acid is added to the second flask before the addition of the arsenazo III. (The oxalic acid masks the zirconium, but not the interfering U, Th, Ti and Ln.) This latter solution serves as a comparison. Photometry is carried out in 2–5 cm cells at 665 nm. The same procedure is followed to produce the calibration plot, with 0.2–10 µg Zr. The sensitivity of the method is 1 ppm Zr.

Notes:

1. In the analysis of steels and cast irons, the customary dissolution with an acid (or acid mixture) is applied, after which zirconium hydroxide is precipitated by the addition of NaOH.

344

2. For ore samples which cannot be digested by fusion with NaOH, other digestion procedures (e.g. fusion with pyrosulphate) can be applied. Here too, the zirconium must be precipitated with alkali.

Determination of zirconium in aluminium and magnesium alloys [6]

The zirconium is extracted from H_2SO_4–KSCN–antipyrine medium, and re-extracted with conc. HCl. The colour is developed with arsenazo III solution. This method is more selective than other common procedures. The determination is disturbed by Hf.

Analytical procedure: 0.1–0.25 g Al/Mg alloy is dissolved in 10–15 cm³ 6 M HCl, and the volume is made up to 50 cm³. An aliquot containing 5–20 μg Zr is taken, 1 cm³ 1.5 M H_2SO_4 and 5 cm³ 30% KSCN are added, and extraction is performed for 1 min with 10 cm³ 1.5% antipyrine in dichloroethane. The organic phase is shaken with 10 cm³ water, and then with 20 cm³ conc. HCl. The aqueous phases are combined in a 50 cm³ volumetric flask, 4 cm³ 0.04% arsenazo III is added, and the volume is made up to the mark with conc. HCl. The absorbance is measured 20–30 min later, at 665 nm. The Beer–Lambert law holds in the interval 0.1–0.5 μg Zr/cm³.

References

[1] Elinson, S. V., Petrov, K. I.: *Analiticheskaya khimiya elementov Zr i Hf.* Nauka, Moscow 1965.
[2] Mukherji, A. K.: *Analytical Chemistry of Zirconium and Hafnium.* Pergamon, Oxford 1970.
[3] Dolgorev, A. V., Palnikova, T. I., Podchainova, V. N.: *Zav. Lab.* **40,** 129 (1974).
[4] Savvin, S. B.: *Arsenazo III.* Atomizdat, Moscow 1966.
[5] Dedkov, Yu. M., Ryabchikov, D. I.: *Organicheskie reagenty v analiticheskoi khimii tsirkoniya.* Nauka, Moscow 1970.
[6] Akimov, V. K., Gvelesiani, L. T., Busev, A. I., Nenning, P.: *Acta Chim. Acad. Sci. Hung.* **97,** 105 (1978).

Appendix

LIST OF THE MOST IMPORTANT REAGENTS

Common name of reagent	Structural formula and chemical name of reagent	Elements determined with reagent
Acetylacetone	$CH_3-C(=O)-CH=C(OH)-CH_3$ 2,4-Pentanedione	Be [2, 9]
6-(p-Acetylphenyl-azo)-2-aminopyrimidine		S [10]
Alizarin	 1,2-Dihydroxyanthraquinone	F [10, 17]

Alizarin fluorine blue (alizarin complexone)	3-Dicarboxymethylaminomethyl-1,2-dihydroxyanthraquinone	F [2, 10, 17]
Alizarin S (alizarin sulphonic acid, Na salt) (alizarin red S)	1,2-Dihydroxyanthraquinone-3-sulphonic acid, Na salt	Ln [1] Zr [1]
Aluminon	Aurintricarboxylic acid, triammonium salt	Al [1, 2] Ga [1]

Common name of reagent	Structural formula and chemical name of reagent	Elemenst determined with reagent
p-Aminoacetophenone	CH_3CO—⟨C₆H₄⟩—NH_2	N [8]
Ammonium pyrrolidine dithiocarbamate	CH₂—CH₂\\N—C(=S)—SNH₄ / CH₂—CH₂ Ammonium 1-pyrrolidinecarbodithioate	Sb [11]
Aniline-4-sulphonic acid	NH_2—⟨C₆H₄⟩—SO_3H	N [6]
AQSA	Anthraquinone-2-sulphonic acid	Co [7]

Arsenazo I	2-(1,8-Dihydroxy-3,6-disulpho-2-naphthylazo) benzene arsonic acid, Na salt	Ln [1] U [2, 6]
Arsenazo II	4,4′-Diarsonobiphenyl-3,3′-bis(2″-azochromotropic acid)	Th [3]
Arsenazo III	2,2′-[1,8-Dihydroxy-3,6-disulpho-2,7-naphthylene-bis(azo)] dibenzenearsonic acid	Cr [11] Ln [1, 2, 4, 8, 22] Sc [5, 7] Th [4, 6–9] U [4, 6–10, 13, 14] Zr [4, 5]

Chemical structure of Arsenazo I: naphthalene ring bearing OH, OH groups, SO$_3$H, HO$_3$S substituents, with an N=N azo linkage to a benzene ring carrying AsO$_3$H$_2$.

Chemical structure of Arsenazo II: two naphthalene units each bearing OH, OH, SO$_3$H and HO$_3$S groups, connected via N=N azo linkages to a biphenyl system carrying AsO$_3$H$_2$ and H$_2$O$_3$As groups.

Chemical structure of Arsenazo III: central naphthalene ring with OH, OH, SO$_3$H and HO$_3$S substituents, linked by two N=N azo groups to benzene rings each bearing AsO$_3$H$_2$ and H$_2$O$_3$As.

351

Common name of reagent	Structural formula and chemical name of reagent	Elements determined with reagent
Azo-azoxy BN	 2-Hydroxy-2′-(2-hydroxy-1-naphthylazo)-5-methylazoxybenzene	Cu [4, 7]
Azomethine H	 N-Salicylidene-(1-amino-2-hydroxynaphthalene-4,7-disulphonic acid)	B [7]
BAPH	 4,7-Bis(phenylazoanilino)-1,10-phenanthroline	Fe [5]

Name	Structure	Determined element
Barbituric acid	(structure)	N [12, 14]
Bathocuproin	2,9-Dimethyl-4,7-diphenyl-1,10-phenanthroline	Cu [14]
Bathophenanthroline	4,7-Diphenyl-1,10-phenanthroline	Fe [7, 8, 14, 15]
5-Benzylidenerhodanine	(structure)	Noble metals [11, 13]

Common name of reagent	Structural formula and chemical name of reagent	Elements determined with reagent
Beryllon II	 4,5-Dihydroxy-3-(8-hydroxy-3,6-disulpho-1-naphthylazo)-2,7-naphthalene-disulphonic acid	Be [2, 3, 6, 7]
Beryllon III	 5-(4-Diethylamino-2-hydroxyphenylazo)-4-hydroxynaphthalene-2,7-disulphonic acid	Be [2]

Beryllon IV	 o-(1-Hydroxy-3-sulpho-6-dicarboxymethylamino-2-naphthylazo)-benzenearsonic acid	Be [2]
8,8′-Biquinoline-disulphide		Cu [10, 12, 13]
Bismuthiol II	 5-Mercapto-3-(2-phenyl)-1,3,4-thiadiazole-2-thione	Te [1–3]

355

Common name of reagent	Structural formula and chemical name of reagent	Elements determined with reagent
4,4'-Bis(N-methyl-N-benzyl-aminophenyl)-antipyryl-carbinol		Sb [15]
Bis(4-Na-tetraazoly-azo-5)-acetic acid, ethyl ester		Ni [6, 13]
BPHA (HBPHA)	\n\nN-Benzoyl-N-phenylhydroxylamine	V [2-4, 6-8]

Br-Benzthiazo	 1-(6-Bromo-Benzthiazolyl-2-azo-1)-naphthol-2	Cd [4]
6-Br-BTAK	 6-Bromo benzothiazolyl-(2'-azo-2)-4-methylphenol	Ni [8]
3,5-Br-MEPADAP	 2-(3,5-Dibromo-4-methyl-2-pyridylazo)-5-diethylaminophenol	V [5]

357

Common name of reagent	Structural formula and chemical name of reagent	Elements determined with reagent
5-Br-PADAP	5-Bromo-2-pyridylazodiethylaminophenol	U [16]
3,5-Br-PADAP	3,5-Dibromo-2-pyridylazo-diethylaminophenol	Co [6]
Brilliant green		Au [15] Ga [1] Sb [6, 7]

Bromo-pyrogallol red	3,3''-Dibromopyrogallol-sulphophthalein	Ln [1] Nb [1] Noble metals [5, 11, 13]
Brucine		N [1]
Butyl-rhodamine B		Te [7]

359

Common name of reagent	Structural formula and chemical name of reagent	Elements determined with reagent
Cadion	4-(Nitrobenzenediazoamino)-azobenzene	Cd [3]
Calcion (calcichrome)		Ca [1, 2]
Carminic acid		B [9, 10]

Reagent	Structure	Determined elements
Catechol violet (pyro-catechol violet)	3,3′4′-trihydroxyfuchsone-2″-sulphonic acid	Al [6] F [13] In [2, 3] Ln [1]; Mo, W [12] Sn [4] Zr [1]
Chloramine T	4-Toluenesulphonic acid-chloroamine	N [12, 14]
Chlorophosphonazo III	2,7-Bis(4-chloro-2-phosphonophenylazo)-1,8-dihydroxynaphthalene-3,6-disulphonic acid	Ba, Sr [2, 3] Ca [3, 6] Mg [1] U [2, 4, 6, 7]

Common name of reagent	Structural formula and chemical name of reagent	Elements determined with reagent
Chrome azurol S		Al [1, 2, 7] Ga [14] In [2–4]
Chrome pyrazole II	 Bis(4-methylbenzylaminophenyl)-antipyrylcarbinol	Cd [5]

Chromotrope 2 R	 2(Phenylazo)-1,8-dihydroxynaphthalene-3,6-disulphonic acid	Mg [1]
Chromotropic acid	 1,8-Dihydroxynaphthalene-3,6-disulphonic acid	Ti [5, 12]
Crystal violet		B [1, 2, 9] Bi [1, 2] Noble metals [4, 11, 13] Sb [14] Tl [6]

363

Common name of reagent	Structural formula and chemical name of reagent	Elements determined with reagent
Cuproin	 2,2'-Biquinolyl	Cu [8, 9]
Curcumin	$CH_2(COCH=CH$ $OH)_2$	B [1, 4]
DDTC	 Diethyldithiocarbamic acid, Na salt	Bi [7] Cu [1–4, 7, 8] Mn [5] Pb [5, 6] Te [1] Ti [7] U [2]
3,3'-Diaminobenzidine	 3,4,3',4'-Tetraaminodiphenyl	Se [1, 3]

2,3-Diaminonaphthalene	NH₂, NH₂ (2,3-diaminonaphthalene structure)	Se [1, 3]
4,5-Diamino-6-thio-pyrimidine	SH, NH₂, NH₂ (pyrimidine structure)	Se [3]
Diantipyrylmethane (DAM)	4,4′-Diantipyrylmethane	Ga [11]; Ni [14]; Sb [12]; Ti [6, 7, 12–14]
1,1′-Dianthrimide	1,1′-Iminodianthraquinone	Se [3]

Common name of reagent	Structural formula and chemical name of reagent	Elements determined with reagent
Dibenzoylmethane		U [2, 4, 6]
Dibromtichromin	 N-Methyl-N,N-bis(methylene-2,2'-dibromochromotropic acid)-amine, Na salt	Ti [4]
Diethyldithiophosphoric acid	$(C_2H_5O)_2 - P \overset{S}{\underset{}{}} SH$	Re [1]
DHNS	 1,8-Dihydroxynaphthalene-4-sulphonic acid	B [8]

Compound	Structure	
Dimercaptothiopyrone	2,6-Dimercapto-7-thiopyran-4-one	Bi [3]
5-(p-Dimethylamino-benzylidene)-rhodanine		Au [16]
Dimethylglyoxime	$H_3C-C-C-CH_3$, HON NOH	Ni [1, 5, 6, 9, 13] Re [1, 3]
2,6-Dimethylphenol		N [7]
N,N-Dimethyl-p-phenylenediamine		S [5]

367

Common name of reagent	Structural formula and chemical name of reagent	Elements determined with reagent
Diphenylcarbazide	$O=C\begin{smallmatrix} NH-NH-C_6H_5 \\ NH-NH-C_6H_5 \end{smallmatrix}$ 1,5-Diphenylcarbohydrazide	Cr [4, 5, 7, 8]
Diphenylcarbazone	$O=C\begin{smallmatrix} N=N-C_6H_5 \\ NH-NH-C_6H_5 \end{smallmatrix}$ Phenylazoformic acid 2-phenylhydrazide	Cl [3]
1,3-Diphenylguanidine	$HN=C\begin{smallmatrix} NH-C_6H_5 \\ NH-C_6H_5 \end{smallmatrix}$	U [10]
Dipicramine	2,2',4,4',6,6'-Hexanitrodiphenylamine	Alkali metals [3]

2,2'-Dipyridyl (2,2'--bipyridyl)		Fe [12]
Disulphophenylfluorone	 2,3,7-Trihydroxy-9-(2',4'-disulphophenyl)-6-fluorone	In [5] Ti [10]
Dithiol	 Toluene-3,4-dithiol	Ge [2] Mo, W [1–7] Sn [5]

Common name of reagent	Structural formula and chemical name of reagent	Elements determined with reagent
Dithiopyrylmethane (DTPM)	4,4'-Dithiopyrylmethane	As [12]
Dithizone	1,5-Diphenylthiocarbazone	Ag [14]; Bi [4] Cd [1]; Cu [1, 2] Hg [4, 8] In [2–4] Pb [1, 2, 4] Noble metals [17] Se [1, 3, 13] Zn [2, 10, 11]
DMNS	3,3'-Dimethylnaphthidinedisulphonic acid	V [14, 15]

	Mg [1]
1-(2-Hydroxy-4 sulphonic acid-6-nitronaphthylazo)-2-hydroxynaphthalene	
	Al [1, 2] F [8, 16, 22] Sc [1]
	Te [7]

Eriochrome black T

Eriochrome cyanine R
(solochrome cyanine R)

Ethylrhodamine B

Common name of reagent	Structural formula and chemical name of reagent	Elements determined with reagent
Formaldoxime	$CH_2 = NOH$	Mn [5]
N-Furoyl-N-phenyl-hydroxylamine		V [9, 10]
2-Furyldioxime	 HON NOH	Ni [11, 12] Ra [1, 3]
Gallein	 4,5-Dihydroxyfluorescein	In [2–4]

Gallion	4-Hydroxy-3-(2-hydroxy-3-chloro-5-nitrophenylazo)-naphthalene-5-amino--2,7-disulphonic acid	Ga [1]
Glycine cresol red	3,3'-Bis(N-carboxymethylaminomethyl)-o-cresolsulphonephthalein	Cr [3]

Common name of reagent	Structural formula and chemical name of reagent	Elements determined with reagent
Glycine thymol blue	3,3'-Bis-(N-carboxymethylaminomethyl)-thymosulphonephthalein	Cr [3]
Glyoxal-bis(2-hydroxyanil)		Ca [5, 7]
5-(2-Hydroxy-3,5-dinitro-phenylazo)-N-oxide-8-hydroxyquinoline		Cu [11] Noble metals [18, 19]

Name	Structure	Elements
H-resorcin	1-(Resorcinazo)-8-hydroxy-3,6-naphthalenedisulphonic acid	B [5, 11]
Indophenol		N [1, 2]
Isobutyldithiopyryl-methane		Te [14, 15]
Lumogallion	4-Chloro-6-(2,4-dihydroxyphenylazo)-1-hydroxybenzene-2-sulphonic acid, Na salt	Nb [1, 4] Sn [3] Ge [4, 11]

Common name of reagent	Structural formula and chemical name of reagent	Elements determined with reagent
Magneson IREA	 2-(2-Hydroxy-5-chloro-3-sulphophenylazo)-1-hydroxynaphthalene	Mg [1, 2, 6]
Malachite green		Ga [1, 3–5, 12] Tl [1]
2-Mercaptobenzimidazole		Se [3]

		Re [1, 3]
8-Mercaptoquinoline	Thiooxine	
4-Methyl-1,2-cyclohexane-dionedioxime		Re [3]
Methylene blue		S [13] Sb [9]
N-Methyl-2-(α-nitroso-α--cyanide)-methyl-benzimidazole		Fe [9]
Methyl orange	4-Dimethylaminoazophenyl-4'-sulphonic acid, Na salt	Ln [1, 2]

377

Common name of reagent	Structural formula and chemical name of reagent	Elements determined with reagent
Methyl thymol blue	$(H_3C)_2CH$ $CH(CH_3)_2$ O $CH_2N(CH_2COOH)_2$ $(HOOCCH_2)_2NH_2C$ CH_3 CH_3 SO_3H	In [2–4]
Methyl violet	$N(CH_3)_2$ $N(CH_3)_2Cl$ $NHCH_3$	Ga [1]

Name	Structure	
Michler's thioketone	4,4'-Bis(dimethylamino)-thiobenzophenone	Noble metals [5, 11–13]
Murexide	H_4ON 5,5'-Nitrilodibarbituric acid, NH_4 salt	Ca [7]
NAAN	5-Nitrophenol-(2-azo-1')-2'-(β-acetylhydrazine)-naphthalene	Zn [2, 10]
1-Naphthylamine		N [6]

Common name of reagent	Structural formula and chemical name of reagent	Elements determined with reagent
N-(1-Naphthyl)-ethylene-diamine dihydrochloride	—NHCH$_2$CH$_2$NH$_2$ · 2HCl	N [10, 11]
Neocuproin	2,9-Dimethyl-1,10-phenanthroline	Cu [8]
Nevazol-NO	2-(2-Hydroxy-5-nitro-3-sulphophenylazo)-1-hydroxy-4-naphthalene-sulphonic acid	V [12]

Reagent	Structure	Determined elements
Nitchromeazo	HO₃S, NO₂, SO₃H, OH, OH, SO₃H, HO₃S, O₂N, N=N, N=N 2,7-Bis(4-nitro-2-sulphophenylazo)-1,8-dihydroxy-3,6-naphthalene-disulphonic acid	Ba, Sr [2, 3] S [15]
Nitroantranilazo	CH₃, N=N, C₆H₅, O, COOH, N=N, O₂N 1-Phenyl-3-methyl(2-carboxy-4-nitrophenylazo)-pyrazolone-5	Alkali metals [1, 2]
1-Nitrozo-2-naphthol	NO, OH	Co [1] Mo, W [11]

Common name of reagent	Structural formula and chemical name of reagent	Elements determined with reagent
Nitroso-R salt	1-Nitroso-2-hydroxy-3,6-naphthalenedisulphonic acid, di-Na salt	Co [1–3, 5]
Nitrosulphophenol S	2,7-(2',2''-Hydroxy-3',3''-disulphonic acid-5',5''-nitrodiphenylazo)-1,8-dihydroxy-3,6-naphthalenedisulphonic acid	Zr [4, 5] Al [10]
Orthanil K		S [14]

Reagent	Structure	Metals
Orthanil S (sulphonazo III)	2,7-Bis(2'-sulphophenylazo)-1,8-dihydroxy-3,6-naphthalenedisulphonic acid	Ba, Sr [2–4]
Oxine	8-Quinolinol (8-hydroxyquinoline)	Al [1, 2] In [2] Mg [7] Nb [1] Zn [1]
PAN	1-(2-Pyridylazo)-2-naphthol	Co [10] In [2–4, 7] Ni [7] Sn [10] Zn [2, 9]

Common name of reagent	Structural formula and chemical name of reagent	Elements determined with reagent
PAR	4-(2-Pyridylazo-)-resorcinol	Co [9] Cr [3, 10] In [2–4, 7] Nb [1] V [11]
1,10-Phenanthroline		In [5] Fe [4, 13] Noble metals [11, 13]
Phenazo		Mg [1, 2]
m-Phenylenediamine		N [8]

Reagent	Structure	Elements determined
o-Phenylenediamine	![NH₂ NH₂ benzene]	Se [1–3]
Phenylfluorone	2,3,7-Trihydroxy-9-phenyl-6-fluorone	Ge [2, 4, 8] Nb [1] Sn [1, 8] Zr [1]
Phenylhydrazine	NH—NH₂	W, Mo [3]
γ-Picoline	4-Methylpyridine	N [12]

Common name of reagent	Structural formula and chemical name of reagent	Elements determined with reagent
Picramine R	 3-Hydroxy-4-(2-hydroxy-3,5-dinitrophenylazo)-naphthalene-2,7-disulphonic acid	Zr [4]
Picramine S	 2,7-Bis(picraminazo)-chromotropic acid	Zr [3]

Fe [6]	SO$_3$NH$_4$ SO$_3$NH$_4$ · H$_2$O 3-(4-Phenyl-2-pyridyl)-5,6-(diphenyl-4,4′-disulphonic acid)-1,2,2-triazine, di-NH$_4$ salt	PPDT-DAS
N [14]		Pyridine
Sn [2]	OH OH HO O 9-(Pyridyl-4)-fluorone (2,3,7-trihydroxy-9-pyridyl-6-fluorone)	Pyridine-3-fluorone

Common name of reagent	Structural formula and chemical name of reagent	Elements determined with reagent
Pyrocatechol violet	(see Catechol violet)	
Pyrogallol red	 Pyrogallolsulphophthalein	Zr [10]
Quinalizarin	 1,2,5,8-Tetrahydroxyanthraquinone	B [6] Ga [1] In [2]

Rhodamine B	$(C_2H_5)_2N$ ... O ... $\overset{+}{N}(C_2H_5)_2Cl^-$ ... COOH Tetraethylrhodaminium chloride	Cd [2] Ga [1, 6–8, 13] In [2–4] Mo, W [9] Sb [13] Tl [1]
Rhodamine 6G	$(C_2H_5)HN$ ... H_3C ... O ... $\overset{+}{N}H(C_2H_5)_2Cl^-$ ... CH_3 ... $COOC_2H_5$	In [4]

Common name of reagent	Structural formula and chemical name of reagent	Elements determined with reagent
p-Rosaniline (p-fuchsin)	p-p'-p''-Triaminotriphenylmethane chloride	S [3, 9]
Salicylic acid	o-Hydroxybenzoic acid	B [1] N [9] Ti [8]

Salicylideneamino-2-thiophenol	2-(o-Hydroxyphenyl)-benzthiazoline	Sn [6]
Silver diethyldithio-carbamate		As [6, 7] Sb [10]
Solochrome cyanine R	(see Eriochrome cyanine R)	
SPADNS	2-(4-Sulphophenylazo)-1,8-dihydroxy-3,6-naphthalene disulphonic acid	Th [2]

Common name of reagent	Structural formula and chemical name of reagent	Elements determined with reagent
Stilbazo	4,4'-Bis(3''-4''-dihydroxyphenylazo)2,2'-stilbenedisulphonic acid	Ga [1] In [2–4]
Sulphanilamide	H_2N—⬡—SO_2NH_2 4-Aminobenzenesulphamide	N [10, 11]
Sulphanilic acid	H_2N—⬡—SO_3H Aniline-4-sulphonic acid	N [6]
Sulphoarsazen		Cd [6] Pb [3, 7]

Sulphochlorophenol S	Nb [1, 7, 8] Zr [3, 4, 5]

2,7-Bis(2-hydroxy-5-chloro-3-sulphophenylazo)-1,8-dihydroxy-3,6-naphthalenedisulphonic acid

Sulphochrome	Be [8]

4-Hydroxy-5,5'-dicarboxy-3,3'-dimethylfuchsone-2'',4''-disulphonic acid

Sulphonazo III

(see Orthanil S)

393

Common name of reagent	Structural formula and chemical name of reagent	Elements determined with reagent
5-Sulphosalicylic acid		Fe [11]
TAN-3,6 S	 1-(2'-Thiazolylazo)-2-hydroxy-3,6-naphthalenedisulphonic acid	Ni [3, 4]
2-Thenoyltrifluoroacetone	 4,4,4-Trifluoro-1-(2-thienyl)-1,3-butanedione	U [4]

Thioglycolic acid	$HSCH_2COOH$ Mercaptoacetic acid	W, Mo [1, 3]
Thionine	 3,7-Diaminophenthiazine	B [5] Sb [8]
Thiourea	$S = C \begin{smallmatrix} NH_2 \\ NH_2 \end{smallmatrix}$	Bi [5] Re [1, 3] Sb [4]
Thorin (Thoron, APANS)	 o-(2-Hydroxy-3,6-disulpho-1-naphthylazo) benzenearsonic acid	Th [1]

395

Common name of reagent	Structural formula and chemical name of reagent	Elements determined with reagent
Tichromin	 N-Methyl-N,N-bis(methylenechromotropic acid)-amine, Na salt	Ti [4]
Tiron	 1,2-Dihydroxybenzene-3,5-disulphonic acid	Ti [11]
Titan yellow		Mg [1]

o-Tolidin	H_2N— ... —NH_2, CH_3, H_3C 3,3'-Dimethyl-4,4'-diaminodiphenyl	Cl [5] Mn [11]
2,6-Xylenol	OH, CH_3, H_3C, CH_3	N [7]
Xylenol orange (XO)	CH_3, CH_3, $CH_2N(CH_2COOH)_2$, SO_3H, HO, $(HOOCCH_2)_2NCH_2$	In [2] Sc [2] Zn [1, 3] Zr [3]
Zincon	COOH, OH, HN—N, N=N—C—C_6H_5, HO_3S 2-Carboxy-2'-hydroxy-5'-sulphoformazylbenzene	Zn [1, 2, 8]

Subject index

The most important reagents are listed in the Appendix.